The Minerals, Metals & Materials Series

Gisele Azimi · Kerstin Forsberg ·
Takanari Ouchi · Hojong Kim ·
Shafiq Alam · Alafara Abdullahi Baba
Editors

Rare Metal Technology 2020

TMS

Springer

Editors
Gisele Azimi
University of Toronto
Toronto, ON, Canada

Kerstin Forsberg
KTH Royal Institute of Technology
Stockholm, Sweden

Takanari Ouchi
The University of Tokyo
Tokyo, Japan

Hojong Kim
Pennsylvania State University
University Park, PA, USA

Shafiq Alam
University of Saskatchewan
Saskatoon, SK, Canada

Alafara Abdullahi Baba
University of Ilorin
Ilorin, Nigeria

ISSN 2367-1181 ISSN 2367-1696 (electronic)
The Minerals, Metals & Materials Series
ISBN 978-3-030-36757-2 ISBN 978-3-030-36758-9 (eBook)
https://doi.org/10.1007/978-3-030-36758-9

This Springer imprint is published by the registered company Springer Nature Switzerland AG
The registered company address is: Gewerbestrasse 11, 6330 Cham, Switzerland

Preface

Rare Metal Technology 2020 is the proceedings of the symposium on Rare Metal Extraction & Processing sponsored by the Hydrometallurgy and Electrometallurgy Committee of the TMS Extraction & Processing Division. The symposium has been organized to encompass the extraction of rare metals as well as rare extraction processing techniques used in metal production and mineral processing. This is the seventh symposium since 2014.

This symposium covers research and developments in the extraction and processing of rare metals from primary and secondary sources such as laterite ore, electric arc furnace (EAF) dusts, and waste electrical and electronic equipment (WEEE) including fluorescent lamps and lithium-ion batteries. Rare metals include strategic metals that are in increasing demand and subject to supply risks (those that are not covered by other TMS symposia). The symposium is focused on primary production as well as secondary production through urban mining and recycling to enable the circular economy. Processing techniques including, but not limited to, hydrometallurgy (solvent extraction, ion exchange, precipitation, and crystallization), electrometallurgy (electrorefining and electrowinning), pyrometallurgy, and aeriometallurgy (supercritical fluid extraction) are discussed in this symposium.

In this proceedings publication, papers are presented on the extraction and processing of rare earth metals including neodymium, dysprosium, scandium, and others; platinum group metals including platinum, palladium, iridium, and others; battery-related metals including lithium, cobalt, nickel, and aluminum; electronics-related materials including copper and gold; and refectory metals including titanium, niobium, zirconium, and hafnium. Other critical materials such as gallium, germanium, indium, and silicon are also included. Several processing techniques including hydrometallurgy, electrometallurgy, pyrometallurgy, and supercritical fluid extraction are discussed.

We acknowledge the efforts of the symposium organizers and proceedings editors: Gisele Azimi, Kerstin Forsberg, Takanari Ouchi, Hojong Kim, Shafiq Alam, and Alafara Abdullahi Baba. The support from TMS staff members, Matt

Baker and Patricia Warren, is greatly appreciated in assembling and publishing the proceedings. We sincerely thank all the authors, speakers, and participants and look forward to continued collaboration in the advancement of science and technology in the area of rare metal extraction and processing.

Gisele Azimi
Lead Organizer

Contents

About the Editors

Gisele Azimi is an Associate Professor cross-appointed between the Departments of Chemical Engineering and Applied Chemistry and Materials Science and Engineering at the University of Toronto. She is also a registered Professional Engineer in Ontario. Her research program is aligned well with the "Sustainability" and "Advanced Materials and Manufacturing" research themes. In her research program, she strives to achieve a sustainable future and mitigates the adverse effects of climate change through (1) advanced recycling and urban mining of waste electrical and electronic equipment (WEEE), utilizing innovative recycling processes based on supercritical fluids; (2) industrial solid waste reduction through waste valorization to produce strategic materials like rare earth elements; (3) development of innovative materials with unique properties with far-reaching applications in structural and energy materials sectors; and (4) energy storage focusing on the development of a new generation of rechargeable batteries made of aluminum. She received her Ph.D. in 2010 from the Department of Chemical Engineering and Applied Chemistry at the University of Toronto. Before returning to the University of Toronto as a faculty member in 2014, she completed two postdoctoral appointments at MIT in the Departments of Materials Science and Engineering and Mechanical Engineering. She has received a number of awards including Emerging Leaders Award in Chemical Engineering, Dean's Spark Professorship, Early

Researcher Award, TMS Light Metals/Extraction & Processing Subject Award–Recycling, and Connaught New Researcher Award.

Kerstin Forsberg is an Associate Professor in Chemical Engineering at KTH Royal Institute of Technology in Sweden. Her research program is focused on separation processes, in particular, crystallization. This knowledge is often applied in projects concerning recovery of resources from waste. She is the Program Director of the Master's program in Chemical Engineering for Energy and Environment at KTH. She is also the Deputy Director for the Research Platform for Industrial Transformation at KTH and she represents the School of Engineering Sciences in Chemistry, Biotechnology and Health as a board member of the Water Centre at KTH and as a member of the management team for the Initiative in Circular Economy at KTH.

Takanari Ouchi is a Research Associate at the Institute of Industrial Science at the University of Tokyo. He received his Ph.D. in Nano-Science and Nano-Engineering from Waseda University in 2011. In this tenure, Dr. Ouchi developed electrochemical deposition processes to fabricate metal nano-structures with both well-controlled crystallinity and uniformity at the single nano-meter scale, and demonstrated the applicability of these processes for the fabrication of bit-patterned magnetic recording media for future hard disk drives. After completing his doctoral degree, Dr. Ouchi joined MIT, where he developed liquid metal batteries, which are, in principle, bi-directional electrolysis (electrorefining) cells, for application in grid-scale energy storage. As a research scientist, Dr. Ouchi led the systematic investigation of the electrochemical properties of liquid metal electrodes in molten salt electrolytes and developed novel lithium, calcium, and sodium liquid metal batteries. Since he began work as a Research Associate at the University of Tokyo in 2017, he has developed new recycling processes for rare metals and precious metals using pyrometallurgical and electrometallurgical methods. As a member of the Hydrometallurgy and Electrometallurgy Committee at The Minerals, Metals & Materials Society (TMS),

Dr. Ouchi has contributed to the development of the vibrant field of metal extraction, organized technical symposia at TMS, solicited papers as a Guest Editor of *JOM*, and has earned several awards and honors, such as the TMS EPD Young Leaders Professional Development Award in 2015, based on his contributions to electrometallurgical processing.

Hojong Kim is an Assistant Professor in Material Science and Engineering at Penn State University. He received a B.S. degree from Seoul National University and Ph.D. degree at MIT in the Uhlig Corrosion Laboratory. Dr. Kim worked as a Senior Researcher at Samsung Corning Precision Glass to improve the process yield for TFT-LCD glass manufacturing by engineering high temperature materials. After 5 years of industrial experience, Dr. Kim returned to MIT as a Postdoctoral Researcher to contribute to the growing need for sustainable technology, with a research focus on molten oxide electrolysis for carbon-free iron production and liquid metal batteries for large-scale energy storage. His current research focuses on electrochemical processes for separation of energy-critical elements and development of corrosion-resistant materials. He is the recipient of the NSF CAREER award and doctoral new investigator award from the American Chemical Society. He served as the Chair (2017–2019) and Vice-Chair (2015–2017) of the Hydrometallurgy and Electrometallurgy Committee of The Minerals, Metals & Materials Society.

Shafiq Alam is an Associate Professor at the University of Saskatchewan, Canada. He is an expert in the area of mining and mineral processing with profound experience in industrial operations, management, engineering, design, consulting, teaching, research, and professional services. As a productive researcher, he has secured two patents and has produced over 170 publications. He is the co-editor of eight books and an Associate Editor of the *International Journal of Mining, Materials and Metallurgical Engineering*. He is the winner of the 2015 Technology Award from the Extraction & Processing Division (EPD) of The Minerals, Metals & Materials Society (TMS).

With extensive relevant industry experience as a registered professional engineer, Dr. Alam has worked on projects with many different mining industries. He is an Executive Committee Member of the Hydrometallurgy Section of the Canadian Institute of Mining, Metallurgy and Petroleum (CIM). During 2015–2017, he served as the Chair of the TMS Hydrometallurgy and Electrometallurgy Committee. He is a co-organizer of many symposia at the international conferences through CIM and TMS. Dr. Alam is one of the founding organizers of the Rare Metal Extraction & Processing Symposium at TMS. He was involved in organizing the International Nickel-Cobalt 2013 Symposium and the TMS 2017 Honorary Symposium on applications of Process Engineering Principles in Materials Processing, Energy and Environmental Technologies. He is also involved in organizing the ninth International Symposium on Lead and Zinc Processing (PbZn 2020), co-located with the TMS 2020 Annual Meeting and Exhibition in San Diego, California.

Alafara Abdullahi Baba is a Professor of Analytical/Industrial and Materials Chemistry in the Faculty of Physical Sciences, University of Ilorin, Nigeria. He holds a Ph.D. degree in Chemistry from the University of Ilorin in 2008. His dissertation titled *Recovery of Zinc and Lead from Sphalerite, Galena and Waste Materials by Hydrometallurgical Treatments* was judged the best in the area of Physical Sciences at University of Ilorin, Nigeria in 2010. Until his current appointment as Head, Department of Industrial Chemistry in 2017, he was Deputy Director—Central Research Laboratories, University of Ilorin (2014–2017). He is a Fellow of the Chemical Society of Nigeria (FCSN) and Materials Science & Technology Society of Nigeria (FMSN); is currently Secretary of the Hydrometallurgy and Electrometallurgy Committee of the Extraction & Processing Division (EPD) of The Minerals, Metals & Materials Society (TMS); is a co-organizer of the Rare Metal Extraction & Processing Symposium and Energy Technologies & Carbon

Dioxide Management Symposium at the TMS Annual Meeting and Exhibition; and is on the TMS Materials Characterization, Education, and EPD Awards committees.

Dr. Baba has a keen interest in teaching, community services, and research covering solid minerals and materials processing through hydrometallurgical routes; reactions in solution and dissolution kinetic studies; and preparation of phyllosilicates, porous, and bio-ceramic materials for industrial value additions. He has 113 publications in nationally and internationally acclaimed journals of high impact, and has attended many national and international workshops, conferences, and research exhibitions to present his research breakthroughs. He has received several awards and honors including 2015 MISRA AWARD of the Indian Institute of Mineral Engineers (IIME) for the best paper on Electro-/Hydro-Bio-Processing at the IIME International Seminar on Mineral Processing Technology–2014 held at Andhra University, Visakhapatnam, India; 2015 MTN Season of Surprise Prize as Best Lecturer in the University of Ilorin—Nigeria category; Award of Meritorious Service in recognition of immense contributions to the Development of the Central Research Laboratories, University of Ilorin, Nigeria (2014–2017); and 2018 Presidential Merit Award in Recognition of Passion, Outstanding and Selfless Service to the Materials Science and Technology Society of Nigeria.

Part I
Lithium, Cobalt, Rare Earth Metals

Development of a Physiochemical Model Combined with an Engineering Model for Predicting Solvent Extraction Performances Within the Context of Lithium-Ion Battery Recycling

Alexandre Chagnes

Abstract Solvent extraction is a mature technology that is used in many industrial applications for the extraction and/or purification of metals contained in aqueous solutions. Such a technology could be advantageously utilized to produce high-grade salts of lithium, cobalt, nickel, and manganese from mine or lithium-ion battery recycling. However, nickel–cobalt–manganese separation is not easy by means of commercial extractants. Such a separation could be achieved providing that process operation would be always optimized. Modelling tools could help achieve this goal. This paper presents the development of a global model implementing a physicochemical model and an engineering model, which can predict the influence of pH, flowrates, and mixers-settlers arrangement on the performances of liquid–liquid extraction.

Keywords Solvent extraction · Modelling · Cobalt extraction · Lithium-ion battery

Introduction

Lithium-ion batteries operate on the reversible exchange of the lithium ion between a positive electrode, usually a lithiated transition metal oxide noted as LiMeO and a graphite negative electrode (Fig. 1):

$$\text{LiMeO} = \text{Li}_{1-x}\text{MeO} + x\text{Li}^{+} + xe^{-} \tag{1}$$

$$\text{C}_6 + x\text{Li}^{+} + xe^{-} = \text{LiC}_6 \tag{2}$$

A. Chagnes (✉)
Université de Lorraine, CNRS, GeoRessources, Vandoeuvre-les-Nancy, France
e-mail: alexandre.chagnes@univ-lorraine.fr

G. Azimi et al. (eds.), *Rare Metal Technology 2020*, The Minerals, Metals & Materials Series, https://doi.org/10.1007/978-3-030-36758-9_1

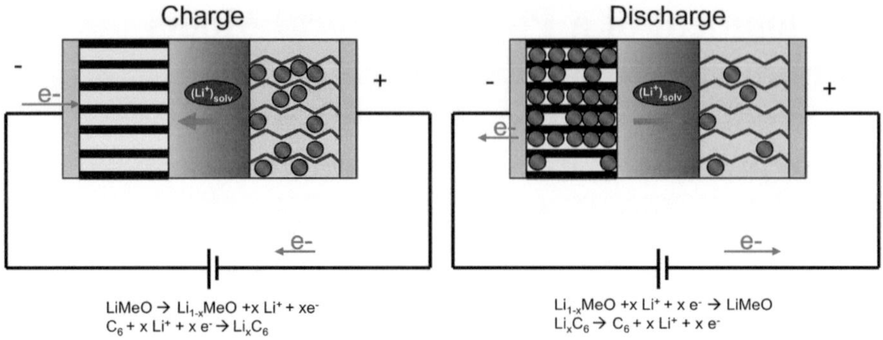

Fig. 1 Principle of lithium-ion battery[1]

Lithium-ion batteries are composed of a mixture of aprotic dipolar organic solvents (mostly alkyl carbonate) in which a lithium salt is dissolved at 1 mol L^{-1} (mainly $LiPF_6$, $LiBF_4$, or LiTFSI) [1]. This electrolyte is impregnated into a porous polypropylene–polyethylene separator which separates the positive electrode and the negative electrode [2]. Lithium-ion battery is the technology of choice for the development of electric vehicles. This technology is now mature; however, there are still many challenges to increase their energy density while ensuring an irreproachable safety of use. The negative electrode has not been changed since the first lithium-ion battery developed by Sony in 1991 and graphite is still mainly used. However, other materials, such as silicon, are presently beginning to find interesting applications. Conversely, many different positive electrodes have been developed and commercialized since the first positive electrode ($LiCoO_2$) implemented in the first Sony's battery. For instance, $LiNiO_2$ or $LiMn_2O_4$ was found as potential candidates for various applications in energy storage in the last decade. Today, polymetallic electrodes such as NMC cathodes, which are lithiated oxides of cobalt, nickel, and manganese at various stoichiometries, are the most used materials, especially for batteries in laptops and also in electric vehicles [3].

It is clear that challenges in lithium-ion battery do not only concern the production of efficient batteries for the electrochemical storage of energy since lithium-ion battery technology relies on the use of critical and/or strategic value resources. As previously stated, cobalt, nickel, manganese, and lithium are the key elements for the production of lithium-ion batteries. Given the significant market for the electric vehicle, a significant growth in demand for these metals is expected. Thus, a multiplication of demand by 31% for cobalt, 69% for nickel, and 46% for lithium is expected between 2017 and 2030 [4]. The supply of cobalt, nickel, manganese, and lithium (but also graphite and fluorine, which are the other key elements for the production of lithium-ion batteries) requires the development of efficient, economical, and sustainable processes capable of extracting and purifying these metals contained in the current resources or those that will be exploited in the near future.

Therefore, it appears crucial to include lithium-ion battery development in a circular economy approach very early. In particular, optimized recycling and reuse of battery components must both minimize their impact on the environment and limit geopolitical issues related to tensions on the mineral resources necessary for lithium-ion battery production. Although recycling will never replace mining, it reduces resource dependence by ensuring the presence of exploitable resources in the territory, which is particularly important for the European countries where exploited or exploitable resources are limited.

Recycling of lithium-ion batteries is not an easy task since there is not only a unique technology, but rather at least one technology for each application. Furthermore, lithium-ion battery technology has changed rapidly in the last decades and it will likely continue changing significantly in the next decades while other technologies will emerge in the near future such as hydrogen fuel cell for electric vehicles. Therefore, it appears difficult to assay the amount and the technology of lithium-ion batteries that will have to be really recycled in the next decades. Within this context, investors prefer to wait and see before investing in recycling units unless recycling processes will be able to treat many types of spent materials without major modification of the process. Robotics and artificial intelligence technology can help develop breakthrough processes capable of adapting to different technologies for recycling. However, artificial intelligence based only on deep-learning algorithms requires a huge amount of data to work efficiently.

Conversely, the amount of data required by artificial intelligence can be drastically reduced by combining deep learning with a mathematical model describing the physiochemistry and the engineering of the process. This paper is not focused on deep learning, but on the development of a global model implementing a physicochemical model and an engineering model that could be further combined with deep-learning technology to build a new process management tool for recycling lithium-ion batteries.

Results and Discussion

Solvent extraction is a mature technology for the recovery of metals from acidic and alkaline media or the purification of solutions containing various metal impurities. For instance, solvent extraction technology is used in many processes for cobalt extraction and cobalt–nickel separation. For this goal, Cyanex®272 (bis (2,4,4-trimethylpentyl)phosphinic acid) is the most widely used extractant [5]. Other extractants belonging to the family of cationic exchangers, such as di-(2-ethylhexyl)phosphoric acid (D2EHPA), can also be used [6].

Cobalt(II) in acidic chloride media is extracted by D2EHPA under the form of Co^{2+}, and $CoCl^{+}$ according to the following equations [7]:

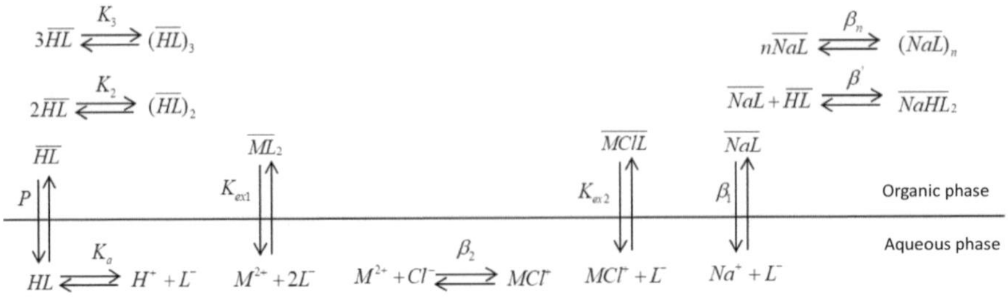

Fig. 2 Equilibria implemented in the physicochemical model

$$Co^{2+} + 2\overline{HL} \rightleftharpoons \overline{CoL_2} + 2H^+ \quad (3)$$

$$CoCl^+ + \overline{HL} \rightleftharpoons \overline{CoClL_2} + H^+ \quad (4)$$

The overbar indicates the species in the organic phase, whereas the absence of overbar indicates the species in the aqueous phase.

Besides the above extraction equilibria, reactions reported in Fig. 2 must be taken into account in the model (dimerization and trimerization of D2EHPA, partition constant of D2EHPA between the organic and the aqueous phase, etc.).

The constants reported in Fig. 3 have been determined in a previous work for various extractants, including D2EHPA [7, 8]. By solving mass balance equations and by using the constants reported in Table 1, it is possible to calculate the extraction isotherm of cobalt(II) by D2EHPA at any pH (Fig. 3). More information about calculations are available in our recent paper published in Ref. [9].

This physicochemical model is used to calculate the extraction isotherms of cobalt(II) by D2EHPA at any pH that is used as input data of an engineering model capable of calculating cobalt concentration in mixer settlers as a function of cobalt concentration in leach solution as well as flowrates of leach solution, stripping solution, and extraction solvent in extraction and stripping stages [10].

The present model has been used to calculate the performances of the flow sheets 4_3, 13_12, 13_21, 22_12, 22_21, 31_12, and 31_21 reported in Fig. 4 for a flowrate of leaching solution of 110 $m^3\ h^{-1}$, a solvent flowrate of 30 $m^3\ h^{-1}$, and a stripping flowrate of 4 $m^3\ h^{-1}$, while pH value in each mixer settler was kept at 5.

The flow sheet performance has been assayed by calculating the residual fraction of metals (f_r) and the concentration factor (f_c) expressed as

$$f_r = \frac{x_{out,\ extr}}{x_0} \quad (5)$$

$$f_c = \frac{x_{out,\ strip}}{x_0} \quad (6)$$

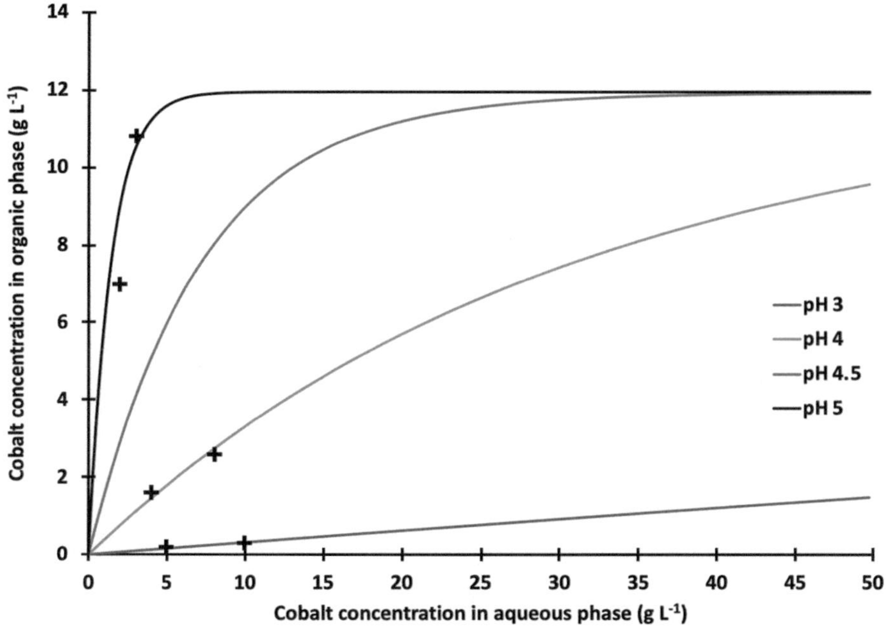

Fig. 3 Extraction isotherms of Co(II) by D2EHPA calculated by solving mass balance equations of equilibria reported in Fig. 1 and corresponding constants reported in Table 1. [HL] = 0.25 mol L^{-1}; $[Cl^-]$ = 1 mol L^{-1}; temperature = 25 °C

Table 1 Parameters used in the physicochemical model to calculate the variation of the extraction efficiency of Co(II) by D2EHPA from acidic chloride media [7, 8]

Parameter	Value
P	211
pK_a	2.6
K_2	184
K_3	8385
β_1	8
β'	230
β_n	10^7
n	7

where x_0, $x_{out,\ extr}$, and $x_{out,\ strip}$ are initial cobalt concentration in the leach solution at the entrance of the flow sheet, cobalt concentration at the outlet of the extraction stage (raffinate), and cobalt concentration at the outlet of the stripping stage (eluate).

Figure 5 shows the residual fraction of cobalt is significantly decreased at high cobalt concentration in the leach solution by using the two-loop flow sheets instead of the classical flow sheet 4_3, while the concentration factor remains high. Furthermore, it is interesting to highlight that the same performance is obtained between the flow sheets 13_12, 13_21, 22_12 and 22_21 as well as between the

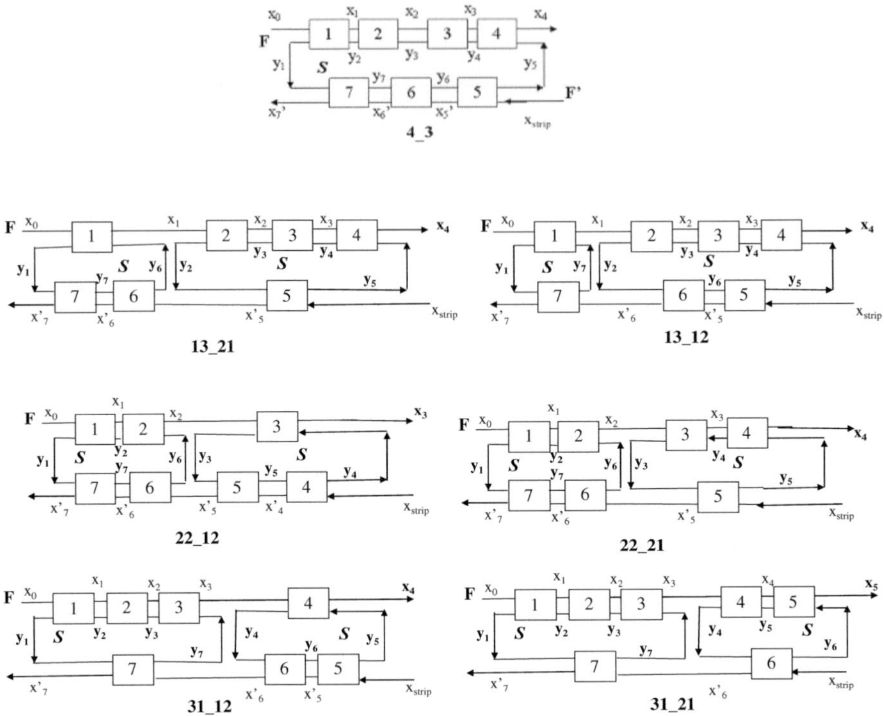

Fig. 4 Arrangement of mixers settlers

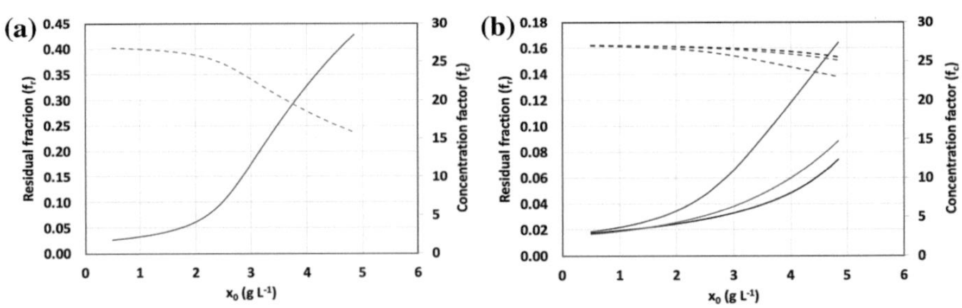

Fig. 5 Influence of cobalt concentration in the leach solution (x_0) on residual fraction (solid line) and concentration factor (dotted line) at pH = 5 for flow sheets (a) 4_3 and (b) 13_12 or 13_21 (black), 22_12 or 22_21 (green), and 31_12 or 31_21 (red). Operating parameters: Leach solution: pH = 5 and $[Cl^-] = 1$ mol L^{-1}; Stripping solution: pH = 1 and $[Cl^-] = 1$ mol L^{-1}; $F = 110$ m^3 h^{-1}; $S = 30$ m^3 h^{-1} and $F' = 4$ m^3 h^{-1}

flow sheets 31_21 and 31_12. It appears that the flow sheets 13_12 and 13_21 exhibit the best performance at pH = 5 since the concentration factor remains high ($f_c \sim 25$) and the residual factor remains low even with the presence of 5 g L^{-1} cobalt in the leach solution.

Conclusions

The combination of a physicochemical model and an engineering model into a global model is a relevant approach to assay the influence of pH, flowrates, and mixers-settlers arrangement on solvent extraction efficiency. This paper demonstrates the interest of such a global model by applying it for the recovery of cobalt (II) from chloride media. It is emphasized to extend this model to the prediction of flow sheet performances for cobalt–nickel–manganese separation since this separation is particularly tricky to implement by using commercial extractants like Cyanex® 272 or D2EHPA (Co–Ni separation can be performed with these extractants but Co–Mn or Ni–Mn is much more difficult since the separation coefficient is not enough high). Thus, the use of such a model combined with deep-learning algorithms and the presence of appropriate sensors inside the mixers settlers (pH and flowrate measurements) could ensure separation under optimized conditions despite the low separation coefficient between Co–Ni and Mn.

References

1. Chagnes A (2015) Lithium battery technologies: electrolytes. In: Chagnes A, Swiatowska J (eds) Lithium process chemistry: resources, extractions, batteries and recycling, 1st edn. Elsevier, Amsterdam, pp 167–189
2. Chagnes, A. (2012) La technologie lithium-ion. La Revue 3EI 69:69–74
3. Chagnes A, Swiatowska J (eds) (2015) Lithium process chemistry: resources, extractions, batteries and recycling. Elsevier, Amsterdam
4. Pillot C (2018) The rechargeable battery market and main trends 2017–2025. Paper presented at R&D stream lithium-ion development & commercialization delivering higher performance at lower cost, Fort Lauderdale, Florida (USA), 28–29 March 2018
5. Chagnes A, Cote G (2010) Séparation du Cobalt et du Nickel à l'aide du Cyanex® 272 par extraction liquide-liquide. L'Actualité Chimique 346:29–347
6. Omelchuk K, Szczepański P, Shrotre A, Haddad M, Chagnes A (2017) Effects of structural changes of new organophosphorus cationic exchangers on solvent extraction of cobalt, nickel and manganese from acidic chloride media. RSC Adv 7:5660–5668
7. Omelchuk K, Chagnes A (2018) New cationic exchangers for the recovery of cobalt(II), nickel(II) and manganese(II) from acidic chloride solutions: modelling of extraction curves. Hydrometallurgy 180:96–103
8. Omelchuk K, Stambouli M, Chagnes A (2018) Investigation of aggregation and acid dissociation of new cationic exchangers for liquid–liquid extraction. J Mol Liq J Mol Liq 262:111–118
9. Chagnes, A. (2019) Simulation of solvent extraction flowsheets by a global model combining physicochemical and engineering approaches—application to cobalt(II) extraction by D2EHPA, accepted in solvent extraction and ion exchange (in Press)
10. Collet S, Chagnes A, Courtaud B, Thiry J, Cote G (2009) Solvent extraction of uranium from acidic sulfate media by Alamine®336: computer simulation and optimization of the flowsheets. J Chem Technol Biotechnol 84:1331–1337

Conclusions

A Fundamental Investigation of Li_2CO_3 Crystallization from Li_2SO_4 System

Hongting Liu and Gisele Azimi

Abstract In this study, a fundamental investigation of the crystallization process of Li_2CO_3 from Li_2SO_4 solution by adding Na_2CO_3 was performed. Experimental data indicated that at optimum conditions, 90% Li from Li_2SO_4 was recovered as Li_2CO_3 solid with 1% impurity in the product and the reaction reached equilibrium within 1 h. The presence of impurities, i.e., $CaSO_4$ and Na_2SO_4, in the initial Li_2SO_4 solution had significant negative impact on both lithium recovery efficiency and purity level of the final Li_2CO_3 product. A feeding rate of Na_2CO_3 solution into Li_2SO_4 solution or adding seed in the initial Li_2SO_4 solution showed minimal effect on the recovery of the product. Seeding properly helped to form the final crystal in desired shape and size with narrow range of particle size distribution.

Keywords Li_2CO_3 · Crystallization · Lithium recovery · Crystal morphology

Introduction

Lithium's high energy density by weight and high electrochemical potential make it a key component of battery technology and production, Lithium Ion Battery (LIB) [1]. By experts' prediction, the demand for lithium will increase by fivefold in less than 10 years, driven majorly by the increasing demand for LIB [2]. Lithium carbonate, as an important cathode material, has attracted significant attention in recent years.

Around 59% of current world's lithium resource is from brine while 25% is from mineral ores. Due to the high expenses associated with lithium extraction from ores,

---

H. Liu · G. Azimi (✉)
Laboratory for Strategic Materials, Department of Chemical Engineering and Applied Chemistry, University of Toronto, 200 College Street, Toronto, ON M5S 3E5, Canada
e-mail: g.azimi@utoronto.ca

G. Azimi
Department of Materials Science and Engineering, University of Toronto, 184 College Street, Toronto, ON M5S 3E4, Canada

G. Azimi et al. (eds.), *Rare Metal Technology 2020*, The Minerals, Metals & Materials Series, https://doi.org/10.1007/978-3-030-36758-9_2

the majority of lithium in the market is extracted from brines [2]. However, because of the explosive growth of lithium demand in recent years, mineral ores have re-entered the market [3].

Spodumene is one of the most common lithium-abundant materials and has been used for lithium primary extraction in industry since 1950 [4]. Natural spodumene, which exists in its α phase, is resistant to many common chemicals. Therefore, it is usually calcinated at 1000–1100 °C to β phase, making it more reactive and less chemical resistant [5]. β-spodumene is then leached by concentrated sulfuric acid at 250 °C to turn lithium into its aqueous phase (Li_2SO_4) followed by precipitation of Li_2CO_3 by adding Na_2CO_3. Due to intrinsic drawbacks of this process, such as high levels of sulfate, heavy metal ions in the product, and complex procedures, many alternative processes to produce Li_2CO_3 were proposed, including chlorine method, hydrofluoric method, sodium carbonate method, etc. [6–8]. However, none of them has been commercialized because of high processing costs, utilization of toxic reagents, special requirement of equipment, and others. Therefore, the sulfuric method for lithium primary extraction still remains the first choice in the industry.

As previously mentioned, the conventional sulfuric method includes two major steps: leaching and precipitation. The leaching process has been well studied and optimized by utilizing fractional factorial design of experiment by Lajoie-Leroux et al. [9]. However, the precipitation process has low lithium recovery efficiency and relatively high impurity level in the product. To the authors' knowledge, there has been no extensive study on this process. Limitations and mechanisms behind this reaction are not well understood. Thus, fundamental and experimental investigations can provide valuable insight into this process, enabling process refinement and optimization.

In this research, precipitation of Li_2CO_3 by adding Na_2CO_3 into Li_2SO_4 was studied in detail by evaluating the effect of different operating variables including initial salt concentration, reaction temperature, impurity level, seeding, feeding rate in both macro-scale (recovery efficiency, product purity, reaction rate, etc.) and micro-scale (particle size, crystal morphology, etc.) in order to elucidate the mechanisms behind the process and achieve optimization.

Experimental

Materials

$Li_2SO_4 \cdot H_2O$ (99.9%), Na_2CO_3 (99.9%), Li_2CO_3 (99.998%), Na_2SO_4 (99.5%), and $CaSO_4$ (99.5%) were purchased from VWR. Solutions were prepared by deionized water. Seed crystals were prepared by Li_2CO_3 powder at different loadings (mass measured by balance) and size (sieved to achieve relatively narrow range distribution).

Experimental Procedure

For batch reactions, Li_2SO_4 and Na_2CO_3 solutions were mixed together in sealed flask. Reaction temperature was controlled by water bath (Fisher Scientific, Inc.). For controlled addition, a peristaltic pump (Cole-Parmer) was used to pump Li_2SO_4 into Na_2CO_3 solution. All the reactions were run for 24 h and the slurry was then filtered through 2.5 μm filter paper and dried at 50 °C in a furnace. For experiments with seed or impurity, initial Li_2SO_4 solution was doped with seeds or impurities and then Na_2CO_3 solution was added into the system.

Sample Preparation and Characterization

Liquid samples were prepared by withdrawing a known amount of solution from the mixture and the concentration of ions was measured by Inductively Coupled Plasma Optical Emission Spectroscopy (ICP-OES) (PerkinElmer Optima 8000). Solid samples were taken from dried filter cake. Morphology of sample was observed under Scanning Electron Microscope (SEM, Hitachi SU 5000). X-ray Diffraction (XRD, Rigaku MiniFlex 600 Diffractometer) was utilized to obtain mineralogical information and particle size distribution of crystals was measured by Diffraction Particle Size Analyzer (Horiba Parica LA-950).

Results and Discussion

Effect of Initial Salt Concentration

The effect of initial salt (Li_2SO_4 and Na_2CO_3) concentration on Li_2CO_3 production reaction rate, recovery efficiency, and product purity was studied in the range of 1 mol/L to 2 mol/L with both salts at equal molar level. In all experiments, reactions were batch and operating conditions were maintained at 25 °C, 300 rpm agitation rate without any seeds or impurities.

Figure 1a shows that recovery efficiency increases dramatically from approximately 44% to 70% with increasing initial salt concentration. As the amount of Li_2CO_3 that can dissolve in water is a fixed value at the same temperature, more Li^+ and CO_3^{2-} in the solution will form more precipitate and thus increase the recovery efficiency. It is also noticed that the impurity level rises as indicated in Fig. 1b, possibly due to the introduction of more Na^+ and SO_4^{2-} in the system, making them easier to be trapped in the precipitate.

As is shown in Fig. 1c, reaction rate of this process is monitored by ICP-OES as a function of Li concentration changing by time (g/L·s). Concentration of Li^+ in the solution decreased as the reaction proceeded because Li^+ precipitated as Li_2CO_3.

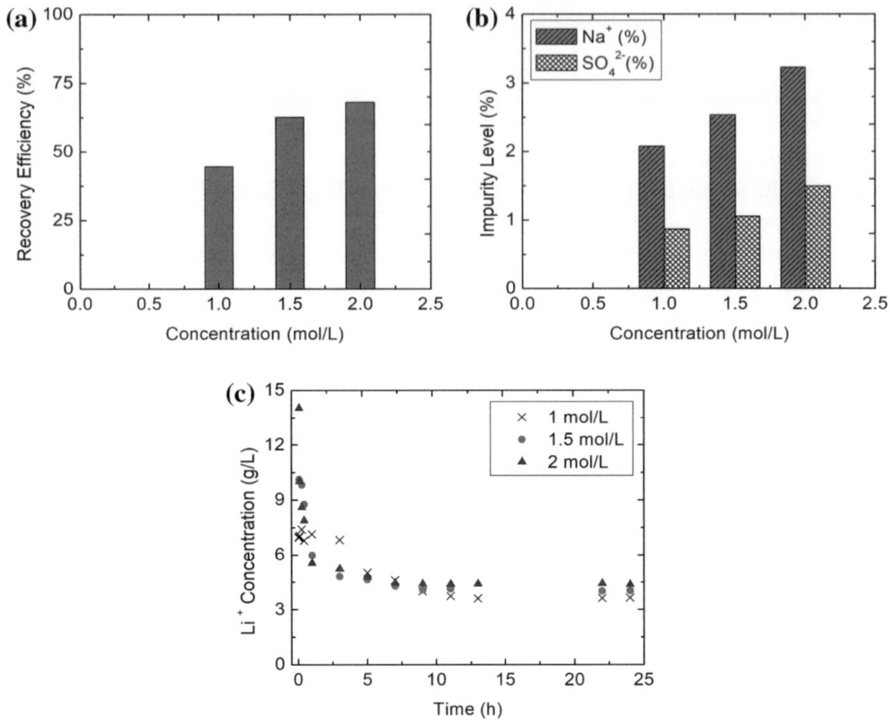

Fig. 1 **a** Recovery efficiency versus initial concentration; **b** Impurity level in the product versus initial concentration; **c** Li concentration change with time at various initial concentrations

With the increase of initial salt concentration, there is a sharp rise in the reaction rate, as can be observed by the slope change in Fig. 1c. This can be explained by the fact that the larger the number of ions in the system, the higher the possibility of collision and thus the faster the reaction.

Effect of Reaction Temperature

The effect of reaction temperature was investigated in the range between 25 and 85 °C with other conditions fixed as following: batch reaction, initial concentration of 2 mol/L, agitation rate of 300 rpm, and no addition of seed or impurities.

As is shown in Fig. 2a, with the increase of temperature from 25 to 85 °C, recovery efficiency increases from 70% to 80% because of the negative solubility trend of Li_2CO_3 versus temperature. In the meantime, as can be seen in Fig. 2b, impurity level, namely, Na^+ and SO_4^{2-}, in the product decreases with increasing temperature, which is possibly because solubility of Na_2SO_4 rises with increasing

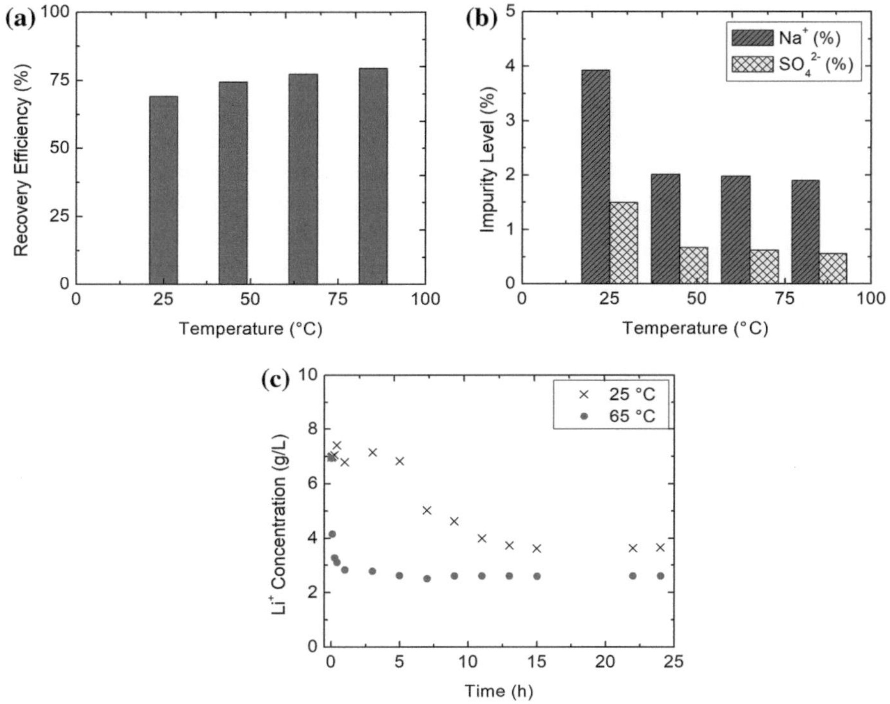

Fig. 2 **a** Recovery efficiency versus reaction temperature. **b** Impurity level in the product versus reaction temperature. **c** Li concentration change with time at various reaction temperatures

temperature. At high temperature, Na^+ and SO_4^{2-} prefer to stay in the solution rather than precipitating together with Li_2CO_3.

Figure 2c compares the reaction rates between 25 and 65 °C at 1 mol/L initial concentration. As reaction at 2 mol/L, 25 °C is already fast, in order to observe a more obvious effect by temperature, this kinetic test was carried out at 1 mol/L. It is noticed that there is a considerable increase in the reaction rate with increasing temperature. Reaction equilibrium time was shortened to less than 2 h from 11 h. This can be explained by the higher kinetics of particles at elevated temperature, enhancing the possibility of molecules collisions, thus increasing the reaction rate significantly [10].

Effect of Impurities

In real industrial processes, Li_2SO_4 solution usually contains impurities such as sodium sulfate and calcium sulfate, where cations originate from the ores while anions come from sulfuric acid. To study the effect of these impurities, before the precipitation reaction, initial Li_2SO_4 solution was mixed with different levels of impurities. Operating condition remained at 25 °C, 1 mol/L initial concentration, 300 rpm agitation rate without seeding.

Figure 3a presents the relation between recovery efficiency and level of Na_2SO_4. As can be seen with increasing Na_2SO_4 concentration, the recovery efficiency of product significantly decreased. Because solubility of Li_2CO_3 in water increases with the addition of Na_2SO_4, which can be explained by the inert ion effect [11]. Comparing the SEM images presented in Fig. 3c (Li_2CO_3 produced without impurities) and Fig. 3d (Li_2CO_3 produced in the presence of 1 mol/L Na_2SO_4) indicates that there is no clear difference between the two cases, because even

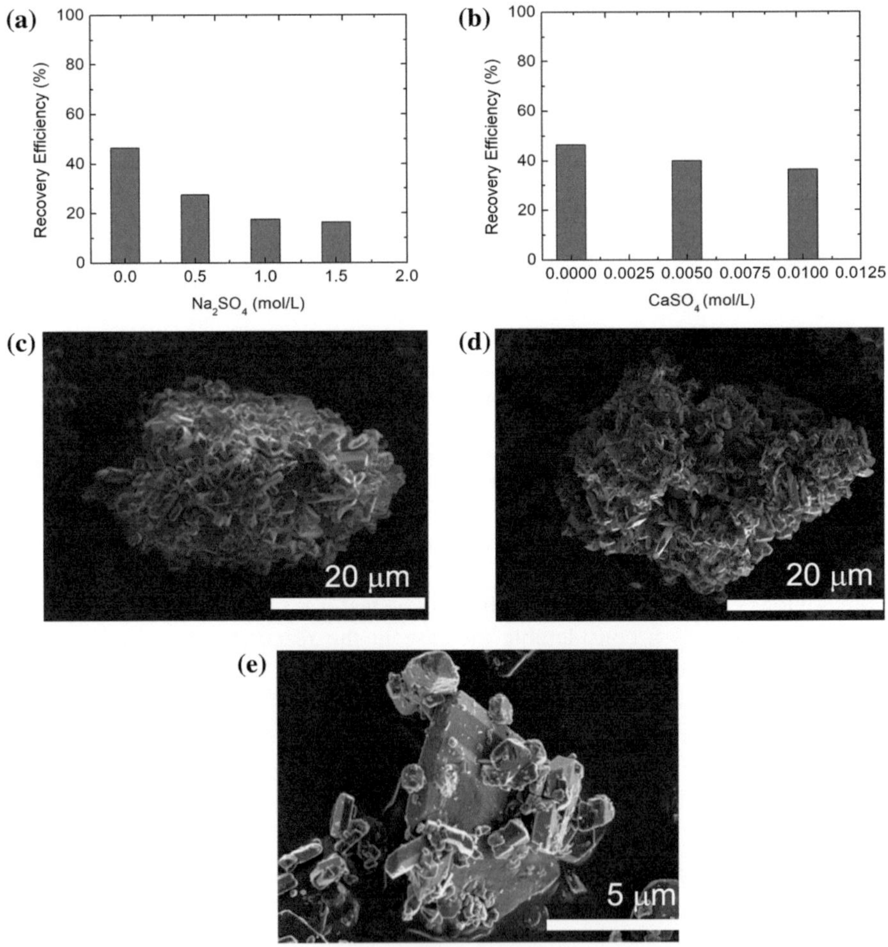

Fig. 3 **a** Recovery efficiency as a function of Na_2SO_4 concentration; **b** Recovery efficiency as a function of $CaSO_4$ concentration; **c** Li_2CO_3 crystal produced without impurities; **d** Li_2CO_3 crystal produced under 1 mol/L Na_2SO_4; **e** Li_2CO_3 crystal produced under 0.005 mol/L $CaSO_4$

without additional Na_2SO_4, there was already Na_2SO_4 produced in the system as a side product, and thus the crystal morphology did not change.

Figure 3b shows that the recovery efficiency also decreases if $CaSO_4$ is present in the system. Before the precipitation reaction, solution contains Li^+ and Ca^{2+}, after introducing Na_2CO_3 to the solution, Li^+ and Ca^{2+} compete to react with CO_3^{2-}. However, the K_{sp} of $CaCO_3$ is about five orders of magnitude less than that of Li_2CO_3, and therefore Ca^{2+} took CO_3^{2-} away from the solution, resulting in a decrease in the Li_2CO_3 yield. Under SEM, particles in Fig. 3e (Li_2CO_3 produced in the presence of 0.005 mol/L $CaSO_4$) show significant reduction in agglomerations and particle size when comparing with Fig. 3c.

Effect of Seeding

To study the effect of seeding, seed crystals at different levels of seed loading and size were put into the mixture of Li_2SO_4 and Na_2CO_3 solution at 1 mol/L, 25 °C, 300 rpm without any impurities.

As is shown in Fig. 4a and b, at different seed loadings, there exist minor changes in the recovery efficiency or purity level in the product. In terms of particle size and uniformity of the products, the information in Table 1 shows that Li_2CO_3 produced without seeding has a wide range of Particle Size Distribution (PSD) with variance of 2171. By adding seed with narrow PSD (Variance 96) and product crystals with 1% or 5% seed loading both have significant decrease in variance, indicating a more uniform PSD, which is desirable in industrial processes. In addition, it is noticed that increasing seed loading results in final product with smaller mean particle size because of the increasing number of crystal growth site. Furthermore, if seed loading is fixed while seed size increases, there is also an increase in the product particle size as shown in Fig. 4d. By observing the particles under SEM in Fig. 4e and f, it is found out that the product without seeding shows severe agglomerations with small crystals packed closely together, while seeded product shows large and single crystals with slight agglomerations. Concluding from these images, seeding is able to facilitate producing large and single crystals, which is beneficial for downstream processes, such as filtering, drying, and further chemical processing.

Effect of Agitation

To study the effect of agitation, experiments were carried out at 1 mol/L, 25 °C without addition of seeds or impurities in the range of agitation rate from 150 rpm to 600 rpm. As shown in Fig. 5, recovery efficiency at 300 rpm, 450 rpm, and 600 rpm is relatively the same while the one at 150 rpm is significantly lower than those at higher rates. As can be seen in Fig. 1c, it took around 11 h for the reaction to

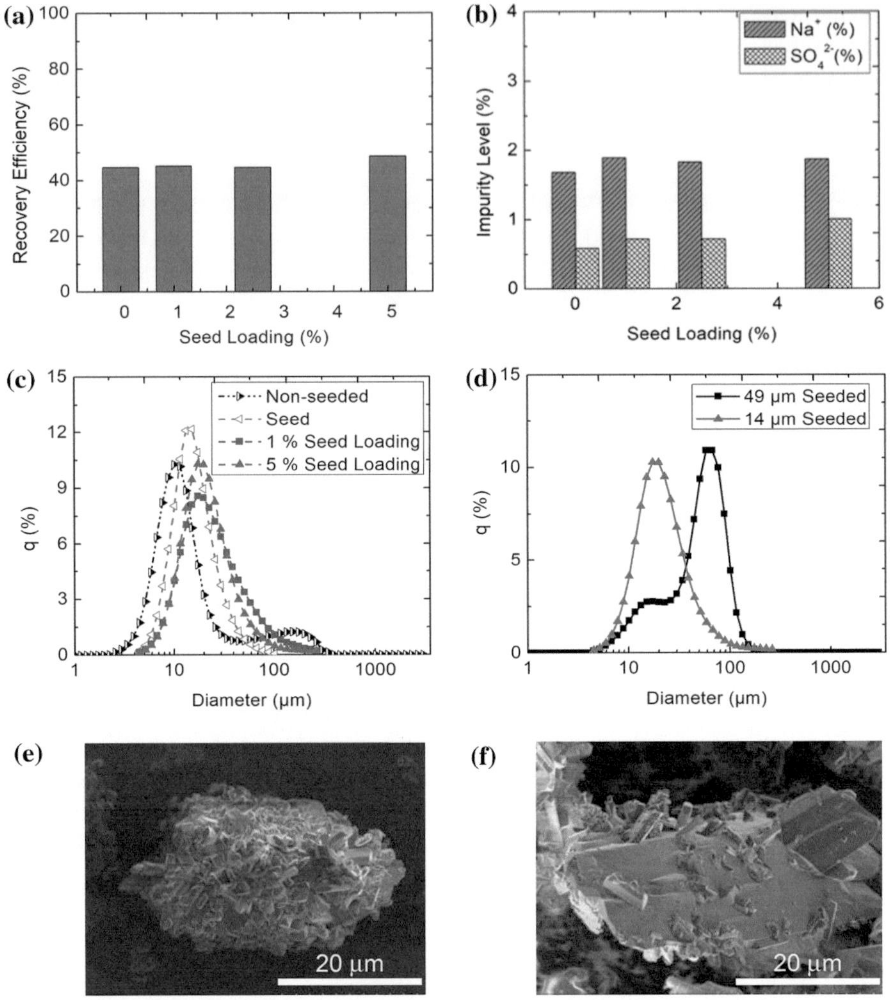

Fig. 4 **a** Recovery efficiency as a function of seed loading; **b** Impurity level of product as a function of seed loading; **c** Particle size distribution of seed, non-seeded, and seeded products; **d** Particle size distribution for seeded products with different seed sizes. **e** Non-seeded Li_2CO_3 crystal; **f** Seeded Li_2CO_3 crystal

Table 1 Mean particle size and variance of seed, non-seeded, and seeded product

	Non-seeded	Seed	1% seeded	5% seeded
Mean size	27 μm	15 μm	29.8 μm	24.9 μm
Variance	2171	96	861	568

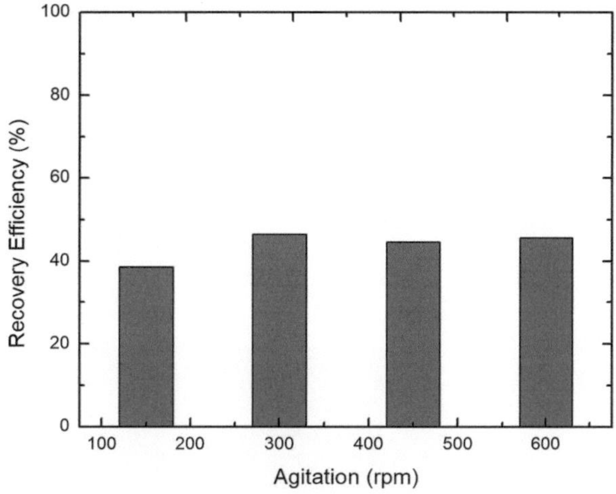

Fig. 5 Recovery efficiency as a function of agitation rate

reach equilibrium at 300 rpm and above conditions. So, the low recovery efficiency is possibly because at low agitation speed, the reaction rate was too slow and did not go to completion when sample was taken. Given a reasonable agitation rate, it is believed that this factor does not have significant effect on recovery of Li_2CO_3.

Effect of Feeding Rate

In this experiment, Na_2CO_3 solution (100 mL) was slowly pumped into Li_2SO_4 solution (100 mL) in the rate range from 0.2 mL/min to 10 mL/min with other conditions fixed at 1 mol/L, 25 °C without seeding or impurities.

Recovery of Li_2CO_3 was not affected by feeding rate as shown in Fig. 6a because the total amount of ions in the system eventually reached the same value, which is the key factor affecting the recovery efficiency. But there exist differences between the crystal's morphology under SEM. Figure 6b and c presents the SEM images of products produced at 0.2 mL/min and 10 mL/min, respectively. It is observed that at 0.2 mL/min, single and large crystal was obtained while at a higher feeding rate, small particles agglomerated together and formed chunk of aggregates. Based on calculation, it took more than 8 h for the entire Na_2CO_3 solution (100 mL) to be pumped into Li_2SO_4 solution, which gave the crystal enough time to grow bigger. However, when feeding was fast, ion concentration increased quickly in the solution, causing high supersaturation levels, which led to more primary nucleation, and thus smaller particles and severe agglomerations.

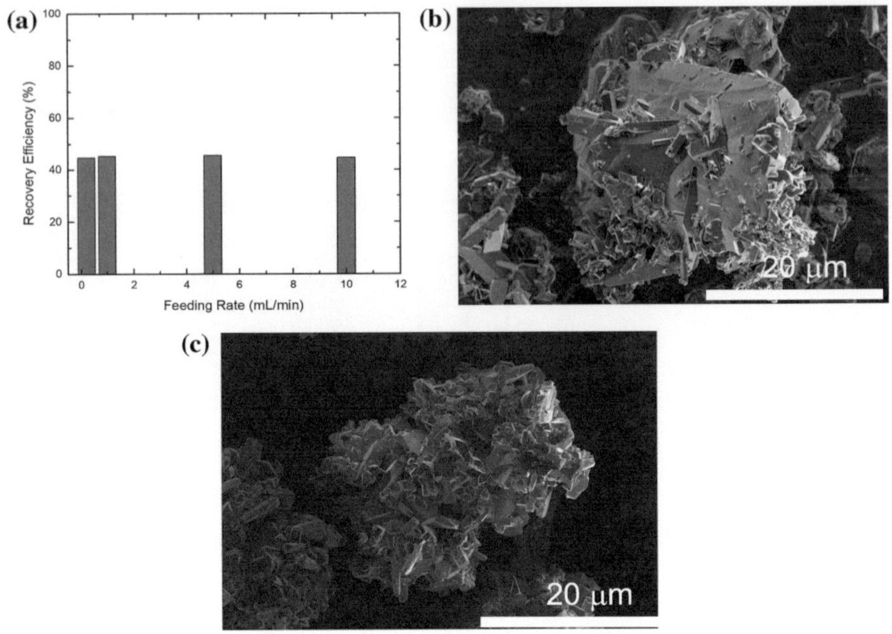

Fig. 6 **a** Recovery efficiency as a function of feeding rate; **b** Li_2CO_3 crystal produced at 0.2 mL/min feeding rate; **c** Li_2CO_3 crystal produced at 10 mL/min feeding rate

Conclusions

A fundamental study of recovering Li from Li_2SO_4 as Li_2CO_3 by adding Na_2CO_3 was performed. It was found that increasing initial salt concentration and reaction temperature both improve recovery efficiency and reaction rate. In the meantime, the former has a negative effect on the product purity while the latter has a positive effect. Common impurities, such as Na_2SO_4 and $CaSO_4$, have a significant negative effect on both recovery efficiency and product purity. Addition of seed with known size and mass helps to produce single crystals with desired size and shape and low level of agglomeration in a narrow range of particle size distribution. Seeding, agitation speed, and feeding rate show minimal effect on the recovery efficiency. Low feeding rate increases single crystal size and reduces agglomeration. On the basis of these results, optimized conditions were determined to be at high initial concentration, high temperature with proper filtration and purification steps (removing of Na^+ and SO_4^{2-}). Under these conditions, 90.1% of Li was recovered from Li_2SO_4 as Li_2CO_3 with less than 1% impurity. Depending on the requirement by the industry or customer, proper seeding and controlled feeding rates can be applied to control the final crystal size and morphology.

References

1. Vikström H, Davidsson S, Höök M (2013) Lithium availability and future production outlooks. Appl Energy 110:252–266. https://doi.org/10.1016/j.apenergy.2013.04.005
2. Kavanagh L, Keohane J, Cabellos GG, Lloyd A (2018) Global lithium sources—industrial use and future in the electric vehicle industry : a review. https://doi.org/10.3390/resources7030057
3. Vieceli N, Nogueira CA, Pereira MFC et al (2018) Hydrometallurgy recovery of lithium carbonate by acid digestion and hydrometallurgical processing from mechanically activated lepidolite. Hydrometallurgy 175:1–10. https://doi.org/10.1016/j.hydromet.2017.10.022
4. Tran T, Luong VT (2015) Lithium production processes. Elsevier Inc
5. Kuang G, Liu Y, Li H et al (2018) Extraction of lithium from β-spodumene using sodium sulfate solution. Hydrometallurgy 177:49–56. https://doi.org/10.1016/j.hydromet.2018.02.015
6. Barbosa LI, González JA, Del Carmen Ruiz M (2015) Extraction of lithium from β-spodumene using chlorination roasting with calcium chloride. Thermochim Acta 605:63–67. https://doi.org/10.1016/j.tca.2015.02.009
7. Rosales G, Ruiz M, Rodriguez M (2016) Study of the extraction kinetics of lithium by leaching β-spodumene with hydrofluoric acid. Minerals 6:98. https://doi.org/10.3390/min6040098
8. Chen Y, Tian Q, Chen B et al (2011) Preparation of lithium carbonate from spodumene by a sodium carbonate autoclave process. Hydrometallurgy 109:43–46. https://doi.org/10.1016/j.hydromet.2011.05.006
9. Lajoie-Leroux F, Dessemond C, Soucy G et al (2018) Impact of the impurities on lithium extraction from β-spodumene in the sulfuric acid process. Miner Eng 129:1–8. https://doi.org/10.1016/j.mineng.2018.09.011
10. Paine DC, Whitson T, Janiac D et al (1999) A study of low temperature crystallization of amorphous thin film indium-tin-oxide. J Appl Phys 85:8445–8450. https://doi.org/10.1063/1.370695
11. Bulavin V, Rushenko I, Blinkov M (2017) Determining a dependence of the effect of inert electrolyte on a difficultly soluble salt under different conditions. Eastern-European J Enterp Technol 4:10–16. https://doi.org/10.15587/1729-4061.2017.108181

Recycling of End-of-Life Lithium-Ion Battery of Electric Vehicles

Ka Ho Chan, Monu Malik, John Anawati and Gisele Azimi

Abstract This study puts the emphasis on developing and optimizing efficient hydrometallurgical processes to recycle a lithium-ion battery of an electric vehicle utilizing systematic experimental and theoretical approaches based on design of experiment methodology. Two leachants, i.e., HCl and $H_2SO_4 + H_2O_2$ were utilized and on the basis of fractional factorial design for the metal leaching efficiency, the most effective leachant was selected as $H_2SO_4 + H_2O_2$. In this case, 1.5 M H_2SO_4 with 1.0 wt% H_2O_2 at a liquid-to-solid ratio of 20 mL g^{-1} and temperature of 50 °C for 60 min resulted in the recovery of 100% lithium, 98.4% cobalt, 98.6% nickel, and 98.6% manganese. Moreover, a process mechanism of $H_2SO_4 + H_2O_2$ leaching of all four metals was proposed. Finally, the Co, Ni, and Mn co-precipitate and Li_2CO_3 precipitate were combined to regenerate a new cathode active material.

Keywords Spent lithium-ion batteries (LIBs) · Cathode active material · Recycling · Hydrometallurgy · Leaching · Kinetics · Lithium · Cobalt

Introduction

Lithium-ion batteries (LIBs) are utilized extensively in various industrial and consumer applications, such as in portable electronic devices and electric transportation [1] because of their high energy density, long storage life, low maintenance, small volume, and lightweight. A recent study estimated that approximately 120 million electric vehicles could be on the road by 2030 in China, the European Union, and the United States [2], and therefore the demands for LIBs for electric

K. H. Chan · M. Malik · J. Anawati · G. Azimi (✉)
Laboratory for Strategic Materials, Department of Chemical Engineering and Applied Chemistry, 200 College Street, Toronto, ON M5S 3E5, Canada
e-mail: g.azimi@utoronto.ca

G. Azimi
Department of Materials Science and Engineering, University of Toronto, 184 College Street, Toronto, ON M5S 3E4, Canada

G. Azimi et al. (eds.), *Rare Metal Technology 2020*, The Minerals, Metals & Materials Series, https://doi.org/10.1007/978-3-030-36758-9_3

vehicles are expected to grow rapidly in the future. Although existing lithium and cobalt reserves can meet such increased demand, more than 90% of the world reserves for lithium and cobalt are located in only a few countries such as China, Australia, and Chile [3], thus causing potential resource scarcity in other countries. In addition, spent LIBs contain many toxic heavy metals, such as Co, Ni, and Mn. Improper disposal of postconsumer LIBs into landfills may allow toxic heavy metals to seep into underground water and soil, resulting in severe environmental contamination. Therefore, considering the increasing demand for LIBs, the resource scarcity of lithium and cobalt, and significant environmental awareness, effective urban mining of postconsumer LIBs has become imperative.

Since the most expensive part of a LIB is the cathode material, traditional recycling methods have mainly focused on recovering valuable metals, such as Li and Co from the cathode material. The state-of-the-art techniques for recycling spent LIBs comprise pretreatment, metal extraction, and product separation [1]. The metal extraction process, in particular, mainly based on pyrometallurgy and/or hydrometallurgy [1], plays an important role in the efficiency of the recycling process. Hydrometallurgy is commonly employed for the recovery of valuable metals via acid leaching because of its high recovery rate, low energy consumption, and limited waste generation [4]. Common inorganic acid leachants that were used in recycling of spent LIBs include HNO_3 [5], H_2SO_4 [4, 6–8], and HCl [9–11], and reducing agents such as H_2O_2 [12, 13] can be added to accelerate the leaching process of metal ions.

Currently, the second generation ($LiNi_xCo_yMn_{1-x-y}O_2$) cathode material has become the most dominant cathode type in LIBs over the first generation ($LiCoO_2$) cathode material due to better performance in electric vehicles [1, 6]. Previous studies have mainly focused on recycling spent LIBs of mobile phones [4, 13, 14] and laptops [7], while there are only a few studies on recycling of spent LIBs from electric vehicles. In addition, most previous studies optimized the recycling process of the first generation cathode material by monitoring the influence of one factor at a time; however, optimization of the second generation cathode material by using multivariate statistic techniques has rarely been reported. Furthermore, relatively few researchers studied the leaching kinetics and mechanism for the $LiNi_xCo_yMn_{1-x-y}O_2$ chemistry.

In the current study, a closed-loop hydrometallurgical process was developed to recover valuable metals from an end-of-life LIB of an electric vehicle. The aim of this study is to find out the most effective leachant and the optimum operating conditions for LIB recycling utilizing systematic experimental and theoretical approaches based on fractional factorial design of experiment methodology and propose the reaction mechanism of leaching LIB cathode material.

Materials and Methods

Chemicals and Materials

The following reagents were employed in this work: concentrated nitric acid (ACS Reagent grade, 68.0–70.0 wt%, VWR), concentrated hydrochloric acid (ACS Reagent grade, 36.5–38.0 wt%, VWR), concentrated sulfuric acid (ACS Reagent grade, 95.0–98.0 wt%, VWR), concentrated sodium hydroxide (ACS Reagent grade, 50 wt%, VWR), hydrogen peroxide (ACS Reagent grade, 30 wt%, VWR), sodium carbonate (Na_2CO_3, ACS Reagent grade, 99.5%, BioShop), and deionized water (0.055 μS, Millipore).

Battery Pack Disassembly and Cathode Material Preparation

The battery pack was first discharged using an electric heater to heat water and then disassembled to cell level using electrical insulated gloves. The electrolyte injector port in the cell casing was opened to remove the electrolyte, and then anode and cathode were unfolded and detached from the separator inside a glove box. Cathode material was effectively peeled off from the aluminum foils using an ultrasonic bath (VWR 97043-964) with deionized water for 2–3 min at room temperature. The cathode active material solution was filtered and dried at 50 °C for 24 h. The dried cathode material was ground into fine powders by a mortar and pestle and mixed evenly with a vortex mixer.

Aqua Regia Digestion and ICP-OES Characterization

Seven independent 0.2 g of samples were digested in approximately 20 mL concentrated aqua regia (3 HCl:1 HNO_3) at 200 °C in a microwave digestion system (MARS6 Xpress). The digested samples were diluted to 50 mL total volume with deionized water, filtered with 0.45 μm nylon syringe filters (Basix^{TM}), and then diluted to the measurement concentration range (0.1–20 mg/L) with 5 wt% HNO_3. The average composition of the cathode material powder feed was measured by Inductively Coupled Plasma Optical Emission Spectrometry (ICP-OES, PerkinElmer Optima 8000) using the following wavelengths: Li 610.362 nm, Co 228.616 nm, Ni 231.604 nm, and Mn 257.610 nm.

Mineralogical and Morphological Analysis

The mineralogical characterization of the cathode material sample was obtained with X-ray diffraction (Rigaku MiniFlex 600). Morphological analysis of the sample was investigated using Scanning Electron Microscopy Energy Dispersive Spectroscopy (SEM-EDS, Hitachi SU5000).

Leaching Process

The leaching experiments were carried out in a water bath (Fisher Scientific ISOTEMP 4100 H21P) inside a fume hood. For all tests, 5 g of powdered cathode active material was added to a 500 mL Erlenmeyer flask containing a pre-heated acid solution. The flask was equipped with a rubber stopper to prevent water loss by evaporation and a magnetic stirrer (Fisher Scientific RT Basic Stirrer 120) to provide continuous agitation. Samples were taken at 60 min and 300 min for H_2SO_4 + H_2O_2 and HCl leaching tests, respectively, and then immediately filtered using 0.45 μm nylon syringe filters. The filtered samples were diluted with 5 wt% HNO_3 to determine the metal concentrations in the leachate by ICP-OES.

Design of Experiment and Empirical Model Building

The effect of the experimental parameters on the extraction of Li, Ni, Mn, and Co in both HCl and H_2SO_4 + H_2O_2 leaching systems were tested using fractional factorial design methodology (H_2SO_4 + H_2O_2: 2_{V}^{5-1}, HCl: 2_{IV}^{4-1}). The following operating parameters were investigated: leaching temperature (X_1), acid concentration (X_2), H_2O_2 concentration (X_3), liquid-to-solid ratio (X_4), and agitation rate (X_5).

The experimental extraction results were fit to an empirical model (Eq. 1 for HCl system and Eq. 2 for H_2SO_4 + H_2O_2 system) with primary factor effects (X_j) and second-order interaction effects (X_jX_k).

$$\hat{y}_i = \hat{\beta}_0 + \hat{\beta}_1 X_1 + \hat{\beta}_2 X_2 + \hat{\beta}_4 X_4 + \hat{\beta}_5 X_5 + \tilde{\beta}_{12} X_1 X_2 + \tilde{\beta}_{14} X_1 X_4 + \tilde{\beta}_{15} X_1 X_5 \quad (1)$$

$$\begin{aligned} \hat{y}_i = {} & \hat{\beta}_0 + \hat{\beta}_1 X_1 + \hat{\beta}_2 X_2 + \hat{\beta}_3 X_3 + \hat{\beta}_4 X_4 + \hat{\beta}_5 X_5 + \hat{\beta}_{12} X_1 X_2 + \hat{\beta}_{13} X_1 X_3 + \hat{\beta}_{14} X_1 X_4 \\ & + \hat{\beta}_{15} X_1 X_5 + \hat{\beta}_{23} X_2 X_3 + \hat{\beta}_{24} X_2 X_4 + \hat{\beta}_{25} X_2 X_5 + \hat{\beta}_{34} X_3 X_4 + \hat{\beta}_{35} X_3 X_5 + \hat{\beta}_{45} X_4 X_5 \end{aligned} \quad (2)$$

The model parameters for each of the empirical models ($\hat{\beta}_j$) were fit by multiple linear least squares regression (mLLSR, Eq. 3).

$$\hat{\beta} = \left(X^T X\right)^{-1}\left(X^T Y\right) \tag{3}$$

Precipitation

A pH curve was plotted by titration of $H_2SO_4 + H_2O_2$ leachate with 2 M NaOH solution. Co, Ni, and Mn were co-precipitated at pH above 11 by adding 2 M NaOH solution to the leachate. The solution was stirred for 1.5 h and filtered by vacuum funnel. The filter cake was washed with deionized water to remove the impurities and dried at 80 °C for 24 h to obtain Co, Ni, and Mn co-precipitate. The remaining filtrate containing lithium was concentrated by water evaporation, and then treated with saturated sodium carbonate (Na_2CO_3) solution at 95 °C to precipitate as Li_2CO_3.

Results

Characterization Results of Cathode Active Material

The chemical composition of the cathode material was analyzed by ICP-OES and mainly contained Co (35.8 wt%), Ni (8.77 wt%), Mn (8.11 wt%), and Li (5.79 wt%), in which 85 wt% was the cathode material solids, with the rest being amorphous carbon. Based on the ICP-OES results, the chemical formula of cathode material was determined as $LiNi_{0.15}Mn_{0.15}Co_{0.70}O_2$. The crystal structure of the cathode active material was analyzed by X-ray Diffraction (XRD). $Li_{0.58}Co_{0.333}Ni_{0.333}Mn_{0.333}O_2$, which is isomorphous with $LiNi_{0.15}Mn_{0.15}Co_{0.70}O_2$, was detected as a single crystal structure. Figure 1 shows the surface morphology and elemental mapping of the cathode active material characterized by SEM-EDS and it was found that the distribution of Co, Ni, and Mn was uniform in the oxide particles.

Extraction Results

It was observed that in both leaching systems, the extraction of Ni, Mn, and Co was almost the same, while the extraction of Li was higher than the other metals. These extraction results were used to construct empirical models to study extraction effects of the parameters.

Based on the overall extraction results, in the case of HCl leachant, only very extreme processing conditions (high temperature (75 °C), high concentration

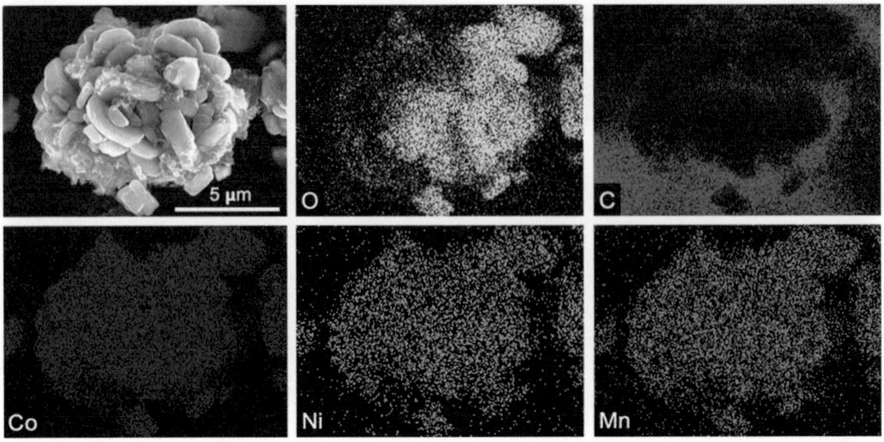

Fig. 1 SEM-EDS characterization of cathode active material powder

(2.0 M HCl), high liquid-to-solid ratio (30 mL g^{-1})) resulted in high leaching efficiencies (y_{Li} = 105.3%, y_{Ni} = 99.3%, y_{Mn} = 97.2%, y_{Co} = 98.4%). In the contrary, the optimal operating conditions in the H_2SO_4 + H_2O_2 system could occur at relatively mild conditions (lower temperature (50 °C), lower concentration (1.5 M H_2SO_4 and 1.0 wt% H_2O_2), lower liquid-to-solid ratio (20 mL g^{-1})), in which 100% extraction efficiencies for all elements could also be achieved (y_{Li} = 100.5%, y_{Ni} = 98.6%, y_{Mn} = 98.6%, y_{Co} = 98.4%). These moderate operating conditions maximize extraction efficiencies while minimizing operating costs compared with leaching at more extreme conditions, which suggests that H_2SO_4 + H_2O_2 is the most effective leachant.

Extraction Parameter Effects

Empirical models in the form of Eqs. 1 and 2 were constructed to study the extraction effects of the parameters (Fig. 2). Experimental results from the fractional factorial design determined that in both leaching systems, agitation rate (X_5) had a negligible effect on extraction, except it had a minor effect on Ni, Mn, and Co extraction in the HCl system.

In the HCl system, all studied primary parameters and most of their interaction parameters have strong positive impacts on extraction. In the H_2SO_4 + H_2O_2 system, the main contributor to the extraction was the concentration of H_2O_2, while other test parameters and interaction parameters were almost insignificant for the extraction.

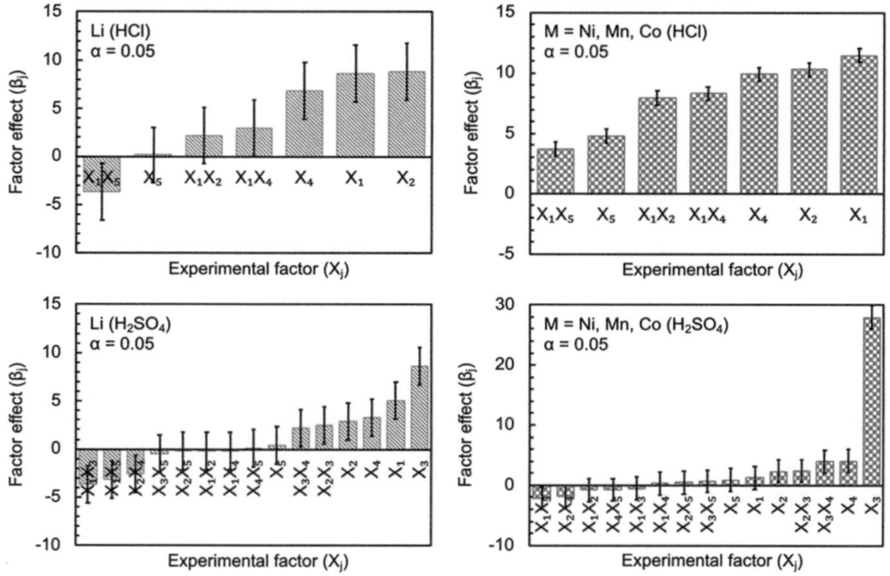

Fig. 2 Ordered factor effect charts for the empirical extraction models. The error bars represent the estimated 95% confidence interval for each factor

Proposed Extraction Mechanism

The $LiNi_xCo_yMn_{1-x-y}O_2$ cathode material has a layered crystal structure with alternating outer Li sheet and inner MO_2 (M = Ni, Mn, Co) sheet. Based on the crystal structure and distinct experimental leaching results, a two-step reaction mechanism for the acid leaching of LIB cathode material was proposed (Fig. 3):

1. Acid reacts with outer Li sheet and breaks down Li–O bonds first, resulting in an aqueous solution of Li^+.
2. Acid then reacts with inner MO_2 sheet and breaks down M–O bonds. Co and Mn are reduced to Co(II) and Mn(II), which become more soluble, by the reducing agent H_2O_2. (Ni is already in its soluble (II) oxidation state).

In reaction 1, 4 mol of HCl are required to dissolve 1 mol of cathode material, producing metal chlorides, water, and Cl_2 gas. In reaction 2, 1.5 mol of H_2SO_4 and 0.5 mol of H_2O_2 are required to dissolve 1 mol of cathode material, producing metal sulfates, water, and O_2 gas.

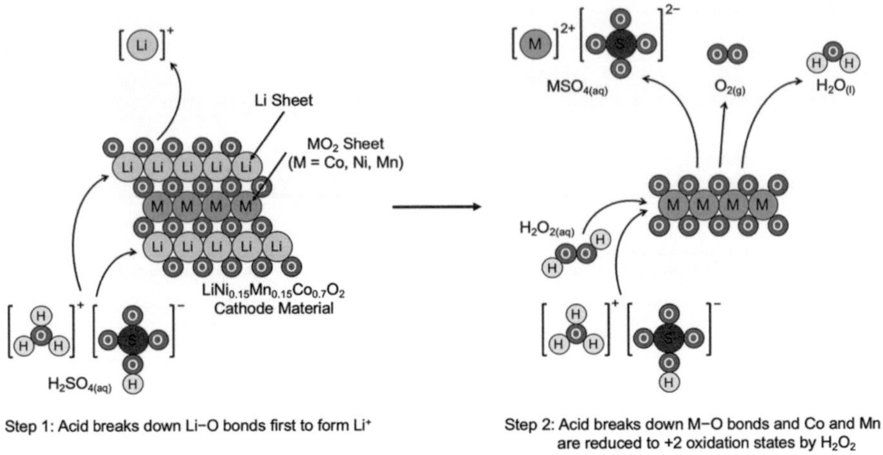

Fig. 3 Illustration of the proposed two-step leaching reaction mechanism

$$2LiNi_{0.15}Mn_{0.15}Co_{0.70}O_{2(s)} + 8HCl_{(aq)} \rightarrow 2LiCl_{(aq)} + 0.3NiCl_{2(aq)} + 0.3MnCl_{2(aq)} + 1.4CoCl_{2(aq)} + 4H_2O_{(l)} + Cl_{2(g)} \quad (1)$$

$$2LiNi_{0.15}Mn_{0.15}Co_{0.70}O_{2(s)} + 3H_2SO_{4(aq)} + H_2O_{2(aq)} \rightarrow Li_2SO_{4(aq)} + 0.3NiSO_{4(aq)} + 0.3MnSO_{4(aq)} + 1.4CoSO_{4(aq)} + 4H_2O_{(l)} + O_{2(g)} \quad (2)$$

Extracted Product Precipitation and Recovery

A pH curve was plotted by titration of H_2SO_4 + H_2O_2 leachate with 2 M NaOH solution to study pH-precipitation phenomenon (Fig. 4). As can be seen, Ni, Co, and Mn were precipitated at pH values of 8, 9, and 10, respectively, while precipitation of lithium was not affected at various pHs.

Since it is difficult to separate Co, Ni, and Mn from one another by pH precipitation due to their similar properties, 100% of Co, Ni, and Mn were co-precipitated by adjusting the pH to above 11, which can be re-introduced into the battery production without separation to synthesize the new cathode active material [15]. The leachate was then concentrated by evaporating water and reacted with saturated Na_2CO_3 solution at 95 °C to form Li_2CO_3. To synthesize the new cathode active material, co-precipitate and Li_2CO_3 were mixed in a puck mill, pressed into pellets and then sintered in a furnace.

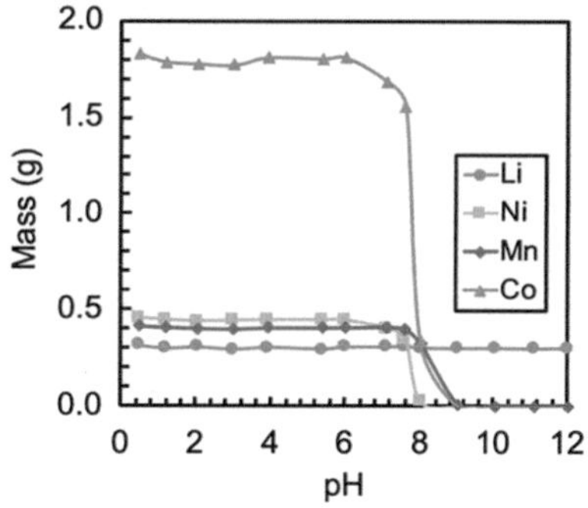

Fig. 4 pH curve by titration of H_2SO_4 + H_2O_2 leachate with 2 M NaOH solution

Conclusions

This work focuses on developing and optimizing a closed-loop hydrometallurgical process to recycle Li, Co, Ni, and Mn from an end-of-life LIB of an electric vehicle. On the basis of design of experimental matrix utilizing a fractional factorial design methodology, H_2SO_4 + H_2O_2 was determined as the most effective leachant because leaching at lower temperature and shorter leaching time with lower reagent consumption could lead to high leaching efficiency of the four metals. H_2SO_4 + H_2O_2 was also considered as a green leachant as H_2O and O_2 were the only by-products in the leaching process. In this case, the highest performing leaching conditions were achieved with 60 min of leaching reaction using 1.5 M H_2SO_4 and 1.0 wt% H_2O_2 at a liquid-to-solid ratio of 20 mL g^{-1} and temperature of 50 °C, resulting in the recovery of 100% lithium, 98.4% cobalt, 98.6% nickel, and 98.6% manganese. Future study of this work will further investigate the optimum operating conditions of both leachants by using response surface methodology and process economics and also the stoichiometric effect of H_2O_2. Finally, the co-precipitate was combined with Li_2CO_3 to regenerate a new cathode material, which provides a closed-loop recycling process for spent LIBs. This simple, green, and economic recycling process not only offers an opportunity to recycle valuable metals from postconsumer LIBs, but also reduces the accumulation of LIB wastes in the landfills.

References

1. Zheng X, Zhu Z, Lin X et al (2018) A mini-review on metal recycling from spent lithium ion batteries. Engineering 4:361–370. https://doi.org/10.1016/J.ENG.2018.05.018
2. Hauke Engel, Russell Hensley, Stefan Knupfer SS (2018) Charging ahead: electric-vehicle infrastructure demand. McKinsey Cent Futur Mobility, McKinsey Co, pp 1–8
3. U.S. Geological Survey (2019) Mineral Commodity Summaries

4. Jha MK, Kumari A, Jha AK et al (2013) Recovery of lithium and cobalt from waste lithium ion batteries of mobile phone. Waste Manag 33:1890–1897. https://doi.org/10.1016/J.WASMAN.2013.05.008
5. Lee CK, Rhee K-I (2002) Preparation of $LiCoO_2$ from spent lithium-ion batteries. J Power Sources 109:17–21. https://doi.org/10.1016/S0378-7753(02)00037-X
6. Gratz E, Sa Q, Apelian D, Wang Y (2014) A closed loop process for recycling spent lithium ion batteries. J Power Sources 262:255–262. https://doi.org/10.1016/J.JPOWSOUR.2014.03.126
7. Meshram P, Pandey BD, Mankhand TR (2015) Hydrometallurgical processing of spent lithium ion batteries (LIBs) in the presence of a reducing agent with emphasis on kinetics of leaching. Chem Eng J 281:418–427. https://doi.org/10.1016/J.CEJ.2015.06.071
8. Swain B, Jeong J, Lee J et al (2007) Hydrometallurgical process for recovery of cobalt from waste cathodic active material generated during manufacturing of lithium ion batteries 167:536–544. https://doi.org/10.1016/j.jpowsour.2007.02.046
9. Huang Y, Han G, Liu J et al (2016) A stepwise recovery of metals from hybrid cathodes of spent Li-ion batteries with leaching-flotation-precipitation process. J Power Sources 325:555–564. https://doi.org/10.1016/J.JPOWSOUR.2016.06.072
10. Barik SP, Prabaharan G, Kumar L (2017) Leaching and separation of Co and Mn from electrode materials of spent lithium-ion batteries using hydrochloric acid: Laboratory and pilot scale study. J Clean Prod 147:37–43. https://doi.org/10.1016/J.JCLEPRO.2017.01.095
11. Zhang P, Yokoyama T, Itabashi O (1998) Hydrometallurgical process for recovery of metal values from spent lithium-ion secondary batteries. 47:259–271
12. Lv W, Wang Z, Cao H et al (2018) A critical review and analysis on the recycling of spent lithium-ion batteries. ACS Sustain Chem Eng 6:1504–1521. https://doi.org/10.1021/acssuschemeng.7b03811
13. Pinna EG, Ruiz MC, Ojeda MW, Rodriguez MH (2017) Cathodes of spent Li-ion batteries: dissolution with phosphoric acid and recovery of lithium and cobalt from leach liquors. Hydrometallurgy 167:66–71. https://doi.org/10.1016/J.HYDROMET.2016.10.024
14. Chen X, Ma H, Luo C, Zhou T (2017) Recovery of valuable metals from waste cathode materials of spent lithium-ion batteries using mild phosphoric acid. J Hazard Mater 326:77–86. https://doi.org/10.1016/J.JHAZMAT.2016.12.021
15. Zou H, Gratz E, Apelian D, Wang Y (2013) Green chemistry. Green Chem 15:1183–1191. https://doi.org/10.1039/c3gc40182k

Optimal Hydrometallurgical Extraction Conditions for Lithium Extraction from a Nigerian Polylithionite Ore for Industrial Application

Kehinde Israel Omoniyi, Peter Ikyernum Agaku and Alafara Abdullahi Baba

Abstract The endowment of Nigeria with solid mineral resources has warranted the present call for economic diversification from petroleum exploration. There is strong demand for lithium and industrial lithium compounds in a wide array of applications such as in the health sector, among others. This study reports the extraction of lithium from polylithionite ore obtained from Keffi, Nigeria in chloride media. The effects of experimental conditions, such as roasting temperature and time, mix ratio, and calcine-to-liquid ratio, were investigated using the Li ore assayed 3.25 wt% Li. The best ratio of polylithionite:NaCl:$CaCl_2$ was 1:1:1 at 900° C and 5 min roasting. Acidic leaching of the residual lithium with defined conditions leached lithium with 83.82% efficiency. Beneficiation of lithium-leach-liquor for industrial value addition shall be reported in due course.

Keywords Polylithionite ore · Roasting · Acidic leaching · Lithium · Extraction

Introduction

The mining of minerals in Nigeria accounts for only 0.3% of its GDP, due to the influence of its vast oil resources. Nigeria currently has over 44 known types of minerals of varying mixes and proven quantities. Diversification of the economy into solid mineral exploration among others is imperative. Lithium (Li) metal, like Tantalum among others, is currently experiencing a global boom as a result of broad spectrum applications. Li is the lightest metal in nature, (^{7}Li) has unique electrochemical properties and the highest specific heat of solid elements. Li and its compounds are used in batteries, glass, lubrication, ceramics, and pharmaceutical industries [1, 2]. Its journey as the "metal of the future" truly began in the late

K. I. Omoniyi (✉) · P. I. Agaku
Department of Chemistry, Ahmadu Bello University, Zaria, Nigeria
e-mail: isrealflourish@yahoo.com; kiomoniyi@abu.edu.ng

A. A. Baba
Department of Industrial Chemistry, University of Ilorin, Ilorin, Nigeria

G. Azimi et al. (eds.), *Rare Metal Technology 2020*, The Minerals, Metals & Materials Series, https://doi.org/10.1007/978-3-030-36758-9_4

1990s where it became greatly popular in portable electronic devices [3]. The consumption of lithium and its related products globally has tremendously increased by more than 20% per year from the past decades.

Li extraction from Li-bearing mineral/pegmatite sources such as lepidolite, spodumene, petalite, zinnwaldite tosuplimate, and natural brine sources of Li for Li battery production has been reported [4]. Keffi pegmatite located in North Central Nigeria is a rare metal type and consists of lepidolite and polylithionite sub-types [5–8].

Polylithionite, chemically known as potassium lithium aluminum silicate fluoride ($KLi_2AlSi_4O_{10}(F,OH)_2$), is named due to allusion to its high Li content. It is a mica group of minerals (though not well known) and is found mostly in pegmatites with other rare minerals [9].

Various extraction techniques including the use of lime and sulphuric acid techniques have been developed to extract lithium from lithium-bearing ores/minerals. However, extraction by sulphuric acid technique involves the use of highly concentrated acid and a complex purification procedure. The lime process ($CaCO_3$), on the other hand, uses too many limestones and a high amount of energy [10, 11]. The drawbacks of the above processes, therefore, limit their further applications. Hydrometallurgical extraction of Li using alkali/and or alkaline chloride salt(s) (NaCl, $CaCl_2$) as reaction additives is, therefore, emerging as an alternative technique, which has proven to be cheaper and efficient for refining precious, base, and refractory metals like Li [12]. Lithium extraction from ores utilizes thermal roasting followed by leaching. If the leaching agent is water (aqueous chemistry), then it is considered hydrometallurgical process.

The study aims to establish the optimal operational conditions for a new hydrometallurgical extraction process of Li from polylithionite under the experimental conditions: roasting temperature and time, mass ratio, leaching temperature/time, and calcine-to-liquid ratio. Maximizing industrial output in solid mineral exploration is paramount in bridging the deficit in increased global demand for metals.

Materials and Methods

Sample Collection and Preparation

The lithium rich polylithionite ore was collected from Keffi pegmatite field, North Central Nigeria in March 2019. The polylithionite ore was first upgraded by crushing and optical sorting treatment. The crushing was carried out by using jaw crusher with open setting at 100 mm of width and 10 mm of gap. The upgraded polylithionite (UP) was subjected to grinding via a ball mill at 60 rmp and sieved with a #200 siever to 100% aperture size of 75 μm for a complete pass, and then assayed using X-ray diffraction (XRD) (Xpert-Pro XRD coupled with PDF-2 software) and chemical composition determined by X-ray fluorescence (XRF) (Oxford Instrument X-supreme 800, UK). Alkali/and or alkaline chloride salts (NaCl, $CaCl_2$) was/were used for roasting for increasing the Li extraction. The leaching agent was water filtered with Whatman filter paper (41).

Roasting of the Ore

Roasting of the ore was carried out according to the method of Lucia et al. [13] in order to obtain water soluble calcine for efficient Li extraction. A specific weighed mass ratio of UP was first mixed with the specific mass ratio(s) of NaCl and/or $CaCl_2$. The mixture was then poured into a graphite crucible and heated in an electrical tubular furnace (Nabertherm programmable furnace, model GmbH, Netherland) at 600–1000 °C. After the specified time, the resulting calcine was removed and cooled to ambient temperature.

Water Leaching of the Roasted Calcine (According to Qunxuan et al. [14])

The roasted calcine sample was leached with water at 60 °C for 0.5 h time, using solid calcine-to-liquid water mass ratios of 1:2.5. A mixture of liquid and solid phases was formed upon standing.

Separation of Liquid–Solid Mixture and Residue Digestion

Filtration was done and the residue was washed with distilled water several times. The leach residue was dried in an oven at 120 °C for 4 h [13].

Chemical Digestion of Residue

The residue was weighed and dissolved with 70% (HF) acid in the ratio 1:2, and the mixture was heated to near dryness and then 100 cm^3 of 1% HCl was added [14]

Determination of Leached Lithium Contents

For each experimental set, Flame Atomic Emission Spectroscopy (FAES) (FAES 280FS, USA) was used to determine the contents of Li in the filtrate and acid digest. XRF was employed to determine the content of other metal elements in the leached liquor.

Effect of Roasting Temperature, Time, and Mass Ratio of Upgraded Ore to $CaCl_2$ or/and NaCl on Lithium Extraction Efficiency

The UP ore was mixed with NaCl or/and $CalCl_2$ at various mass ratios and then roasted for 30 min to form the solid calcine soluble state. The effect of roasting temperature on Li extraction was considered from 600 to 1000 °C. Roasting time of 5–90 min was considered for the extraction. Series of mass ratios of polylithionite ore to NaCl to $CaCl_2$ was considered. The optimized condition was then used for determining the lithium extraction efficiency from the ore.

Leachability of Lithium

The content of Li in both the filtrate and leached residue after chemical digestion was determined by FEAS. The leachability of Li in both cases is expressed in percentage as follows:

$$L(\%) = \frac{M_f Li_2O}{M_f Li_2O + M_r Li_2O} \times 100 \tag{1}$$

$$R(\%) = \frac{M_r Li_2O}{M_r Li_2O + M_f Li_2O} \times 100 \tag{2}$$

$$V_{Li}(\%) = 100 - (L + R) \tag{3}$$

where:
L = Lithium extraction efficiency by water (filtrate),
R = Residual ratio of lithium in leached residue,
$M_f Li_2O$ = Mass concentration of lithium in the filtrate,
$M_r Li_2O$ = Mass concentration of lithium in the leached residue after chemical digestion, and
V_{Li} = Volatilization ratio of lithium during roasting/calcination.

Results and Discussion

Chemical Analysis of the Ore

From Table 1, the order of the composition (%) of oxides in the polylithionite ore is $SiO_2 > Al_2O_3 > K_2O > Cs_2O > Li_2O > Na_2O > F > MnO$. The Li content (3.25%) in the ore is greater than the contents reported in many other Li-bearing minerals (lepidolite, spodumene, petellite) and is within the Li content of the global Li-bearing minerals (1.55–7.03% w/w). The result of the bulk chemical composition obtained agrees with the findings of Akinola et al. for pegmatitic minerals. It was stated that the Li contents of these ores may vary greatly as their grades largely depend on the composition and degree of fractionation of the host pegmatitic fluid [5].

Mineralogical Analysis of the Ore

The mineralogical composition of the first upgraded raw ore (UP) by XRD depicted in Fig. 1 indicated that the ore is mainly polylithionite; $KLi_2\ AlSi_4O_{10}\ (F,OH)_2$. This result, on one hand, is not in close agreement with what was obtained by most previous researchers whose XRD patterns of the first UP ore were

Table 1 Bulk chemical composition analysis (%) of the first upgraded raw polylithionite ore analysed by XRF and FEAS

Oxide composition	Value (%) w/w
SiO_2	58.49
TiO_2	<0.001
Al_2O_3	14.52
F	1.78
Fe_2O_3	0.439
CaO	0.06
MgO	0.01
Na_2O	3.11
K_2O	10.68
MnO	1.55
V_2O_5	<0.001
Cr_2O_3	0.045
CuO	0.082
Li_2O	3.25
ZnO	0.254
Cs_2O	4.14
Nb_2O_5	0.80
Ta_2O_5	0.29
L.O.1	0.50

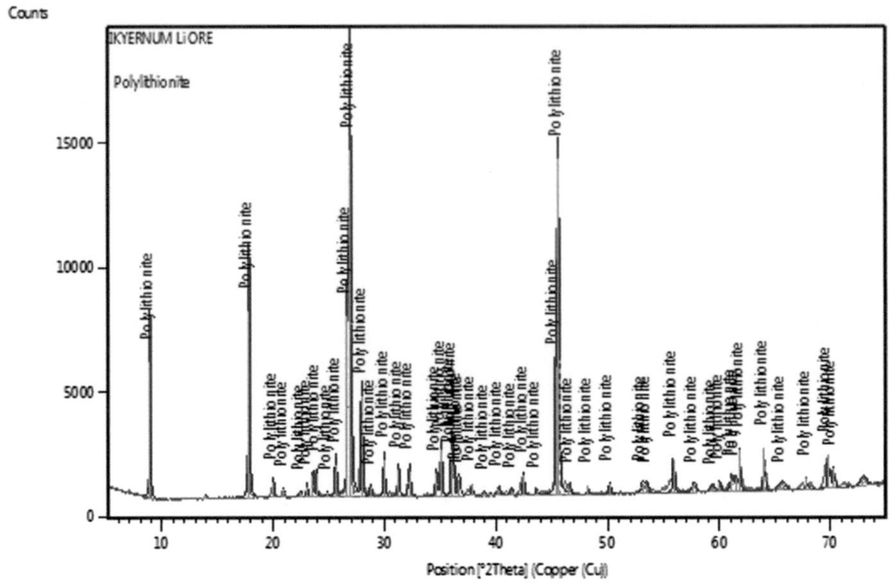

Fig. 1 X-ray diffraction (XRD) pattern of the first upgraded raw ore

poly-mineralogical [14, 15]. This difference may be attributed to the differences in environmental geochemistry. On the other hand, the result obtained herein is in strong association with what was obtained by Lucia et al. in which the XRD pattern was mono-mineralogical [13]. This may be attributed to similar environmental geochemistry but different compositions and degrees of fractionation of the host pegmatite.

Effect of Roasting Temperature on Lithium Extraction Efficiency

As presented in Table 2, Li extraction efficiency increases significantly from the roasting temperature 600–900 °C. Above 900 °C, the sample begins to fuse, so Li extraction decreased with increased temperature due to loss of Li (LiCl) by

Table 2 Effect of roasting temperature on Li extraction efficiency based m (UP ore):m ($CaCl_2$) = 1:1; 30 min roasting time

Roasting temperature (°C)	L (%)	R (%)	V_{Li} (%)
600	19.87	80.13	0
700	28.47	71.53	0
800	35.33	64.67	0
900	71.41	26.21	2.38
1000	57.98	29.17	12.85

volatilization. So, further experiments were carried out at 900 °C, the optimum roasting temperature with Li extraction efficiency of 71.41% and low Li volatilization of 2.38% during roasting. The result is in concomitance with the findings obtained by Qunxuan et al. [14].

Effect of Roasting Time on Lithium Extraction Efficiency

We use 5–90 min of roasting time and 1:1:1 mass ratio of UP ore to NaCl to $CaCl_2$ at the optimum temperature of 900 °C. Table 3 indicated that Li extraction efficiencies generally increased with roasting time investigated. However, Li extraction efficiency is more favorable at 30 min roasting time (71.41%) due to the low volatilization of LiCl with value of 2.38% compared to high volatilization of LiCl obtained at 60–90 min with values of 10.88–11.14%. The result obtained herein is similar to that obtained by Lucia et al. for lepidolite ore [13].

Effect of Mass Ratio of Upgraded Polylithionite Ore to $CaCl_2$ or NaCl on Lithium Extraction Efficiency

The use of different mass ratios of polylithionite ore to $CaCl_2$ or NaCl at the optimum conditions (900 °C and 30 min roasting time) on Li extraction efficiency (Table 4) indicated that 2:1 of polylithionite ore:$CaCl_2$ had Li extraction efficiency of 57.51%. Polylithionite ore:$CaCl_2$ at 1:1 had the Li extraction efficiency being 71:41%. Similarly, polylithionite ore:NaCl at mass ratio of 1:1 gave Li extraction efficiency of 61.98%. For polylithionite ore:NaCl at 1:2 mass ratio, the Li extraction efficiency obtained was 72.49%. The study, therefore, infers that $CaCl_2$ additive is more efficient than NaCl for extracting Li from polylithionite. This finding is corroborated with the report of Barbosa and co-workers in 2014 on Li extraction from spodumene and lepidolite [16].

Table 3 Effect of roasting time on Li extraction efficiency on UP ore:$CaCl_2$:(NaCl) = 1:1:1; 900 °C optimized roasting temperature

Roasting time (min)	L (%)	R (%)	V_{Li} (%)
5	16.18	83.82	0
15	44.62	55.38	0
30	71.41	26.21	2.38
60	75.32	13.80	10.88
90	76.12	12.74	11.14

Table 4 Effect of mass ratio of polylithionite ore to chloride salt additives on Li extraction efficiency based on m(Up ore):m($CaCl_2$) = 1:1, 2:1; m(p-ore): m(NaCl) = 1:1–1:2; at 900 °C, 30 min

Chloride salt additive	Mass ratio of polylithionite ore to chloride salt additive	Lithium extraction efficiency%
$CaCl_2$	2:1	57.51
	1:1	71.41
NaCl	1:2	72.49
	1:1	61.98

Effect of Fractional Mixed Mass Ratio of NaCl and $CaCl_2$ to Upgraded Polylithionite Ore on Lithium Extraction Efficiency

The result from using various mass ratios of $CaCl_2$ and NaCl roasted with a fixed mass ratio (1 g) of UP ore at 900 °C and 30 min roasting time presented in Table 5 shows that Li extraction efficiencies significantly increase with increase in the proportion of NaCl reaching the peak with 88.84% extraction efficiency with 1:0.6:0.4 mass ratio of UP ore/NaCl/$CaCl_2$. This is because high proportion of NaCl acts as a partial substitute of $CaCl_2$ augmenting high Li extraction. In addition, $CaCl_2$ has more extraction strength than NaCl, and hence it is required in less proportion though close in proportion to NaCl. Hence, the mixed mass ratio of 1:0.6:0.4 (UP ore:NaCl:$CaCl_2$) gave an optimum Li extraction efficiency of 88.84% and is recommended as the best mix ratio of chlorinating additives for Li production from the ore.

The stoichiometric reaction for the roasting of UP ore with NaCl and/or $CaCl_2$ for the Li extraction may be proposed, thus (M = Li, K)

$$CaCl_{2(s)} + M_2O_{(s)} + Al_2O_{3(s)} + 2SiO_{2(s)} \xrightarrow{heat} 2MCl_{(s)} + CaAlSi_2O_{8(s)} \tag{4}$$

$$2NaCl_{(s)} + M_2O_{(s)} + Al_2O_{3(s)} + 6SiO_{2(s)} \xrightarrow{heat} 2MCl_{(s)} + NaAlSi_3O_{8(s)} \tag{5}$$

Table 5 Effect of fractional mixed mass ratio of NaCl and $CaCl_2$ to polylithionite ore (P-ore) on lithium extraction efficiency based on optimized 900 °C roasting temperature; 30 min roasting time

Ore:NaCl:$CaCl_2$	Lithium extraction efficiency (%)
1:0.1:0.9	73.82
1:0.2:0.8	75.34
1:0.3:0.7	79.14
1:0.4:0.6	84.21
1:0.5:0.5	86.98
1:0.6:0.4	88.84
1:0.7:0.3	71.72
1:0.8:0.2	63.31

The chemical stoichiometric ratio of Al_2O_3 and Si_2O in Eqs. (4–5) proved that $CaCl_2$ is more efficient than NaCl in the decomposition of the ore and subsequent Li extraction even in its less proportion. However, roasting the UP ore with a chloride salt mixture (NaCl and $CaCl_2$) gives a better Li extraction yield than either of the single chloride salts. This is because the melting point of NaCl and $CaCl_2$ mixture is lower than that of any of the chloride salts alone, which increases the fluidity of the chloride melt. This then allows rapid diffusion of the chloride salt to the surface of the polylithionite facilitating selective Li extraction. This assertion is similar to the report on Li extraction from spodumene and lepidolite, in which Li extraction efficiency was greatly enhanced when the roasting additives were mixed and induced activated Li extraction activity [14].

Effect of Roasting Time and Temperature on the Optimized Fractional Mixed Mass Ratio of $CaCl_2$ and NaCl to Polylithionite Ore

Using the optimized fractional mix mass ratio 1:0.6:0.4 for UP ore:NaCl:$CaCl_2$ with various roasting times and temperatures, Fig. 2 shows that Li extraction efficiency rapidly increased following increase in roasting time from 15 to 30 min; and decreased appreciably from 35 to 45 min. Li extraction efficiency also increased with increase in roasting temperature from 800 to 880 °C and then declined. The maximum extraction efficiency was 89.90% at 880 °C and 30 min roasting temperature and time, respectively. So, 880 °C and 30 min are the optimized roasting temperature and time for the fractional mixed mass ratio 1:0.6:0.4 UP ore, NaCl,

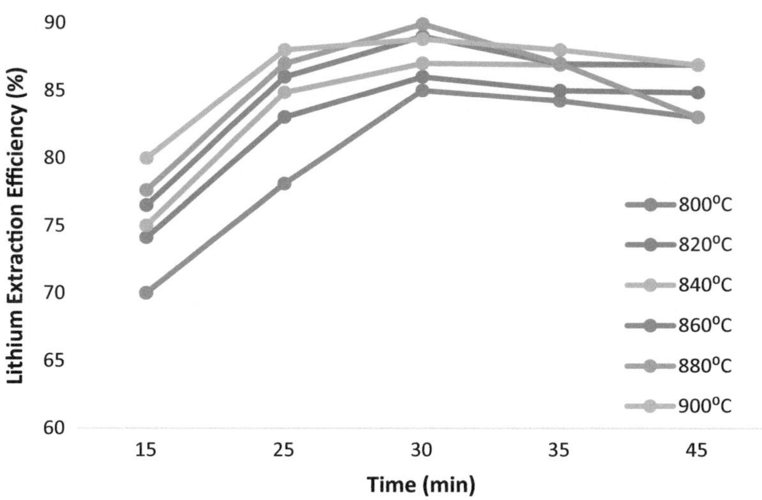

Fig. 2 Effect of roasting time and temperature on lithium extraction efficiency based on optimized fractional mixed mass ratio (p-ore):(NaCl):($CaCl_2$) = 1:0.6:0.4

and $CaCl_2$. This outcome closely aligns with the findings of other studies, in which Li extraction efficiency is significantly favoured with an increase in roasting temperature and time, respectively. However, the increase in both roasting reaction temperature and time has a limit of maximum extraction beyond which the Li extraction efficiency has no corresponding increase [17, 18].

Effect of Leaching Temperature and Time on Lithium Extraction Efficiency

The result for the effect of leaching reaction temperature and time on Li extraction efficiency performed using the optimized fractional mix mass ratio of 1:0.6:0.4 UP ore to NaCl to $CaCl_2$, at roasting temperature 880 °C, time 30 min, and solid-to-liquid water ratio 1:2.5, respectively, in Fig. 3 indicated that the leaching reaction temperature is 30 °C. The Li extraction efficiency increases progressively from 10 to 50 min leaching time, though it did not reach the maximum Li extraction efficiency.

When the leaching reaction temperature was increased further to 60 and 90 °C, the Li extraction efficiency increased remarkably at 10–30 min and reached an optimal value of 89.90% at 30 min. Further increase in leaching reaction time from 30 to 50 min led to a steady extraction efficiency of 89.90%. Although the Li extraction efficiencies at 60 and 90 °C increased simultaneous and reached the peak value (89.90%) at 30 min leaching time, the Li extraction efficiency at 60 °C leaching temperature increases more rapidly than that at 90 °C before maintaining a steady peak value of 89.90% beyond that at 30 min leaching time. Therefore, 60 °C

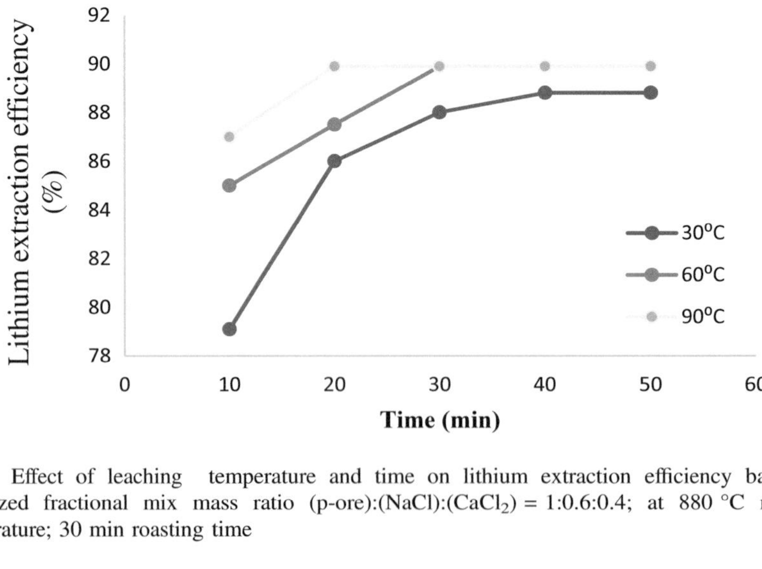

Fig. 3 Effect of leaching temperature and time on lithium extraction efficiency based on optimized fractional mix mass ratio (p-ore):(NaCl):($CaCl_2$) = 1:0.6:0.4; at 880 °C roasting temperature; 30 min roasting time

and 30 min leaching temperature and time are considered the best optimal leaching temperature and time that favour optimal Li production. The result of the leaching temperature and time obtained in this study is similar to that obtained by Yan and co-workers in 2012 for lepidolite, petalite, and spodumene. Li extraction efficiency is a function of leaching temperature and time; however, the longer the leaching time, the lower the leaching temperature, and thus the lower is the Li extraction efficiency [16].

Effect of Solid Calcine-to-Liquid Water (S/L) Ratio on the Leaching Process and Lithium Extraction Efficiency

The result of solid calcine-to-liquid water ratio on Li extraction efficiency was investigated from 1.0 to 5.0% (w/v). The leaching conditions were as follows: pre-optimized fractional mixed mass ratio 1:0.6:0.4 UP ore, NaCl, $CaCl_2$, roasting temperature 880 °C, roasting time 30 min, leaching temperature 60 °C, and leaching time 30 min, respectively. The result shows that the S/L ratios of 1:1.0, 1:1.3, and 1:1.6, respectively, resulted to slight increase in the Li extraction efficiency of values 85.0%, 87.0%, and 88.0%, respectively. When the S/L ratios were 1:2.5 and 1:5.0, respectively, Li extraction efficiencies remarkably increased and reached the maximum values of 89.84% and 89.90%, respectively. Therefore, the S/L ratio 1:2.5 is the most efficient, favorable, and economical condition for industrial Li production. This is in agreement with the work of Thi et al. in which S/L mass ratio affects Li extraction [19].

Plausible Route to Recover Li from the Leached Solution

The recovery of Li from the solution can be achieved by addition of (NaOH) to adjust the pH of the solution between 13 and 14, and then precipitate and remove silicate and aluminate by filtration. Then, Na_2CO_3 solution would be added to the resulting filtrate to lower the pH down to 6.5 and the resulting solution would be concentrated by heating to precipitate Li_2CO_3.

Conclusion

An improved optimization hydrometallurgical process with alkali and/or alkaline chloride salts (NaCl, $CaCl_2$) roasting followed by water leaching was used to extract lithium from polylithionite ore. The results demonstrated that roasting temperature, roasting time, leaching temperature, leaching time, and mass ratio of chloride salt additives (NaCl and $CaCl_2$) affected the lithium extraction efficiency significantly. The optimum operating process parameters with lithium extraction efficiency of 89.90% were established as follows: fractional mixed mass ratio of polylithionite ore/NaCl/$CaCl_2$ of 1:0.6:0.4, roasting temperature of 880 °C, roasting time of 30 min, leaching temperature of 60 °C, and leaching time of 30 min, and solid calcine-to-liquid (S/L) mass ratio of 1:25, respectively.

References

1. Moon KS, Fuerstenau DW (2003) Surface crystal chemistry in selective flotation of spodumene from other aluminosilicates. Int J Miner Process 72(1–4):11–24
2. Ebensperger A, Maxwell P, Moscoso C (2005) The lithium industry: its recent evolution and future prospects. Resour Policy 30(3):218–231
3. Ferro R, Pipoli R (2010). The future of lithium. Mining intelligence series, p 134
4. Mohr SH, Mudd GM, Giurco D (2012) Lithium resources and production: a critical global assessment and global projections. Mineral 2:65–84
5. Akinola OO, Okunlola OA, Obasi RA (2014) Physico-chemical characteristics and industrial potentials of Lepidolite from Ijero—Aramoko pegmatite field, South Western Nigeria. Intern J Scientific Tech Res 3(3):278–283
6. Okunlola OA (2005) Metallogeny of Ta–Nb mineralization of Precambrian pegmatite of Nigeria. Mineral Wealth 137
7. Okunlola OA, Akinola OO (2010) Petrochemical characteristics of the Precambrian rare metal pegmatite of Oke-Asa area, South Western Nigeria: implication for Ta–Nb mineralization. RMZ Mater Geoenviron 57(4):525–538
8. Tanko IY, Adam M, Dambring PD (2015) Field features and mode of emplacement of pegmatites of Keffi area, North-Central, Nigeria. Int J Sci Tech Res 4:215–216
9. www.galleries.com.polylithionite. Polylithionite: potassium lithium aluminum silicate fluoride. Accessed 6 May 2019
10. Brandt F, Haus R (2010) New concept of Lithium minerals processing. Mineral Eng. 23:659–661
11. Chen Y, Tian Q, Chen B, Shi X, Liao T (2011) Preparation of lithium carbonate from spodumene by a sodium carbonate autoclave process. Hydrometallurgy 109:43–46
12. Brocchi EA, Moura FJ (2008) Chlorination methods applied to recovery of refractory metals from tin slags. J Minerals Eng 21(2):150–156
13. Lucia IB, Jorge AG, Mariadel CR (2015) Extraction of lithium from B-spodumene using chlorination roasting with in $CaCl_2$. Mineral Eng 19:53–66
14. Qunxuan Y, Xinhai L, Zhixing W, Xifei W, Jiexi W, Huajun G, Qiyang H, Wenjie P (2012) Extraction of lithium from lepidolite by sulphation roasting and water leaching. Int J Miner Process 23:110–111
15. Gustavo DR, Mariadel CR, Mario HR (2014) Novel process for the extraction of lithium from β-spodumene by leaching with HF. Hydrometallurgy 1–6:147–148
16. Yan Q, Li X, Wang Z, Wang JX, Guo H, Hu Q, Peng W, Wu XF (2012) Extraction of Lithium from lepidolote using chlorination roasting—water leaching process. Trans Nonferrous Metals Soc China 22(7):1753–1759
17. Chan CY, Kirks DW (1999) Behaviour of metals under the conditions of roasting incinerator for flash ash and chlorinating agent. J Hazardous Mat B64:75–89

18. Dunn WE, Van JJ (2005) Cyclical vacuum chlorination process, including lithium extraction. https://patentimages.storage.googleapis.com/23/9b/77/7da27e93eca812/US20050220691A1.pdf. Accessed May 2019
19. Thi Thu H, Van Tri L, RetoGiere TT (2015) Extraction of lithium from lepidolite via iron sulphide roasting and water leaching. Hydrometallurgy 153:154–159

Selective Lithium Recovery from Brines Using Hydrothermally Treated Titania Slag

Rajashekhar Marthi and York R. Smith

Abstract Ion-exchange adsorbents such as delithiated lithium titanium oxides (LTOs) are highly effective for selective lithium adsorption from brines resources. In this work, we have synthesized LTO from waste titania slag and immobilized on a diatomaceous earth (DE) support. Titania slag was hydrothermally treated in alkaline solution to remove slag impurities. Acidic leaching followed by hydrolysis was performed to dissolve impurities and immobilize TiO_2 on DE. Subsequent solid-state synthesis with Li_2CO_3 resulted in the formation of LTO. Batch adsorption studies show that around 99% of lithium was adsorbed from a buffered 50 ppm Li. Thermodynamic and kinetic studies show the lithium adsorption to be an endothermic, chemisorption process.

Keywords Titania slag · Hydrothermal treatment · Lithium titanium oxide

Introduction

Due to the growing market of electric vehicles (EVs), there has been an increasing demand for lithium used in battery and energy storage applications. Around 56% of the total lithium in use is being used for battery applications [1]. The lithium consumption for battery applications is expected to increase from 47.6 kt in 2018 [1] to 184–986 kt by 2050 [2]. Lithium enriched brine is one of the promising resources for lithium with less environmental impacts, compared to ore-based resources such as spodumene [3]. Lithium from brines is usually extracted using precipitation techniques which are completely dependent on the brine chemistry. However, precipitation techniques are not feasible for lithium extraction from brines having a low lithium concentration (<100 ppm) and high Mg^{2+}/Li^{+} molar ratio [4]. High magnesium content in the brine results in magnesium contamination

R. Marthi · Y. R. Smith (✉)
Department of Materials Science and Engineering, University of Utah, Salt Lake City, UT, USA
e-mail: york.smith@utah.edu

G. Azimi et al. (eds.), *Rare Metal Technology 2020*, The Minerals, Metals & Materials Series, https://doi.org/10.1007/978-3-030-36758-9_5

and lower recovery of lithium [5]. Recently, lithium titanium oxide (LTO) based ionic sieves are gaining popularity due to their high selectivity of lithium over other alkali and alkaline earth metal ions. Unlike lithium manganese oxide (LMO) ionic sieves, which suffer from manganese dissolution during delithiation, LTOs have higher stability of titanium in the LTO matrix [6–8], and are considered to be better alternatives for lithium adsorption compared to LMO.

Generally, LTO can be synthesized by solid-state [6] or sol–gel techniques [9] by using a TiO_2 and a lithium salt precursor. Titania slag is a good source for producing titanium oxide [10] or its derivative products such as lithium titanium oxides [11]. Alkaline treatment of titania slag is gaining popularity over traditional chloride or sulfate processes [12]. However, the focus of these methods has been mainly on the production of pigment-grade TiO_2. Nano-sized TiO_2 can be obtained from titania slag by controlling the hydrolysis parameters. However, nano-sized TiO_2 is difficult to recover by simple filtration or sedimentation techniques. For practical purposes, nano-sized TiO_2 can be grown on a stable and porous support such as diatomaceous earth (DE). The nano-sized TiO_2 is not only immobilized on DE, but DE can also control the distribution and size of the nano-size TiO_2 immobilized [13]. Such composites can be easily separated by either sedimentation or filtration [14].

In this work, we have synthesized nano-sized LTO and immobilized on DE, which acts as a stable support. Titania slag was hydrothermally treated in alkaline solution to remove impurities such as iron and silicon. The hydrothermal product was leached in acidic solution to dissolve titanium and iron impurities. The titanium-rich leachate was hydrolyzed on the surface of DE to obtain TiO_2@DE, where TiO_2 acts as a titanium precursor for the formation of LTO@DE. Solid-state reaction of TiO_2@DE was performed with Li_2CO_3 to obtain LTO@DE. Subsequent delithiation with dilute acid results in delithiated lithium titanium oxide (DLTO@DE) which is used as an ionic sieve for lithium adsorption. The adsorbent parameters and the optimization of the adsorbent composite materials for lithium adsorption were studied in a LiCl buffer solution. After optimizing the synthesis parameters, the adsorbent composites will be later tested using Great Salt Lake water [15].

Experimental Method

Synthesis of the Adsorbent Composite

The DLTO@DE adsorbent composite was synthesized from upgraded titanium slag as shown in Fig. 1. In the first step, 2 g of titania slag was stirred in 30 ml of 10 M NaOH solution for 30 min. The mixture was then transferred into a Teflon lined stainless steel container. Alkaline hydrothermal treatment was performed at 200 °C for 24 h to remove the impurities such as silicon and iron present in the slag.

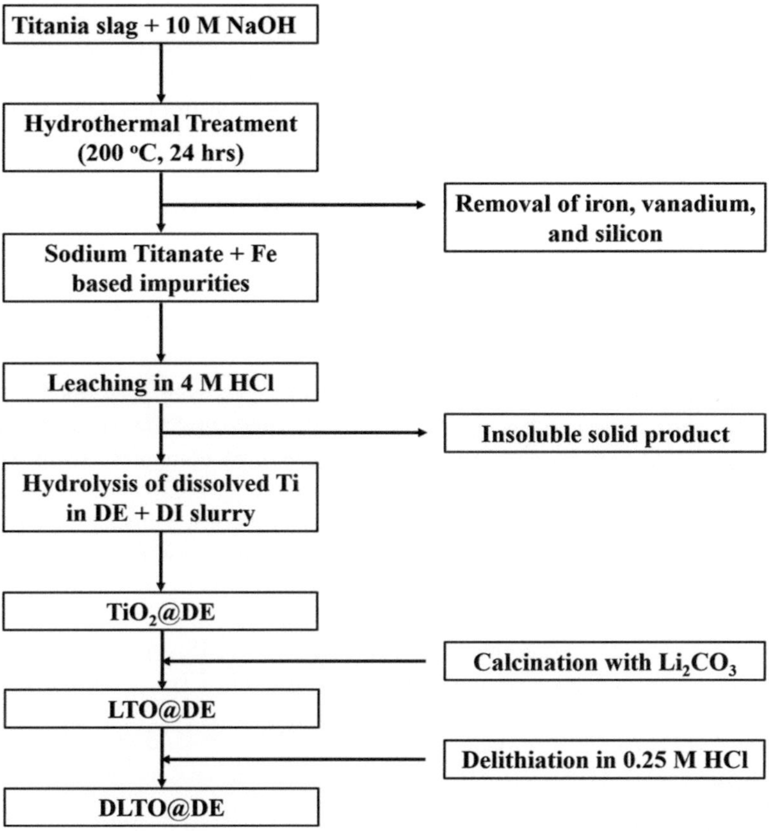

Fig. 1 Synthesis route of DLTO@DE from titania slag

The obtained white precipitate was collected and washed with plenty of deionized (DI) water. The precipitate was dried in a vacuum oven at 60 °C overnight. It was observed that the precipitate still has some greyish shade, indicating the presence of minor impurities. To remove the minor impurities, the precipitate was pulverized in mortar using pestle and was leached in 4 M HCl acid at 50 °C for 4 h (S: L = 20 ml: 1 g) to dissolve titanium and minor impurities that are present in the hydrothermal product. The titanium-rich leachate was separated from the undissolved solids using a vacuum filter. The leachate was later transferred into a slurry of DE in deionized water to obtain a final solid-to-liquid ratio of 1 g: 500 ml. Hydrolysis of the above slurry was performed at 90 °C for 4 h to obtain TiO_2@DE. The obtained TiO_2@DE composite material was filtered, washed with plenty of DI water, and dried at 60 °C overnight. The amount of nano-TiO_2 immobilized on DE was calculated by measuring the mass difference of DE before and after the hydrolysis.

The TiO_2@DE adsorbent composite material was mixed with Li_2CO_3 (molar ratio of Li/Ti = 2.05) in a mortar and was calcined at temperature 500–700 °C for 4 h to obtain LTO@DE composite material. Delithiated lithium titanium oxide composite (DLTO@DE) was obtained by eluting lithium from LTO@DE by stirring LTO@DE in 0.25 M HCl for 24 h (S:L, 1 g: 200 ml). Lithium adsorption studies were performed by stirring DLTO@DE in 50 ppm Li^+ buffered solution (pH = 10.20, adjusted by using NH_4OH and NH_4Cl) for 24 h.

Characterization

X-ray diffraction (XRD, Rigaku MiniFlex 6000) studies were performed by using Cu K_α radiation source (step size = 0.1°, dwell time = 0.5°/min). The software PDXL2 was used to determine the crystalline phases by comparing them with the standard index cards. The morphology of the adsorbent composite materials was characterized using scanning electron micrography (SEM, Hitachi S-4800). The elemental composition of the titania slag and the hydrothermally treated products was obtained using energy dispersive X-ray analysis (EDX, Oxford Instruments). Lithium concentrations before and after the adsorption were analyzed using inductively coupled plasma-mass spectrometry (ICP-MS, Agilent 7900). The standards and the analytes were prepared and analyzed in 2 vol.% nitric acid matrix solutions.

Kinetic Studies

Approximately, 0.5 g of DLTO@DE was stirred in 200 ml of 50 ppm LiCl buffer (pH = 9.50, adjusted by 0.1 M NH_4OH and 0.1 M NH_4Cl) for 24 h. At certain time intervals, ~5–7 ml of the slurry was collected and separated using a syringe filter. The liquid was analyzed for lithium using ICP-MS. To understand the adsorption mechanism, non-linear forms of pseudo-first-order (PFO) and pseudo-second-order (PSO) kinetic models were applied. Lithium uptake was calculated by

$$q_t = (C_0 - C_t) * \frac{V}{m * f} \tag{1}$$

$$q_e = (C_0 - C_e) * \left(\frac{V}{m * f}\right) \tag{2}$$

where q_t and q_e represent the amount of lithium adsorbed in milligrams per gram of LTO at time t and the equilibrium, respectively, C_0 and C_t represent the initial lithium concentration and concentration of lithium at time t, m represents the mass of the adsorbent, and f represents the mass fraction of LTO in LTO@DE. The percentage lithium adsorption is given as

$$\%\text{Adsorption} = \frac{C_0 - C_e}{C_0} * 100 \quad (3)$$

Results and Discussion

Figure 2a shows the XRD patterns of raw titania slag and the hydrothermal product. It can be observed that the raw titania slag consists of highly crystalline rutile titania (JCPDS: 76-1940) and pseudobrookite ($FeTi_2O_5$, JCPDS: 89-8065) phases. After hydrothermal treatment of the raw titania slag, it can be observed that unreacted rutile titania with poor crystallinity is present, and a broad peak between 25 and 30° represents the amorphous sodium titanate phase [16]. The raw titania slag is orange in color (Fig. 2b), which might be due to the presence of impurities such as iron and silicon (<1 wt%) (Table 1) [17]. Upon alkaline hydrothermal treatment of the raw titania slag, a white precipitate with greyish shade is obtained (Fig. 2c). Impurities such as silicon converts into sodium metallates, which are soluble in the aqueous solution, whereas the titanium in the slag converts into sodium titanate, which is insoluble in aqueous solution [16]. However, the hydrothermal product is still grey in color, which indicates the presence of some minor impurities which cannot be detected by EDX (Table 1). Techniques such as ICP-MS are required to identify these impurities quantitatively. Sodium titanate is insoluble in alkaline and neutral aqueous solution, but it can easily dissolve in acidic solutions [16]. Hence, the hydrothermal product was leached in 4 M HCl acid solution to dissolve titanium and other possible impurities.

The titanium-rich leachate is hydrolyzed in a DE-DI water slurry for 4 h at 90 °C to obtain TiO_2@DE. The XRD pattern of TiO_2@DE (Fig. 3a) shows the presence of both anatase (JCPDS: 71-1168) and rutile phase (JCPDS: 76-0318). The immobilization of the anatase or rutile phase depends upon the surface area and the amount of DE present in the slurry [18]. Quantitative analysis using reference intensity ratio (RIR) shows that around 30% of the immobilized titania is in the form of rutile phase. The immobilized TiO_2@DE was later calcined with Li_2CO_3 at temperatures varying from 500 to 600 °C. Upon calcination, it can be observed that TiO_2 is transformed into LTO. The LTO peaks correspond to Li_2TiO_3 (JCPDS: 33-0831) which has a monoclinic crystal structure (Fig. 3b). It can be observed that as the calcination temperature increases there is an increase in the peak intensities of LTO indicating an increase in the crystallinity of immobilized LTO with increasing temperature. However, it can also be observed that the presence of unreacted anatase TiO_2 and by-products is minimum when calcination was performed at 600 °C.

The percentage adsorption of lithium by DLTO@DE at different calcination temperatures is given in Table 2. It can be seen that LTO@DE obtained from

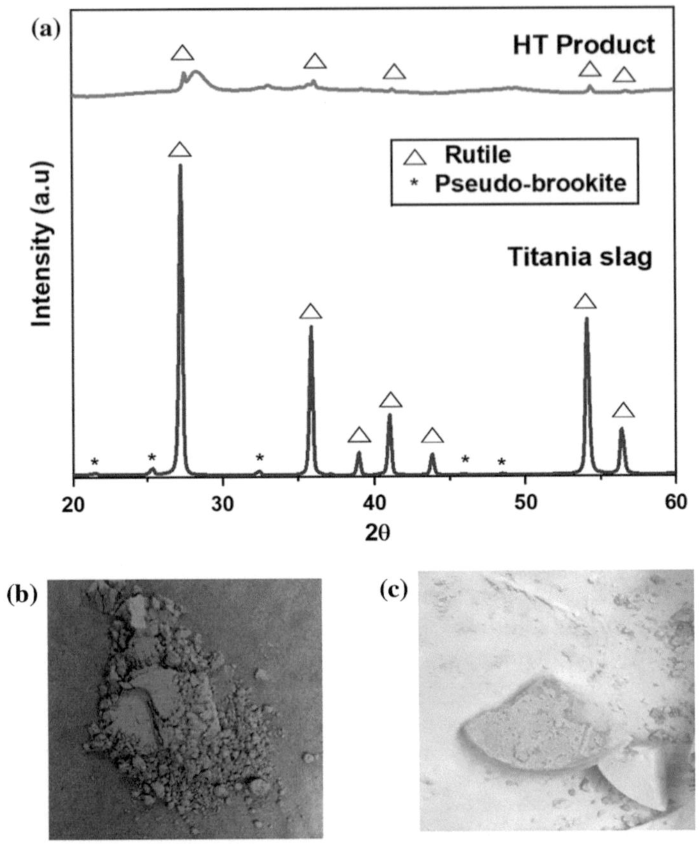

Fig. 2 **a** XRD patterns of raw titania slag and the hydrothermal (HT) product. Images of **b** raw titania slag and **c** titania slag after the hydrothermal treatment. It can be observed that after hydrothermal treatment, a white precipitate is obtained due to the removal of impurities from the slag

Table 1 EDX composition of the raw titania slag and the hydrothermal product

	Ti	Fe	Si	O	Na
Titania slag	44.29 ± 1.20	0.78 ± 0.21	0.65 ± 0.14	54.29 ± 1.25	–
Hydrothermal product	29.22 ± 0.74	–	–	57.17 ± 1.00	13.61 ± 0.56

calcination at 600 °C adsorbs around 99% of lithium from the buffered solution and has a better performance than DLTO@DE obtained from calcination at other temperatures. This can be attributed to more lithium specific active sites present in LTO@DE synthesized at 600 °C, due to better conversion of TiO_2 into LTO and smaller crystallite size which increases the amount of lithium eluted from LTO.

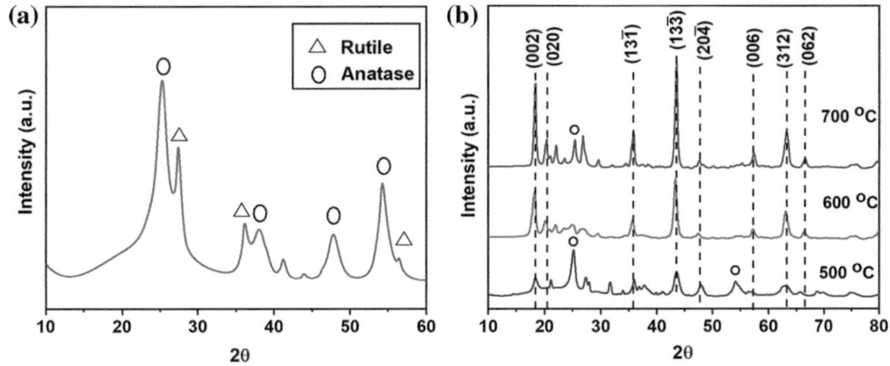

Fig. 3 X-ray diffraction patterns of **a** TiO_2@DE and **b** LTO@DE calcined at different temperatures. It can be observed that with increasing calcination temperature there is an increase in the crystallinity of the LTOs

Table 2 % Lithium adsorption by adsorbent composites synthesized at various calcination temperatures (pH = 10.20, S:L = 1 g: 200 ml)

Calcination temperature (°C)	C_0 (ppm)	C_e (ppm)	% Adsorption
500	49.3	23.0	53.3
600	48.2	0.3	99.4
700	48.2	2.8	94.2

As a porous material, DE has macroporous channels of various pore sizes (Fig. 4a–b). Figure 4c–d shows the immobilized nano-DLTO on the surface of the DE. It can be observed that the immobilized DLTO is distributed uniformly on the surface of DE and does not block the macroporous channels present on the surface of DE which can allow better contact between the lithium rich brine and DLTO@DE [19].

Figure 5 shows that the normalized lithium uptake increases rapidly with time initially and then follows a slower equilibrium stage. It can be observed that the equilibrium is achieved within 4 h of the adsorption. Non-linear PFO and PSO kinetic models were applied to determine the better fit for lithium sorption. The non-linear form of PFO is given as [20]

$$q_t = q_e\left(1 - e^{-k_1 t}\right) \tag{4}$$

The PSO kinetic model is usually associated with chemisorption mechanism, which involves covalent bonding or bond breakage. Its non-linear form is given as [21]

Fig. 4 SEM micrographs of **a-b** DE and **c-d** DLTO@DE. It can be observed that DLTO has been immobilized on the surface of DE uniformly without blocking the macroporous channels

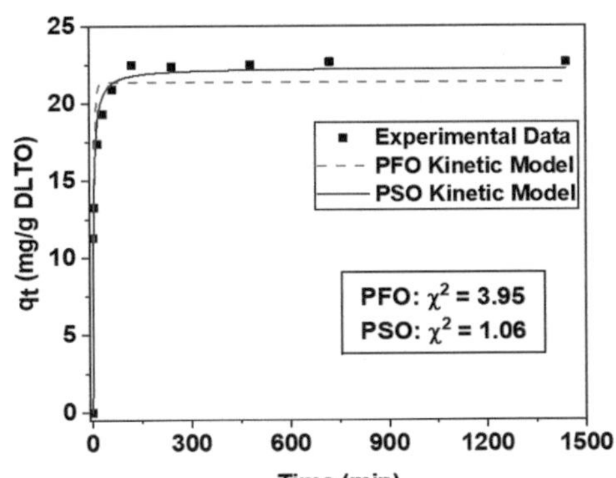

Fig. 5 PSO kinetic model for lithium adsorption shows a better fit with experimental data indicating chemisorption as the main mechanism for lithium adsorption (pH = 9.50, S:L = 1 g: 400 ml)

$$q_t = \frac{q_e^2 k_2 t}{1 + k_2 q_e t} \tag{5}$$

The better fit for the PFO or PSO is determined by comparing the χ^2 values. It can be observed that the χ^2 value for the PSO model is lower than the PFO kinetic model indicating that lithium sorption on DLTO@DE involves a chemisorption mechanism. The value of PSO rate constant is 0.016 g • mg^{-1} • min^{-1} and the equilibrium lithium sorption capacity is 23.11 mg/(g • LTO).

Conclusion

A lithium selective ionic sieve was successfully synthesized from titania slag by alkaline hydrothermal treatment. Immobilization of LTO on DE was successful without blocking the macroporous channels of DE. The crystalline structure of immobilized LTO is monoclinic Li_2TiO_3, and the crystallinity of the LTO increased with increasing calcination temperature. The synthesis parameters for LTO@DE calcined at 600 °C for 4 h with Li/Ti molar ratio of 2.05 were shown to give the highest performance in this study. Lithium adsorption follows a pseudo-second-order kinetic model indicating that lithium adsorption involves covalent bond formation or bond breakage. It was demonstrated that the equilibrium adsorption capacity at pH = 9.50 and S:L (1 g: 400 ml) is 23.11 mg/(g • LTO). An optimized LTO@DE composite will be later tested with a real brine sample such as The Great Salt Lake for its feasibility in practical applications.

References

1. Production D, Columbia B, Relations NT (2018) U.S. Geological Survey, no 703, February 2019. Mineral Commodity Summaries, pp 2018–2019
2. Speirs J, Contestabile M, Houari Y, Gross R (2014) The future of lithium availability for electric vehicle batteries. Renew Sustain Energy Rev 35:183–193
3. Choubey PK, Seuk Kim M, Srivastava RR, Lee J, Lee J-Y (2016) Advance review on the exploitation of the prominent energy-storage element Lithium. Part I: from mineral and brine resources. Miner Eng 89:119–137
4. Munk LA, Bradley D (2016) Lithium brines: a global perspective. January 2016:339–365
5. Meshram P, Pandey BD, Mankhand TR (2014) Extraction of lithium from primary and secondary sources by pre-treatment, leaching and separation: a comprehensive review. Hydrometallurgy 150:192–208
6. Shi XC, Zhang ZB, Zhou DF, Zhang LF, Chen BZ, Yu LL (2013) Synthesis of Li+ adsorbent (H_2TiO_3) and its adsorption properties. Trans Nonferrous Met Soc China (English Ed.) 23 (1):253–259
7. Wang S et al (2016) Hydrothermal synthesis of lithium-enriched β-Li_2TiO_3 with an ion-sieve application: excellent lithium adsorption. RSC Adv 6(104):102608–102616
8. Moazeni M, Hajipour H, Askari M, Nusheh M (2015) Hydrothermal synthesis and characterization of titanium dioxide nanotubes as novel lithium adsorbents. Mater Res Bull 61:70–75
9. Zhang L-Y, Liu Y-W, Huang L, Li N (2018) A novel study on preparation of H_2TiO_3-lithium adsorbent with titanyl sulfate as titanium source by inorganic precipitation–peptization method. RSC Adv 8(3):1385–1391
10. Middlemas S, Fang ZZ, Fan P (2013) A new method for production of titanium dioxide pigment. Hydrometallurgy 131–132:107–113

11. Tang D, Zhou D, Zhou J, Zhang P, Zhang L, Xia Y (2015) Preparation of H_2TiO_3-lithium adsorbent using low-grade titanium slag. Hydrometallurgy 157:90–96
12. Dong H, Jiang T, Guo Y, Chen J, Fan X (2012) Upgrading a Ti-slag by a roast-leach process. Hydrometallurgy 113–114:119–121
13. Li A, Wang J, Zhang W, McNaughton R, Anderson S, Zhang X (2016) Sol-gel based TiO_2 thin film deposition on frustules towards facile and scalable manufacturing. J Phys Conf Ser 773(1)
14. Wang B, de Godoi FC, Sun Z, Zeng Q, Zheng S, Frost RL (2015) Synthesis, characterization and activity of an immobilized photocatalyst: natural porous diatomite supported titania nanoparticles. J Colloid Interf Sci 438:204–211
15. Marthi R, Smith YR (2019) Selective recovery of lithium from the Great Salt Lake using lithium manganese oxide-diatomaceous earth composite. Hydrometallurgy 186:115–125
16. Zhang Y et al (2016) A novel chemical pathway for energy efficient production of Ti metal from upgraded titanium slag. Chem Eng J 286:517–527
17. Zhang W, Zhu Z, Cheng CY (2011) A literature review of titanium metallurgical processes. Hydrometallurgy 108(3–4):177–188
18. Bouna L et al (2011) Synthesis, characterization and photocatalytic activity of TiO_2 supported natural palygorskite microfibers. Appl Clay Sci 52(3):301–311
19. Hong HJ et al (2018) Highly porous and surface-expanded spinel hydrogen manganese oxide (HMO)/Al_2O_3 composite for effective lithium (Li) recovery from seawater. Chem Eng J 337 (December 2017):455–461
20. Lagergren SK (1898) About the theory of so-called adsorption of soluble substances. Sven Vetenskapsakad Handingarl 24:1–39
21. Blanchard G, Maunaye M, Martin G (1984) Removal of heavy metals from waters by means of natural zeolites. Water Res 18(12):1501–1507

Molecular Recognition Approach to REE Extraction, Separation, and Recycling

Gulaim A. Seisenbaeva

Abstract Rare earth elements (REE) are highly requested by modern industry, being indispensable for construction of electric vehicles and environmentally friendly energy sources. Challenge in their production lies in their joint occurrence and similar chemical properties. It can be answered by using solid phase extraction, applying highly selective nanostructured adsorbents. The latter are bearing organic ligands with high specific affinity to particular REE grafted on their surface via covalent bonding. This work is a comparative study to demonstrate the ligand properties in relation to distinct REE as a function of various operating conditions, such as concentration of solutions and pH. Mechanisms of adsorption are elucidated with a variety of techniques including FTIR (Fourier-transform infrared spectroscopy), solid-state NMR spectroscopy (Nuclear Magnetic Resonance), and EXAFS spectroscopy (extended X-ray absorption fine structure) in combination with analysis of particles bearing adsorbed cations by SEM-EDS (scanning electron microscopy/energy dispersive X-ray spectroscopy) and of solutions by complexometric titration. Recommendations are elaborated for specific choice of ligands and adsorption–desorption conditions for improved separation of single REE. The results are verified in separation of REE from complex leachates both by magnetic nanoadsorbents and by mesoporous microparticles applied as a sorbent in High Performance Chelation Ion Chromatography (HPCIC).

Keywords Rare earth elements · Amino polycarboxylic acids · Nanoadsorbent · Uptake and release mechanism

---

G. A. Seisenbaeva (✉)
Department of Molecular Sciences, BioCenter SLU, Box 7015, 75007 Uppsala, Sweden
e-mail: gulaim.seisenbaeva@slu.se

G. Azimi et al. (eds.), *Rare Metal Technology 2020*, The Minerals, Metals & Materials Series, https://doi.org/10.1007/978-3-030-36758-9_6

Introduction

Rare earth elements (REE) are a group of 17 elements, including scandium, yttrium, lanthanum, and 14 lanthanide elements. They actually are not rare in the earth's crust with the contents comparable with widely used 3d-transition metals such as cobalt, nickel, or copper. The challenges in producing them at large scale are because they rarely form separate minerals, and are caused by strong similarity between the radii of their cations, resulting in rather similar properties. These factors lead to difficulties in producing REE from the ores and in their separation from each other, as the industrial applications usually demand obtaining these elements in high purity with very low levels of admixtures of other REEs.

Today, the extraction and purification of REE exploit either solvent extraction, relying on use of large volumes of hazardous organic solvents [1], or chromatographic separation, which is a rather slow technique operating with quite small volumes of products at a time [2].

An attractive alternative to these conventional methods is solid-state extraction exploiting magnetic nanoadsorbents, easily distributed and retracted from solutions, exploiting magnetic field [3].

Magnetic Nanoadsorbents—The Concept

The idea of using nanomaterials as adsorbents is quite well established. Early studies dealing with adsorption of REEs were focused on irreversible adsorption with REE considered as models for radioactive components of nuclear fission products and employed non-magnetic silica [4]. Magnetic silica has been introduced originally for grafting of catalysts in organic catalysis to permit their separation from reaction mixtures [5]. In our studies, we proposed a concept of magnetic adsorbents, where the magnetic properties could be introduced via implanting superparamagnetic particles of iron oxide into either mesoporous matrices or as cores for dense core-shell nanoparticles (see Fig. 1).

Extraction and separation of REEs set rather distinct requirements on the stability of the adsorbent when subjected to harsh conditions of subsequent desorption of REEs from the adsorbent. The latter usually employs 0.1–1.0 M strong acids (nitric or hydrochloric acid). This kind of stability can be demonstrated when the highly reactive iron oxide particles are covered by a dense silica shell with thickness of at least about 20 nm according to our previous studies [6]. This challenge limits the use of mesoporous matrices, requiring to protect the magnetic particles first, before their inclusion into the mesoporous matrix. Another obstacle in using mesoporous adsorbents in the approaches exploiting magnetic extraction is the slow diffusion of solvent through nanostructured matrices when no mechanically supported liquid flow is applied (which is the case of high-pressure liquid

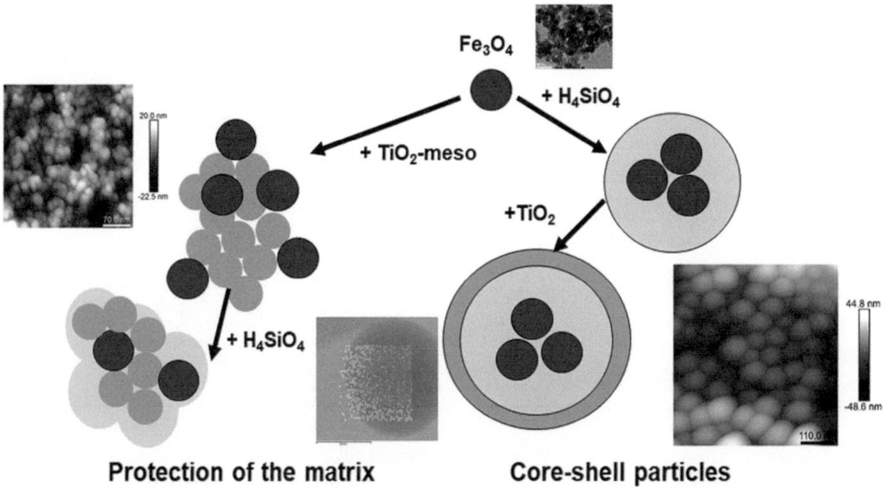

Fig. 1 General concept in creation of functional magnetic nanoadsorbents

chromatography applications). Mesoporous adsorbents revealed extremely long equilibration times, i.e., 17 h or longer [7], which forced us to focus our attention on dense core-shell particles.

Grafting of Functional Ligands

Silica surface is negatively charged at neutral and near-neutral acidic pH, which makes it attractive for use in adsorption of metal cations. However, to achieve selectivity, it is more attractive to graft onto it more specific ligand functions, offering stronger affinity to distinct group or even to single metal cation among those present in solutions. There are two principal approaches, which can be used for building up monolayers bearing complex functions. In one possible option, a function is built up through organic synthetic approaches, and attached then to an alkoxy-silane molecule. The latter is in its turn grafted on the surface of the pre-produced silica particle (see Fig. 2a).

The challenge in this approach lies in the need to separate and purify each reactant at each intermediate step of the synthesis. This can be a time-demanding and expensive procedure, especially in the view that the final alkoxy-silane cannot be purified by column chromatography that, otherwise, can result in its hydrolysis. Individual approaches involving crystallization or solvent extraction will thus be required to make the final reactant pure enough [8]. An alternative is provided via step-by-step construction of a ligand on the surface of the particle (Fig. 2b). It is important in this latter case to keep in mind that some fraction of each intermediate remains unreacted on the surface of the particle and potentially may interfere in

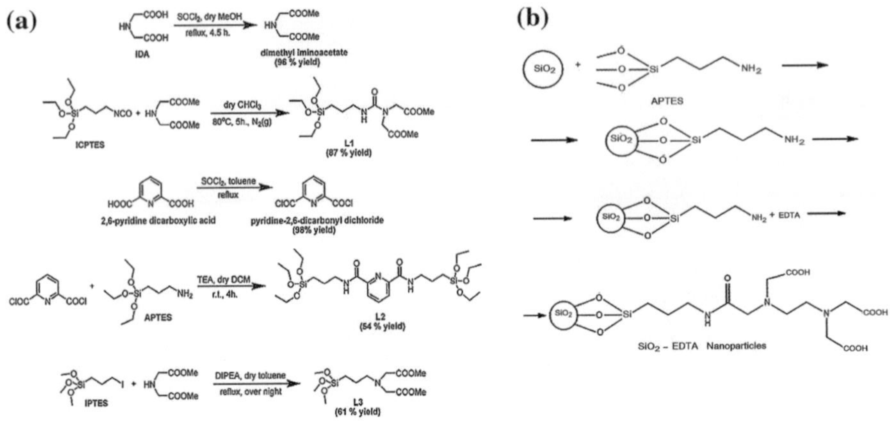

Fig. 2 Synthetic approaches to single-source functional alkoxy-silane precursors (**a**) (Reprinted with permission from [8]) and step-by-step construction of a ligand function (**b**)

surface reactivity [9]. Commonly, only a minor fraction of silanol OH-groups remains unreacted, while only about one-fourth of the alkyl amino groups grafted on the surface gets further converted into amide links to a complexonate ligand. Typical ligand density on magnetic silica nanoparticles produced by step-by-step approach is not exceeding 0.5 mmol/g, while almost double as much can be grafted in optimized single-source approach.

Surface Complexation of Metal Cations

The interaction of metal cations with functionalized particles occurs via relatively complex mechanisms dependent on the nature of the cations and ligands themselves. It is very common that binding on adsorption occurs originally as outer-sphere complexation through electrostatic interaction of the cations with negatively charged carboxylic groups of the ligand functions (see Fig. 3). On storage or on the

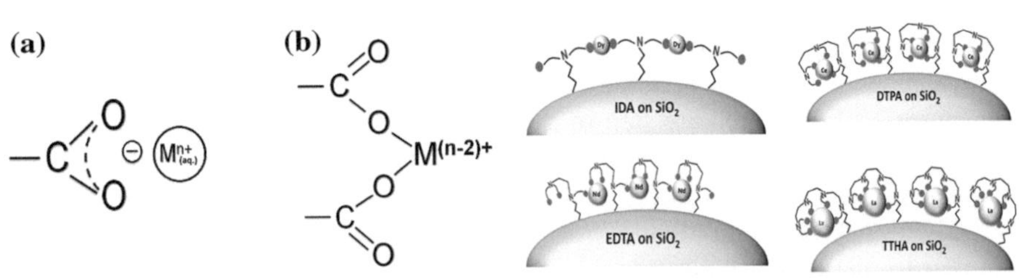

Fig. 3 Binding of metal cations in surface complexes: outer-sphere mode (left, **a**), inner-sphere mode (left, **b**), and structure motifs in inner-sphere complexes of REE with amino carboxylate ligands (right, reprinted with permission from [12])

action of increased acidity (lowering of pH), which improves mobility of the ligands and kinetics of ligand exchange, the outer-sphere complexes transform into inner-sphere ones, where the cations interact directly with oxygen atoms of the ligands. The selectivity in formation of outer-sphere surface species is quite reasonably very low, related to differences in surface charge between cations that are only minor between REE. In contrast, the structure and reactivity of inner-sphere complexes can differ quite pronouncedly between different REEs that can be in complexes formed at lower pH featuring different coordination numbers and different coordination geometries [10, 11].

These differences can cause considerable effects on selectivity and affinity of functional adsorbents towards distinct REE cations. These effects are caused by geometry of the surface complex layers, being more or less compatible for coordination of a given REE. This can be explained by taking into account that the distance between alkyl silane functions on the surface of a nanoparticle is a relatively well-defined value. In primary grafting of an alkoxy-silane, i.e., the case in the use of a single-source precursor or in grafting of an alkyl silane function in the first step of a step-by-step procedure, it constitutes about 6.3 Å in average, according to the data obtained from molecular models.

As a comparison of the possible binding modes of REE cations with IDA ligands reveals the coordination on the surface should differ considerably (see Fig. 4). For La^{3+}-cations, having largest cation radius, the optimal CN = 10 and the distance between ligands should be much shorter (4.983 Å) and the ligand density should be much higher than that available on a real silica surface. For Nd^{3+}, being slightly smaller, the distance still is too short (5.638 Å), compared with the really observed one of about 6.3 Å. In contract, for the octa-coordinated Dy^{3+}-cations, the required and observed inter-ligand distances are matching perfectly. This explains why the silica with IDA-ligand grafted via single-source precursor approach has considerably better capacity for dysprosium compared to lanthanum and why lanthanum cations are much more readily desorbed from the surface at moderate pH = 3, leading to strongly improved selectivity on desorption (Dy:La = 81:1, compared to 4.2:1 ration observed for adsorption at pH = 6.5) [10]. It can thus be concluded that

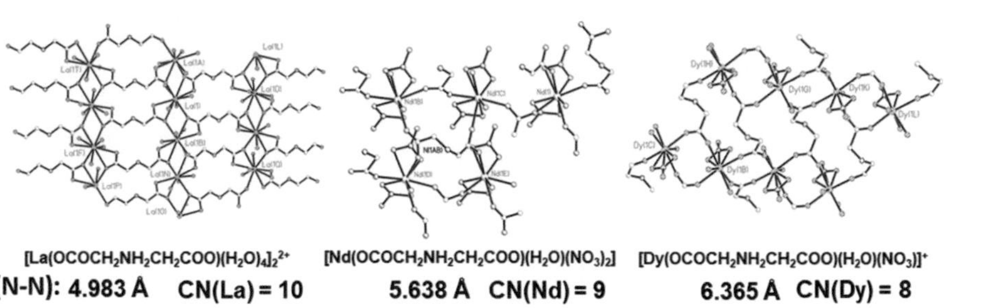

Fig. 4 The structure motifs and optimal inter-ligand distances and coordination numbers (CN) in surface complexes of REE with imino diacetic acid (IDA) surface layers as revealed by molecular models, derived from single crystal studies [10, 11]

the adsorbent, bearing functional ligands, exerts in formation of inner-sphere complexes with the molecular recognition of REE, opening for their efficient separation in few steps. After publishing our first paper on this topic, grafting of functional ligands on silica-shell nanoadsorbents has attracted considerable attention as an approach for extraction and separation of REE [13–15]. However, the understanding of the most pronounced effects achieved on desorption in acidic medium could only be obtained from the advanced analysis of the molecular structures of the surface complexes [10–12] and their confirmation by solid-state NMR and EXAFS spectroscopic investigation of the surface complexes [16].

It is important to mention that practical application of magnetic nanoadsorbents requires development of equipment suitable for their use, namely, special reactors with magnetic distribution and then retraction of the adsorbent, insuring completeness of removal for nanomaterial. Development of such reactors is on the way and models for enhanced lab-scale separation are already available [17].

Nanoadsorbents in High Performance Chelation Ion Chromatography

In our previous study, we investigated the feasibility of a larger scale valorization of functional nanoadsorbents and possibility to compare them with presently commercially available ion-exchange adsorbent resins. In this study, we chose to apply an industrially produced mesoporous adsorbent Kromasil^{TM} functionalized with aminopropyl groups as matrix for further grafting of functional ligands. Kromasil^{TM} is a mesoporous adsorbent composed of small silica nanoparticles (less than 20 nm in diameter), assembled into spherical beads of about 10 μm in size (see Fig. 5).

Chemical modification does not change morphology nor size of the beads, leading, however, to some losses of the active surface area from 263 to 200 m^2/g as determined with the nitrogen adsorption experiments. The latter is quite logical reflecting that some space within the pores is consumed by introduction of a monolayer of amino carboxylic acid on the surface. The amount of introduced ligand is comparable to that of the surface of free nanoparticle carrier, i.e., 0.44 mmol/g (compared with approximately 0.5 mmol/g in non-aggregated nanoparticle material). The produced material demonstrated attractive characteristics. Its capacity towards single REE constituted 0.85–0.95 mmol/g indicating that adsorption occurs as expected through formation of outer-sphere complexes. This value corresponds to capacity of 140–150 mg REE per gram of adsorbent, which is considerably higher than common values for commercially available organic ion-exchange resins. The adsorption kinetics was rather quick with equilibrium achievable within few minutes in dynamic regime. Desorption could be carried out in quantitative fashion through elution with solutions of nitric acid. The efficiency of separation was related to the load of the REE, concentration of the eluent and its gradient, liquid flow, etc. It could be demonstrated that mixtures containing 6 REEs

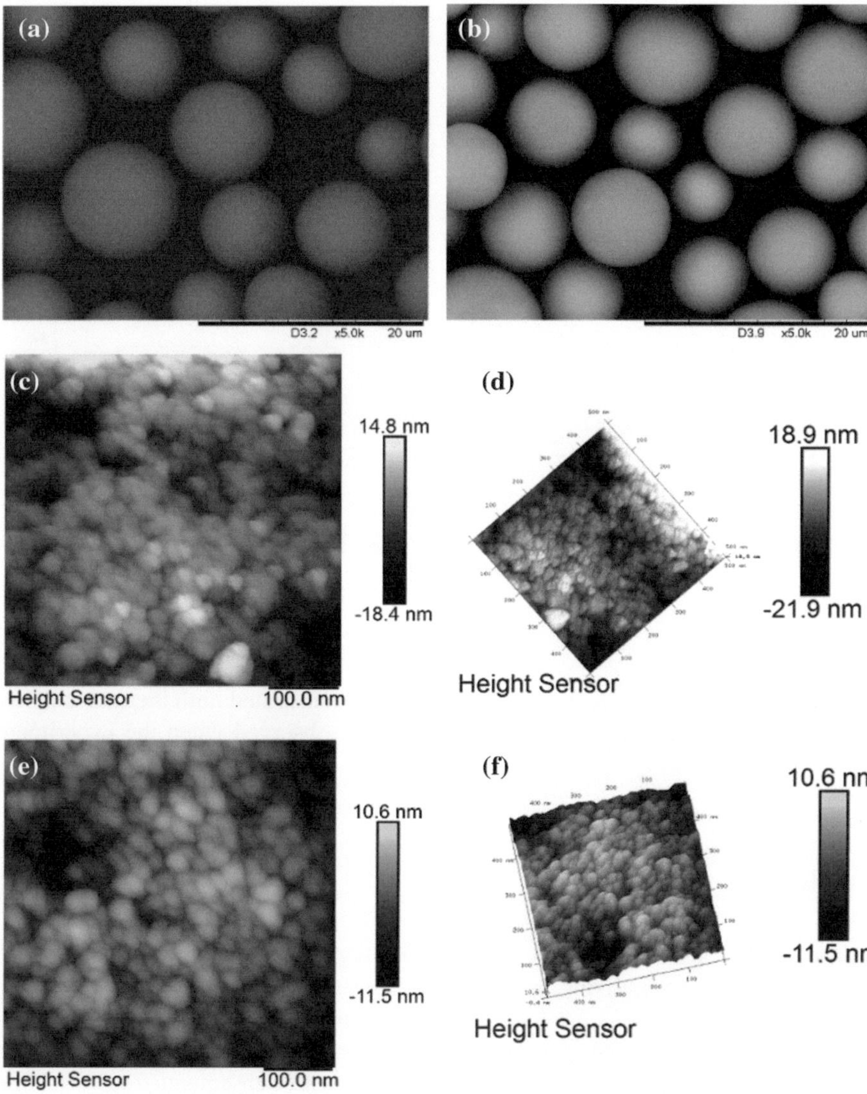

Fig. 5 SEM (**a**, **b**) and AFM (**c**–**f**) of the Kromasil microparticles (**a**, **c**, **d**: pure particles; **b**, **e**, **f**: particles after functionalization by DTPA). Reprinted with permission from [18]

in comparable concentration could be separated in one run with over 90% purity of fractions containing individual REEs dependent on the experimental conditions (see Fig. 6).

Produced adsorbent demonstrated impressive stability under the experimental conditions, showing no detectable change in its properties for at least 50

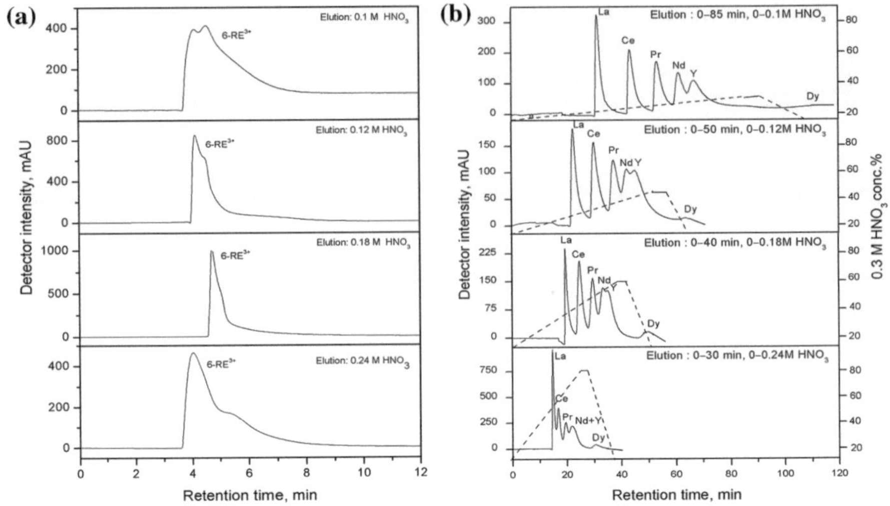

Fig. 6 Chromatograms from **a** isocratic and **b** linear gradient elution studies at different concentrations of HNO_3 in the eluent. Eluent flow rate = 1 mL min^{-1}, column temperature = 55 °C, injected volume = 50 μL, and total REE concentration = 1000 mg L^{-1}. Reprinted with permission from [18]

adsorption–desorption cycles. The material also demonstrated high thermal stability in the presence of nitric acid. Best separation was observed when the experiment was carried out at 55 °C.

Conclusions

Nanostructured adsorbents bearing amino carboxylate ligand functions have been demonstrated to be promising materials for extraction and separation of REEs in both magnetic extraction and when used as ion-exchange adsorbents in chromatography. The revealed mechanism of selective action involves adsorption occurring initially via outer-sphere complexation with subsequent transformation into inner-sphere complexes on desorption in acidic medium. This can be denoted as molecular recognition phenomenon. Produced nanomaterials demonstrate high capacity, selectivity, and stability making them potentially attractive alternatives to present technologies both for treatment of leachates produced from products of primary mining and, potentially, for recycling of REE-based materials.

Acknowledgemnet The author is grateful to EuroStars EU Program for support of the ERA-MIN MetRecycle Project (supported via VinnovaSwedish Research Council grant No. 2018-00739).

References

1. Xie F, Zhang TA, Dreisinger D, Doyle F (2014) A critical review on solvent extraction of rare earths from aqueous solutions. Mater Eng 56:10–28
2. Max-Hansen M (2014). Modeling and optimization of rare earth element chromatography. Lund University, Faculty of Engineering, Department of Chemical Engineering
3. Balomenos E, Davris P, Deady E, Yang J, Panias D, Friedrich B, Binnemans K, Seisenbaeva G, Dietrich C, Kalvig P, Paspaliaris I (2017) The EURARE project: development of a sustainable exploitation scheme for Europe's rare earth ore deposits. Johnson Matthey Technol Rev 61(2):142–153
4. Melnyk IV, Goncharyk VP, Stolyarchuk NV, Kozhara LI, Lunochkina AS, Alonso B, Zub YL (2012) Dy(III) sorption from water solutions by mesoporous silicas functionalized with phosphonic acid groups. Porous Mater 19(5):579–585
5. Shirina ZB, Bronstein L (2018) Magnetically recoverable catalysts: beyond magnetic separation. Front Chem 6(1–6):298
6. Pogorilyi RP, Melnyk IV, Zub YL, Carlson S, Daniel G, Svedlindh P, Seisenbaeva GA, Kessler VG (2014) New product from old reaction: uniform magnetite nanoparticles from iron-mediated synthesis of alkali iodides and their protection from leaching in acidic media. RSC Adv 4:22606–22612
7. Seisenbaeva GA, Melnyk IV, Hedin N, Chen Y, Eriksson P, Trzop E, Zub YL, Kessler VG (2015) Molecular insight into the mode-of-action of phosphonate monolayers as active functions of hybrid metal oxide adsorbents. Case study in sequestration of rare earth elements. RSC Adv 5:24575–24585
8. Demirel Topel S, Polido Legaria E, Tiseanu C, Rocha J, Nedelec JM, Kessler VG, Seisenbaeva GA (2014) Hybrid silica nanoparticles for sequestration and luminescence detection of trivalent rare-earth ions (Dy^{3+} and Nd^{3+}) in solution. J Nanopart Res 16 (1–17):2783
9. Pogorilyi RP, Pylypchuk I, Melnyk IV, Zub YL, Seisenbaeva GA, Kessler VG (2017) Sol-gel derived adsorbents with enzymatic and complexonate functions for complex water remediation. Nanomaterials 7(1–17):298
10. Polido Legaria E, Demirel Topel S, Kessler VG, Seisenbaeva GA (2015) Molecular insights into the selective action of a magnetically removable complexone-grafted adsorbent. Dalton Trans 44:1273–1282
11. Polido Legaria E, Saldan I, Svedlindh P, Wetterskog E, Gunnarsson K, Kessler VG, Seisenbaeva GA (2018) Coordination of rare earth element cations on the surface of silica-derived nanoadsorbents. Dalton Trans 47:1312–1320
12. Polido Legaria E, Samouhos M, Kessler VG, Seisenbaeva GA (2017) Toward molecular recognition of REEs: Comparative analysis of hybrid nanoadsorbents with the different complexonate ligands EDTA, DTPA, and TTHA. Inorg Chem 56:13938–13948
13. Dupont D, Luyten J, Bloemen M, Verbiest T, Binnemans K (2014) Acid-stable magnetic core–shell nanoparticles for the separation of rare earths. Ind Eng Chem Res 53:15222–15229
14. Callura JC, Perkins KM, Noack CW, Washburn NR, Dzombak DA, Karamalidis AK (2018) Selective adsorption of rare earth elements onto functionalized silica particles. Green Chem 20:1515–1526
15. Callura JC, Perkins KM, Baltrus JP, Washburn NR, Dzombak DA, Karamalidis AK (2019) Adsorption kinetics, thermodynamics, and isotherm studies for functionalized lanthanide-chelating resins. J Colloid Interface Sci. https://doi.org/10.1016/j.jcis.2019.08.097

16. Polido Legaria E, Rocha J, Tai CW, Kessler VG, Seisenbaeva GA (2017) Unusual seeding mechanism for enhanced performance in solid phase magnetic extraction of Rare Earth Elements. Sci Rep 7(1–13):43740
17. https://www.labonabead.se/loabeads-magsep/
18. Ashour RM, Samouhos M, Polido Legaria E, Svärd M, Högblom J, Forsberg K, Palmlöf M, Kessler VG, Seisenbaeva GA, Rasmuson ÅC (2018) DTPA-functionalized silica nano- and microparticles for adsorption and chromatographic separation of rare earth elements. ACS Sustain Chem Eng 6:6889–6900

Production of Energy Saving Materials from the Waste Mixtures of REEs

Moufida Mansouri, Cristian Tunsu, Burcak Ebin, Lucy Ajakaiye Jensen and Martina Petranikova

Abstract This work details the procedures and steps for the synthesis of new magnetocaloric materials starting from the outputs generated by recycling for re-use in magnetic refrigeration application. The outputs utilized were a rare earth element-rich product obtained during the hydrometallurgical processing of nickel–metal hydride (NiMH) batteries. This stream contained a mix of rare earth elements (REEs), mainly lanthanum and cerium. After removing some impurities from this feed, especially aluminium, we have used the obtained product in the manufacturing of advanced REEs-based magnetocaloric materials, especially manganites-, orthoferrites-, and REEs-based alloys. The composition of the output from hydrometallurgical processing of NiMH batteries and typical compositions of magnetocaloric materials was determined using inductively coupled plasma-optical emission spectroscopy. The magnetocaloric oxides were successfully synthesized by using conventional solid-state reaction method. Carbothermic and calciothermic reduction methods were used for the synthesis of the as-cast alloy. The X-ray diffraction analysis shows that the magnetocaloric oxides are well crystallized with presence of secondary phases. The effect of temperature on the crystal structure is briefly described.

Keywords Hydrometallurgy · Precipitation · Solid-state reaction · Carbothermic reduction · Calciothermic reduction

Introduction

Rare earth elements (REEs) are the core of many future-sustainable technologies, including permanent magnets, energy efficient lighting (fluorescent and LED lamps), and NiMH batteries. Recently, a great attention has been paid to the use of

M. Mansouri · C. Tunsu · B. Ebin · L. A. Jensen · M. Petranikova (✉)
Department of Chemistry and Chemical Engineering, Nuclear Chemistry and Industrial Materials Recycling, Chalmers University of Technology, Gothenburg 412 96, Sweden
e-mail: martina.petranikova@chalmers.se

G. Azimi et al. (eds.), *Rare Metal Technology 2020*, The Minerals, Metals & Materials Series, https://doi.org/10.1007/978-3-030-36758-9_7

REEs in magnetic refrigeration application, which is based essentially on the magnetocaloric effect (MCE). Magnetic refrigeration is gaining more and more interest due to better efficiencies and being more environmentally friendly than gas-compression refrigeration [1, 2].

Mining, leaching, pre-concentration, and the numerous separation stages required to produce virgin REEs lead to high amounts of secondary wastes and unwanted emissions [3]. Given the scarcity and high demand for REEs, and the environmental issues associated with their mining, it is important to look at additional and cleaner sources of such elements. Substitution of certain REEs in products is desired, but problematic since elements, which have the same desired properties as the targeted REEs, are not available [4].

Hydrometallurgical processing of NiMH batteries generates a feed containing several REEs, notably lanthanum and cerium. This feed has low market value and has no current industrial applicability. The majority of spent NiMH batteries generated in the EU are processed pyrometallurgically by Umicore (BE). The REEs are not recovered in the process and end up in the slag. The EU commission emphasised maximizing material recovery through the battery directive (2006/66/EC), which is currently not achievable using pyrometallurgical treatment. Thus, it is expected that hydrometallurgical treatment will become a leading approach in battery processing. In the majority of the developed processes, REEs are recovered as a mixture [5, 6]. To increase their potential, the utilization of REE mixtures should be explored. Otherwise, the demand for individual separation of REEs will limit the feasibility of current technologies, as it happened with many developed recycling processes. The REEs in the feed are the main REEs used in commercial magnetocaloric materials with general formula La $(Fe, Si)_{13}$. Recently, the recovery of REEs from magnetocaloric materials has been investigated by C.Tunsu et al. [7]. The $La(Fe,Si)_{13}$-based compounds attracted much interest due to the high MCE, tunable T_C, tunable transition nature, non-toxicity, and the low cost [8, 9]. Manganites and orthoferrites for magnetocaloric purposes are, in general, FM perovskites, with the formula $La_{1-x}RE_xMn_{1-x}M_xO_3$, $La_{1-x}RE_xFe_{1-x}M_xO_3$ (where RE is a rare earth which partially or totally substitutes lanthanum, M is a transition element, which partially substitutes Mn) [10, 11]. Thus, one of the options is to use the REE mixtures for magnetocaloric applications. The synthesis of new magnetocaloric materials starting from output of NiMH batteries is investigated in this paper.

Experimental Details and Methods

Materials: NiMH batteries, HCl, Cyanex 923, tributyl phosphate diluted (TBP) in kerosene, decanol, $NaNO_3$, HNO_3, NaOH, Cl_3Al, $6H_2O$, Cl_3La, $7H_2O$, Cl_3Ce, $7H_2O$, Cl_3Nd, $6H_2O$, Cl_3Pr, xH_2O, Cl_3Y, $6H_2O$, Cl_2Co, $6H_2O$, $Cl_2Mn,4H_2O$, $NiCl_2$, $ZnCl_2$, and oxalic acid were purchased from Merck. Cyanex 923 is purchased from Cytec. Erasteel provided the magnetocaloric material that we used as a reference.

Instrumentation: For elementary analysis, Inductively Coupled Plasma-Optical Emission Spectroscopy (iCAP 6500, Thermo Fischer) was utilized. The phase purity and the crystal structure were determined using X-ray diffraction measurements at room temperature using an X-ray diffractometer Bruker D8 Advance.

Generation of the REEs Stream by Processing Spent NiMH Batteries

The dissolution of spent NiMH waste is the method that we used to produce an aqueous solution, which will be subsequently treated using separation and recovery methods. Black mass containing active battery material was obtained by manual dismantling of car NiMH batteries. Hydrochloric acid (8 M) was used as a leaching medium and leaching temperature was 30 °C. The extraction system consisted of the solvating extractant trialkylphosphine oxide mixture Cyanex 923 and tributyl phosphate diluted (TBP) in kerosene. Pre-main extraction process (8% Cyanex 923, 10% TBP, 82% kerosene) was used to remove Fe and Zn from the leachates. Zn and Fe were extracted in four extraction stages followed by three scrubbing stages to remove co-extracted metal ions and four stripping stages to remove extracted Zn and Fe (99.9%). Al and REEs were separated from Ni, K, and Mg in three main extraction stages (70% Cyanex 923, 10% TBP, 10% kerosene, 10% 1-Decanol). High purity nickel (>99.9%) in a raffinate was obtained after main extraction. Loaded organic was stripped with $NaNO_3$ and HNO_3 to remove co-extracted Co, Mn, and Ni. Al and REEs were recovered by stripping process using 1 M HCl after previous scrubbing with 1 M $NaNO_3$ which was used to remove residual Co and Mn.

The final stripping product collects all REEs present in the black mass together with aluminum. Low concentration of cobalt and manganese is also present.

Composition of the leachate is given in Table 1 and composition of the separated streams is given in Table 2.

Table 1 Concentration of the metal ions in the leachate

Element	Content [g/L]/type of material	Element	Content [g/L]/type of material
Al	1.9 ± 0.2	Mn	6.0 ± 0.1
Ce	8.2 ± 0.1	Nd	3.0 ± 0.1
Co	12.0 ± 0.1	Ni	69.0 ± 0.3
Fe	0.3 ± 0.01	Pr	3.0 ± 0.1
La	21.0 ± 0.2	Y	2.0 ± 0.2
Mg	0.07 ± 0.01	Zn	0.5 ± 0.02

Table 2 Concentration of the metal ions in the stripping product

Element	Content [g/L]/type of material	Element	Content [g/L]/type of material
Al	1.9 ± 0.2	Mn	6.0 ± 0.1
Ce	8.2 ± 0.1	Nd	3.0 ± 0.1
Co	12.0 ± 0.1	Ni	69.0 ± 0.3
Fe	0.3 ± 0.01	Pr	3.0 ± 0.1
La	21.0 ± 0.2	Y	2.0 ± 0.2
Mg	0.07 ± 0.01	Zn	0.5 ± 0.02

Removal of the Impurities Using Oxalate Precipitation and Solvent Extraction

The separation is defined by two main parameters, which are the distribution ratio and the separation factor. The distribution ratio, D, is defined as the ratio between the concentration of metal (X) in the organic phase and the concentration of metal in the aqueous phase at equilibrium [12].

The composition of the output from processing of NiMH batteries is given in Table 3. As mentioned before, the solution consists of HCl with dissolved REEs (lanthanum, cerium, neodymium, praseodymium, and yttrium) and aluminum. A commercial magnetocaloric material containing lanthanum, cerium, iron, manganese, and silicon was investigated and its composition is given in Table 3. In addition, the composition of commercial magnetocaloric materials belonging to the family $La(FeSi)_{13}$, according to a European manufacturer, is given. The predominance of lanthanum and cerium in the solution after processing of NiMH batteries and their ratio makes this feed attractive for manufacturing of similar magnetocaloric alloys.

Table 3 Composition of the output from hydrometallurgical processing of NiMH batteries and typical compositions of magnetocaloric materials

Element	Output solution from processing of NiMH batteries (g/L)	Investigated commercial magnetocaloric material (%)	Composition of magnetocaloric material according to a European manufacturer (%)
Al	1.2	–	–
La	13.5	12	0–15
Ce	5.2	4.5	0–8
Nd	1.6	–	–
Pr	2.1	–	–
Y	0.8	–	–
Fe	–	80	Balance
Mn	–	1	0–5
Si		2.5	3–10

Precipitation Separation

Precipitation is one of the methods to separate elements from aqueous solution. Rare earth ions in solution can be precipitated in various forms, e.g., as hydroxides (REE $(OH)_3$), fluorides ($REEF_3$), sulfates ($REE_2(SO_4)_3$), phosphates (REE PO_4), etc. Of notable importance is the low solubility of REE oxalates, $REE_2(C_2O_4)_3$. Oxalic acid precipitation is commonly used to precipitate REEs according to Eq. 1. The oxalates are insoluble in water, diluted acids, and can be easily separated using filtration.

$$2\,REE^{3+}_{(aq)} + 3\,C_2O_{4\,(aq)}^{2-} \rightarrow REE_2(C_2O_4)_{3(s)} \quad (1)$$

Since REEs have similar chemical behavior, precipitation with oxalic acid will lead to a mixed REE precipitate.

Results and Discussion

We have prepared a solution similar to the feed obtained from hydrometallurgical processing of NiMH batteries starting from stoichiometric amount of Cl_3Al, $6H_2O$, Cl_3La,7 H_2O, Cl_3Ce, $7H_2O$, Cl_3Nd, $6H_2O$, Cl_3Pr, xH_2O, Cl_3Y, $6H_2O$, Cl_2Co, $6H_2O$, Cl_2Mn, $4H_2O$, $NiCl_2$, and $ZnCl_2$. Different amounts were dissolved in 0.1 M HCl solution. The prepared solution was mixed with a precise amount of NaOH (99 wt%, Merck) under magnetic stirring for 24 h. Then, it was filtered and treated with oxalic acid. A mixture of REEs oxalate was obtained. The pH of the resulting solution was calculated for each step. The ICP-OES analysis was performed for three different oxalates 'A', 'B', and 'C' and the calculated composition is given in Table 4.

Synthesis of Manganites and Orthoferrites

Among the many possible synthetic procedures, we exploited the solid-state reaction method. This method has been widely used for the synthesis of manganites and orthoferrites for magnetic refrigeration application. High purity powders of oxalate

Table 4 Calculated composition of the prepared oxalates

Composition	% Co	% Ni	% Zn	% Al	% La	% Ce	% Y	% Mn	% Nd	% Pr	Na
Oxalate A	0.009	0	0.007	0	69.8	15.8	1.23	0.03	6.73	6.46	0
Oxalate B	0.009	0	0.006	0	69.3	15.9	1.36	0.04	6.86	6.53	0
Oxalate C	0.001	0.004	0.007	0	71.24	14.52	1.34	0.05	6.29	6.51	0.05

'A' or oxalate 'C', Mn $(C_2O_4)_3 \cdot 2H_2O$, Fe $(C_2O_4)_3 \cdot 2H_2O$ were mixed in stoichiometric proportions according to the following reactions:

$$\begin{aligned} &REE_2(C_2O_4)_{3(s)} + 2(1-x)Mn(C_2O_4)_3.H_2O_{(s)} + \delta' O_{2(g)} \\ &\rightarrow 2\ REE'Mn_{1-x}M_xO_{3(s)} + \delta'' CO_{2(g)} + \delta''' H_2O \end{aligned} \tag{2}$$

$$\begin{aligned} &REE_2(C_2O_4)_{3(s)} + 2(1-x)Fe(C_2O_4)_3 \cdot 2H_2O_{(s)} + \delta' O_{2(g)} \\ &\rightarrow\ 2\ REE'Fe_{1-x}M_xO_{3(s)} + \delta''\ CO_{2(g)} + \delta''' H_2O \end{aligned} \tag{3}$$

where M = Zn, Mn, Co and REE′ = REEs − M.

The mixtures were first grinded by hand, calcined, and finally sintered at different temperatures with intermediate grinding and pelleting under an axial pressure. The preparation was realized under air atmosphere. The annealing temperatures were 500 °C for 12 h, 800 °C for 24 h, and 1000 °C for 24 h for the orthoferrite 'A', and 500 °C for 24 h, 800 °C for 24 h, and 900 °C for 24 h for the orthoferrite 'C'. For the manganite 'C', the calcination and sintering were realized at 500 °C for 24 h, 800 °C for 24 h, 900 °C for 24 h, 1000 °C, and 1200 °C for 24 h.

The X-ray diffraction patterns in Fig. 1a and b show that the orthoferrite 'A' and orthoferrite 'C' are crystallized in a well-defined orthorhombic structure with the presence of rare earth oxide and Fe_2O_3 as secondary phases. The manganite 'C' is

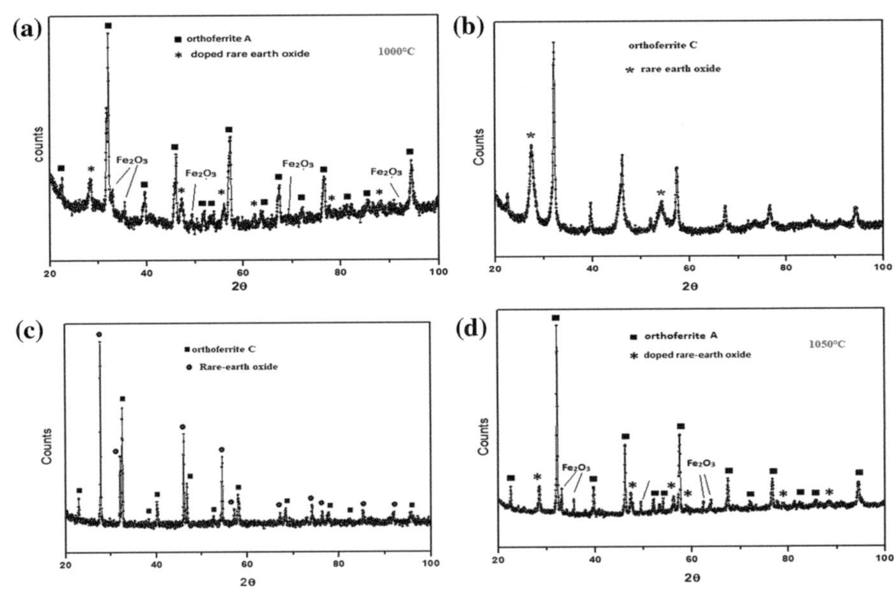

Fig. 1 **a** X-ray diffraction pattern of orthoferrite 'A' prepared by solid-state reaction, **b** X-ray diffraction pattern of the orthoferrite 'C' prepared by solid-state reaction at 1000 °C, **c** X-ray diffraction pattern of manganite 'C' prepared by solid-state reaction, **d** Orthoferrite 'C' sintered at 1050 °C

crystallized with presence of high intense peaks of rare earth oxide. We can note that this compound is a composite (Fig. 1c).

For the orthoferrite 'A', the effect of sintering temperature on structural properties is studied (Fig. 1d). A zoom of the main peaks indicates that the crystal structure has been transformed from rhombohedral to orthorhombic structure after annealing of the sample obtained at 1000 °C for 1 h at 1050 °C in air.

Synthesis of the As-Cast Alloy REEs $(Fe_{1-x}M_x)_{11.5}Si_{1.5}$

The cubic $NaZn_{13}$-type La $(Fe_{1-x}Si_x)_{13}$ compounds have attracted a great attention owing to the sizable MCE that they are showing at low field. An important feature of these compounds is the strong dependence of its T_C and nature of phase transition on element substitution or external perturbations such as magnetic field and pressure [13, 14]. The transition temperature can be tuned by substituting Fe with Co or Mn [15, 16]. Moreover, by substituting La with rare earths (Ce, Nd, and Pr), a slight enhancement of MCE was observed [17–19]. A previous study showed that multi-replacement of La by magnetic atoms Ce, Pr, and Nd enhances the Curie temperature T_C, and La–Ce–Pr–Nd mischmetal was used to produce $La_{0.7}$(Ce, Pr, $Nd)_{0.3}Fe_{11.6}Si_{1.4}$ and its hydrides [20]. The importance of substitution in La and Fe sites makes the output obtained from hydrometallurgical processing of NiMH batteries a very good candidate that can be used instead of purified REE metals for the synthesis of magnetocaloric alloys. Our objective from re-use of outputs is the cost reduction.

The conventional arc and induction melting processes followed by a long annealing treatment, which can take up to 7 days at 1323–1523 K, have been used. To minimize the time of annealing, new alternatives such as melt spinning for producing thin ribbons, spark plasma sintering, ball milling, strip casting, and powder metallurgy method were employed [21–24]. Teixeira et al. [25] have applied the reduction diffusion process to produce $La(FeSi)_{13}$ using La_2O_3 as raw material.

To synthesize the as-cast alloy, carbothermic followed by calciothermic reduction and hydrogen reduction followed by calciothermic reduction were applied in this work. For the carbothermic reduction method, a stoichiometric mixture of orthoferrite 'A', Fe_2O_3, and SiO_2 was grinded with ball miller for 30 min and annealed in air for 2 h at 1000 °C. This composite was milled with carbon and treated for 4 h at 1300 °C under argon atmosphere. Then the resulting compound was mixed with Ca and annealed for 3 h. During this process, CaO was also formed. The reduced samples were thoroughly washed with deionized water and a little amount of dilute acetic acid (0.5 M) was added to remove the CaO and then washed with deionized water several times. Figure 2 shows the diffraction pattern for the as-cast sample REE′ $Fe_{11.5-x}M_xSi_{1.5}$. The sample contains a majority of α-Fe (Si) phase and a minor amount of rare earth cobaltite phase and carbon.

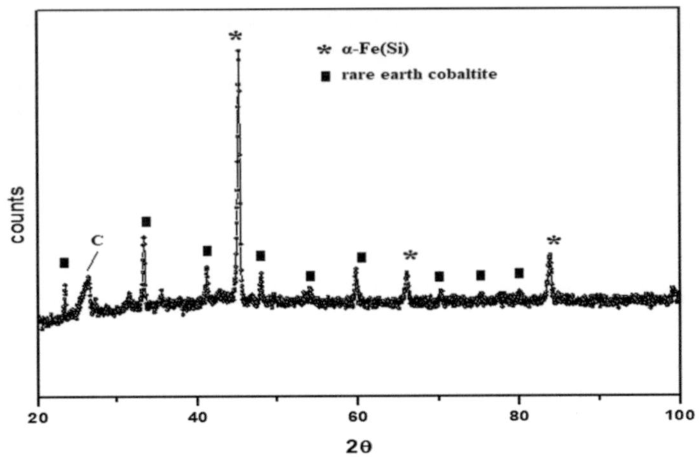

Fig. 2 X-ray diffraction pattern of the as-cast alloy REE'$Fe_{11.5-x}M_xSi_{1.5}$ produced by carbothermic followed by calciothermic reduction

For calciothermic reduction, oxalate 'B' treated for 3 h at 300 °C, Fe (99.99%), Si (99.99%), and Ca (99.99%) were milled for 1 h with ball miller. 10% excess of oxalate amount and 100% excess of calcium were added to compensate the loss of REEs during the reaction. The mixture was put into porcelain combustion boat and placed inside the horizontal quartz tube. The experiments were performed at 800 °C under 500 mL/min constant 5% H_2 (g)–N_2 (g) mixture. After 1 h, the hydrogen gas flowing is stopped. The reduction by calcium was achieved at 1250 °C for 10 h under 250 mL/min flowing argon gas. The reduced sample was washed with

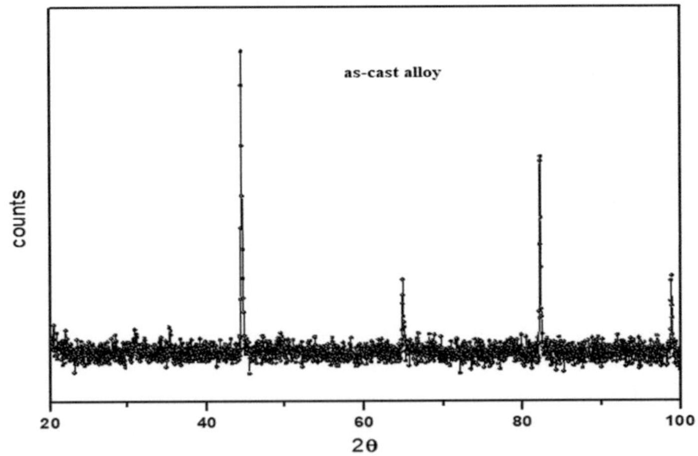

Fig. 3 X-ray diffraction pattern of the as-cast alloy REE'$Fe_{11.5-x}M_xSi_{1.5}$ produced by hydrogen reduction followed by calciothermic reduction

deionized water and dilute acetic acid (0.5 M) to remove CaO. The XRD pattern in Fig. 3 shows the formation of major phase α−Fe (Si), and LaFeSi phase with the presence of La_5Si_4 phase is detected after the washing step.

Conclusions

A mixture of rare earth oxalate was obtained after the hydrometallurgical processing of spent NiMH batteries was successfully carried out in this study. This mixture was successfully used to synthesize manganites and orthoferrites. The X-ray diffraction measurements showed the presence of impurities on the samples. Carbothermic and calciothermic reductions were used to prepare the as-cast alloy which will be treated at high temperature to obtain the REE FeSi-13-based alloys. Further studies will be performed to determine the magnetic and magnetocaloric properties of the obtained samples.

Acknowledgements This research was supported by FORMAS (Grant No: 2017-00637) and ÅForsk (Grant No: 16-406).

References

1. Shah N, Park WY, Bojda N, McNeil M, Waide, P (2014) Superefficient refrigerators: opportunities and challenges for efficiency improvement globally
2. Bruck E (2005) Developments in magnetocaloric refrigeration. J Phys D Appl Phys 38 (23):381–391
3. Tunsu C, Petranikova M, Gergoric M, Ekberg C, Retegan T (2015) Reclaiming rare earth elements from end-of-life products: a review of the perspectives for urban mining using hydrometallurgical unit operations. Hydrometallurgy 156:239–258
4. USGS (2002) Rare earth elements—critical resources for high technology
5. Larsson K, Ekberg C, Ødegaard-Jensen A (2013) Using Cyanex 923 for selective extraction in a high concentration chloride medium on nickel metal hydride battery waste: Part II: mixer–settler experiments. Hydrometallurgy 133:168–175
6. Zhang P, Yokoyama T, Itabashi O, Wakui Y, Suzuki TM, Inoue K (1998) Hydrometallurgical process for recovery of metal values from spent nickel-metal hydride secondary batteries. Hydrometallurgy 50:61–75
7. Tunsu C, Petranikova M (2018) Perspectives for the recovery of critical elements from future energy-efficient refrigeration materials. Clean Prod 197(1):232–241
8. Hu FX, Shen BG, Sun JR, Cheng ZH, Rao GH, Zhang XX (2001) Influence of negative lattice expansion and metamagnetic transition on magnetic entropy change in the compound $LaFe_{11.4}Si_{1.6}$. Appl Phys Lett 78:3675
9. Fujita A, Fujieda S, Hasegawa Y, Fukamichi K (2003) Itinerant-electron metamagnetic transition and large magnetocaloric effects in $La(Fe_xSi_{1-x})_{13}$ compounds and their hydrides. Phys Rev B 67:104416
10. M'nassri R, Cheikhrouhou-Koubaa W, Chniba Boudjada N, Cheikhrouhou A (2013) Effect of barium-deficiency on the structural, magnetic, and magnetocaloric properties of $La_{0.6}Sr_{0.2}Ba_{0.2-x}\square_xMnO_3$ ($0 \leq x \leq 0.15$). J Appl Phys 113:073905

11. Chihi I, Baazaoui M, Mahjoub S, Cheikhrouhou-Koubaa W, Oumezzine M, Farah K (2019) Study of the magnetic and magnetocaloric properties of new perovskite-type materials: $La_{0.6}Ba_{0.2}Sr_{0.2}Mn_{1-x}Fe_xO_3$. Appl Phys A
12. Petranikova M, Ebin B, Tunsu C (2019) Selective recovery of cobalt from the secondary streams after NiMH batteries processing using Cyanex 301. Waste Manag 83:194–201
13. Lyubina J, Nenkov K, Schultz L, Gutfleisch O (2008) Multiple metamagnetic transitions in the magnetic refrigerant La (Fe, Si) (13) H−x. Phys Rev Lett 101:177203
14. Chen YF, Wang F, Shen BG, Wang GJ, Sun JR (2003) Magnetism and magnetic entropy change of $LaFe_{11.6}Si_{1.4}C_x$ ($x = 0 - 0.6$) interstitial compounds. J Appl Phys 93:1323–1325
15. Chen X, Chen YG, Tang YB, Xiao DQ (2014) The system study of 1:13 phase formation, the magnetic transition adjustment, and magnetocaloric property in La(Fe, $Co)_{(13-x)}Si_x$ alloys. J Magn Magn Mater 368:155–168
16. Basso V, Kupferling M, Curcio C, Bennati C, Barzca A, Katter M et al (2015) Specific heat and entropy change at the first order phase transition of La(Fe–$MnSi)_{(13)}$–H compounds. J Appl Phys 118(5):053907
17. Fujieda S, Fujita A, Fukamichi K (2004) Enhancements of magnetocaloric effects in La $(Fe_{0.90}Si_{0.10})_{(13)}$ and its hydride by partial substitution of Ce for La. Mater Trans 45:3228–3231
18. Anh DTK, Thuy NP, Duc NH, Nhien TT, Nong NV (2003) Magnetism and magnetocaloric effect in $La_{1-y}Nd_y(Fe_{0.88}Si_{0.12})_{(13)}$ compounds. J Magn Magn Mater 262:427–431
19. Fujieda S, Fujita A, Fukamichi K (2007) Reduction of hysteresis loss in itinerant-electron metamagnetic transition by partial substitution of Pr for La in La $(Fe_xSi_{1-x})_{(13)}$. J Magn Magn Mater 310:E1004–5
20. Bao LF, Hu FX, Chen L, Wang J, Sun JR et al (2012) Magnetocaloric properties of La(Fe, $Si)_{13}$-based material and its hydride prepared by industrial mischmetal. Appl Phys Lett 101:162406
21. Liu XB, Liu XD, Altounian Z (2005) Phase formation and magnetocaloric effect in rapidly quenched $La(Fe_{1-x}Co_x)_{11.4}Si_{1.6}$. J Appl Phys 98:113904
22. Zhang MX, Liu J, Zhang Y, Dong JD, Yan AR, Skokov KP et al (2015) Large entropy change, adiabatic temperature change, and small hysteresis in La(Fe, $Mn)_{11.6}Si_{1.4}$ strip-cast flakes. J Magn Magn Mater 377:90–94
23. Dong JD, Yan AR, Liu J (2014) Microstructure and magnetocaloric properties of melt-extracted La–Fe–Si microwires. J Magn Magn Mater 357:73–76
24. Patissier A, Fast P-BV (2015) Synthesis of $LaFe_{13-x}Si_x$ magnetocaloric compounds by reactive Spark Plasma Sintering. J Alloys Comp 645:143–150
25. Teixeira CS, Caron L, Anastasopol A, Eijt SWH, Lozano JA, Brück E, Wendhausen PAP (2012) A new feature of the reduction–diffusion process applied for the synthesis of magnetocaloric $LaFe_{13-x}Si_x$ compounds. J Alloys Comp 541:84–87

Selective Recovery of Scandium from Nickel Laterite Ore by Acid Roasting–Water Leaching

John Anawati, Runlin Yuan, Jihye Kim and Gisele Azimi

Abstract Scandium is a rare and expensive metal that is in increasing demand because of its unique properties. Laterite ore is known to contain scandium in appreciable amounts. In this study, a two-stage process called "acid roasting–water leaching" was developed to extract scandium from laterite ore with minimal co-extraction of iron. In this process, the ore was mixed with concentrated sulfuric acid (98 wt%) and roasted in a furnace at or above 600 °C. The roasted ore was then leached in water at ambient conditions. During acid roasting, $Fe_2(SO_4)_3$ is decomposed to Fe_2O_3, which is insoluble in water, and thus it can be separated during the leaching process. A fractional factorial design methodology was utilized to investigate the effect of various operating parameters during the roasting and leaching processes and to optimize the processes. After optimization, 80% of scandium was recovered with less than 15% co-extraction of iron.

Keywords Acid roasting–water leaching · Selective sulfation · Pyro-hydrometallurgy · Primary scandium extraction

Introduction

In an effort to achieve and surpass performance, efficiency and sustainability goals, various industries in the global technological economy are increasingly relying on advanced specialty materials with unique properties. One such material is scandium (Sc), a rare and expensive metallic element which is primarily used to produce high-performance aluminum and magnesium alloys with high strength, good

J. Anawati · R. Yuan · J. Kim · G. Azimi (✉)
Laboratory for Strategic Materials, Department of Chemical Engineering and Applied Chemistry, 200 College Street, Toronto, ON M5S 3E5, Canada
e-mail: g.azimi@utoronto.ca

G. Azimi
Department of Materials Science and Engineering, University of Toronto, 184 College Street, Toronto, ON M5S 3E4, Canada

G. Azimi et al. (eds.), *Rare Metal Technology 2020*, The Minerals, Metals & Materials Series, https://doi.org/10.1007/978-3-030-36758-9_8

weldability, corrosion resistance, and heat tolerance for the aerospace and sports equipment industries [1, 2]. This metal is also employed in specialty applications, such as dopant in high intensity lamps and garnets for lasers and solid oxide fuel cells. Scandium is expensive and has a low global production rate, with a current Sc_2O_3 (99.99%) market price of US$4600/kg and an estimated annual production of 10–15 tonnes [3].

Geologically, scandium exists as a minor element in ores of other materials, primarily rare earths, titanium, apatite, and uranium, and does not concentrate into its own mineral ores, because of its large ionic radius-to-charge ratio; [4] as a result of its scarcity and high value, resources with a scandium grade between 20 and 50 mg/kg can be considered to be scandium ores [5]. Typically, scandium is produced as a by-product of other industrial processes, mainly the extraction and refining of uranium [3]. Work has also been conducted to recover Sc from secondary waste sources [6]. Dedicated scandium mines are not common; in 2011, only three mines operated worldwide, and work is currently underway to open the first primary Sc mine; however, to address projected increase in demand, more scandium-specific extraction and refining is required [6, 7]. One potential scandium resource is nickel laterite ore, which can have scandium grades of up to 50–60 mg/kg [6, 8]. In this work, a nickel laterite ore from New Caledonia was investigated as a potential scandium source.

Previous work on scandium recovery from low grade sources has typically focused on direct acid leaching as it is generally a simple and economical approach; [9] however, direct acid leaching has several disadvantages, such as the production of large volumes of acidic waste, large consumption of reagents, and the co-extraction of undesired materials. In particular, the co-extraction of Fe is particularly problematic because of the chemical similarities between Fe and Sc, which makes their separation complex [10, 11].

One alternative process which has been proposed for the extraction of Sc or other valuable materials from iron-rich materials is acid roasting–water leaching, a multi-step hybrid pyro-hydrometallurgical process which combines a dry sulfation step with a heat treatment step to selectively produce a solid mixture of soluble Sc species and insoluble Fe species, allowing the extraction of Sc by leaching with water at ambient conditions (1 atm, 25 °C) [12–15]. Most of the research in the selective sulfation-roasting-leaching field has focused on the extraction of scandium from bauxite residue; however, the concept of exploiting the differences in the thermal stability of different metal sulfates to selectively roast and leach ores has been previously explored; [16–18] in particular, selective fixed bed sulfation using SO_3 gas and sulfuric acid roasting has been used to selectively leach nickel and cobalt from nickel laterite ore while rejecting Fe [19–21].

The main driving mechanism for this selective process is the differences in thermal decomposition temperatures of different water-soluble metal sulfates: the onset of $Fe_2(SO_4)_3$ decomposition is reported as 545 °C and the onset of $Sc_2(SO_4)_3$ decomposition is 700 °C [12]. In the first step of the process, the ore is mixed with concentrated H_2SO_4 and incubated at 50 °C for 24 h to convert all the Fe and Sc species to their sulfate forms (Reaction 1 and Reaction 2).

$$2Fe(OH)_{3(s)} + 3H_2SO_{4(l)} \rightarrow Fe_2(SO_4)_{3(s)} + 6H_2O_{(l)} \quad \text{(Reaction 1)}$$

$$2Sc(OH)_{3(s)} + 3H_2SO_{4(l)} \rightarrow Sc_2(SO_4)_{3(s)} + 6H_2O_{(l)} \quad \text{(Reaction 2)}$$

The sulfated solids are then roasted at 600–800 °C to convert the soluble contaminant sulfates ($Fe_2(SO_4)_3$), which have lower thermal stability to water-insoluble oxide forms, while driving off the sulfate groups as SO_3 vapors (Reaction 3). The process conditions, such as roasting temperature, duration, and the addition of water to the mixture during the sulfation process, must be tuned to avoid the conversion of soluble $Sc_2(SO_4)_3$ to insoluble Sc_2O_3 (Reaction 4).

$$Fe_2(SO_4)_{3(l)} \rightarrow Fe_2O_{3(s)} + 3SO_{3(g)} \quad \text{(Reaction 3)}$$

$$Sc_2(SO_4)_{3(l)} \rightarrow Sc_2O_{3(s)} + 3SO_{3(g)} \quad \text{(Reaction 4)}$$

This roasting step has the benefit of selectively allowing Sc to be extracted by leaching with water while minimizing co-extraction of Fe. A side benefit of this process is that some of the H_2SO_4 used in the process can potentially be recovered and regenerated by collecting and condensing the SO_3 fumes produced in the roasting process. Temperatures above the onset of $Sc_2(SO_4)_3$ decomposition (Reaction 4) were investigated to assess whether the benefit of the accelerated rate of $Fe_2(SO_4)_3$ decomposition (Reaction 3) at higher temperatures would outweigh Sc losses.

The ARWL process produces a leachate which is enriched in Sc, Co, and Ni, which can be subjected to solvent extraction (SX) with D2EHPA (Di-(2-ethylhexyl) phosphoric acid) for the purification and recovery of Sc, and other valuable metals,

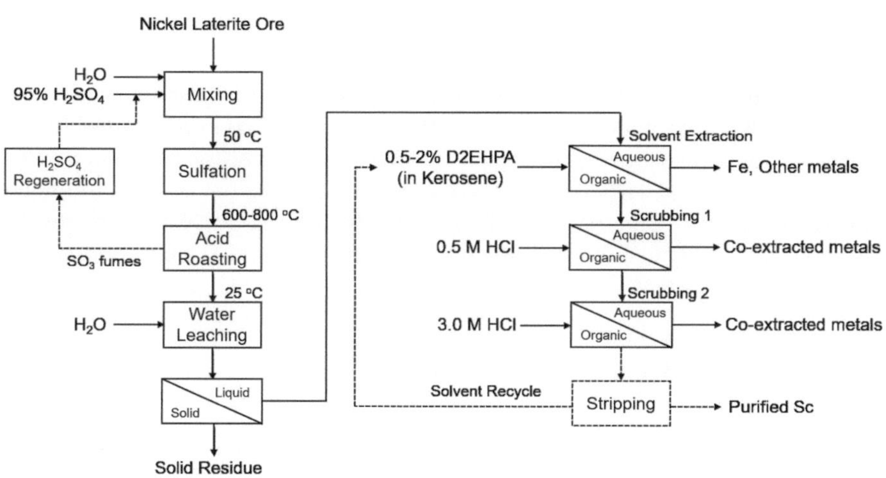

Fig. 1 Proposed process flow diagram for the ARWL-SX process for the recovery of Sc, Co, and Ni from nickel laterite ore. The solid lines represent steps described in this study and dashed lines represent the potential final process flow at industrial scale

which are simplified by the reduced Fe content of the leachate, and a solid residue, comprising primarily Fe_2O_3 which can be disposed of or recycled. The process flow diagram of the proposed process is shown in Fig. 1.

This work outlines initial screening trials to understand the different parameters which can affect the ARWL process and provides a preliminary optimization of these parameters to maximize Sc extraction from laterite ore, while minimizing Fe co-extraction. The overall goal of this work is to assess the feasibility of using ARWL as a technique to efficiently recover Sc from this primary resource to address global demands for this rare and expensive metal. The extractions of Ni and Co were also measured in this work; however, the extraction of these metals from laterite ores by ARWL has been previously described in the literature [20, 21] and was not the focus of this study.

Materials and Methods

Materials

A nickel laterite ore (LO) from New Caledonia was used as the process feedstock. 95.0–98.0 wt% H_2SO_4 (ACS Reagent grade), 68.0–70.0 wt% HNO_3 (ACS Reagent grade), and 36.5–38.0 wt% HCl (ACS Reagent grade) were purchased from VWR. 95% Bis-(2-ethylhexyl)-phosphate (D2EHPA) was purchased from Alfa Aesar. Deionized water (0.055 μS, Millipore) was used for dilutions and water leaching.

Laterite Ore Characterization

The characterization of the laterite ore feed was carried out by microwave-assisted aqua regia digestion followed by ICP-OES (inductively coupled plasma optical emission spectrometry—Perkin Elmer Optima 8000) analysis. The concentrated aqua regia solution (3 HCl:1 HNO_3) was used to digest the ore at a solid-to-liquid ratio of 7.0 g_{LO}/L. Digestion was carried out in a MARS6 Xpress microwave digestion system (9 °C/min ramp up, 20 min dwell at 200, 5 °C/min ramp down).

Acid Roasting–Water Leaching (ARWL)

Dry laterite ore (LO) was mixed with 0.5–2 $g_{H_2SO_4}/g_{LO}$ and 0–1 g_{H_2O}/g_{LO} in a porcelain crucible in ratios prescribed by the experimental design. The acid ore–water mixtures were heated at 50 °C for 24 h to ensure sulfation. The sulfated ore was then roasted at 600–800 °C for 30–90 min in a muffle furnace. The acid-roasted ore samples (ARLO) were cooled, and then mixed with deionized water at a liquid-to-solid ratio of 4–16 mL H_2O/g_{ARLO}, in a magnetic stirrer-equipped flask with an agitation rate of 300–500 rpm. The leachate solution was sampled at 30 min and 120 min using 0.45 μm nylon syringe filters (Fisher Scientific). The filtered leachate solutions were immediately diluted with 5 wt% HNO_3 to allow ICP-OES elemental quantification.

Solvent Extraction

The leachate solutions obtained after the ARWL process were mixed with 0.5–2 vol.% of extractant, D2EHPA, diluted in kerosene at an organic-to-aqueous ratio of 1:1 (10 mL each). The mixture was then shaken for 20 min at room temperature. After phase separation, the organic solutions were mixed with 0.5 M and 3 M HCl for scrubbing. Aqueous solutions obtained after each step were analyzed by ICP-OES.

ICP-OES Analysis

The leachate solutions produced in the water leaching step and the aqueous solutions produced in the solvent extraction steps were analyzed by ICP-OES for their elemental concentrations.

SEM-EDS and XRD Characterization

Sample characterizations were performed with scanning electron microscopy and energy dispersive X-ray spectroscopy (Hitachi SU3500). Mineralogical analysis was performed with X-ray diffraction (XRD) (Rigaku MiniFlex 600 Diffractometer).

Experimental Design

The effects of the experimental parameters (X_1: acid-to-LO ratio, X_2: water-to-ARLO ratio, X_3: roasting duration, X_4: water-to-LO ratio, X_5: leaching agitation rate, and X_6: roasting temperature) were assessed with a series of tests following a 2_{IV}^{6-2} fractional factorial design of experiments. The leaching duration (X_7) was additionally incorporated into the experimental design by sampling the leachate at both 30 and 120 min. The full experimental design matrix is given in Table 1. For analysis, the extraction parameters (X_i) were coded between −1 (low) and +1 (high) according to Eq. 1 (x_i corresponds to the uncoded factor effect level).

Table 1 Experimental design matrix for ARWL. The uncoded parameter levels for each trial are given. The trials were conducted in random order

Test ID	X_1: Acid–LO ratio ($g_{H_2SO_4}/g_{LO}$)	X_2: Water–ARLO ratio (mL H_2O/g_{ARLO})	X_3: Roast duration (min)	X_4: Water–LO ratio (g_{H_2O}/g_{LO})	X_5: Agitation rate (rpm)	X_6: Roast temperature (°C)	X_7: Leach time (min)
2_{IV}^{6-2} *fractional factorial trials*							
A1	2.00	4	90	1.00	300	600	30, 120
A2	0.50	16	30	1.00	500	600	30, 120
A3	2.00	4	30	0	500	600	30, 120
A4	0.50	16	90	0	300	600	30, 120
A5	0.50	4	30	0	300	600	30, 120
A6	2.00	16	30	1.00	300	600	30, 120
A7	2.00	16	90	0	500	600	30, 120
A8	0.50	4	90	1.00	500	600	30, 120
A9	2.00	4	90	0	300	800	30, 120
A10	2.00	4	30	1.00	500	800	30, 120
A11	0.50	16	30	0	500	800	30, 120
A12	0.50	4	90	0	500	800	30, 120
A13	0.50	4	30	1.00	300	800	30, 120
A14	2.00	16	90	1.00	500	800	30, 120
A15	0.50	16	90	1.00	300	800	30, 120
A16	2.00	16	30	0	300	800	30, 120
A17	1.25	10	60	0.50	400	700	30, 120
A18	1.25	10	60	0.50	400	700	30, 120
A19	1.25	10	60	0.50	400	700	30, 120
Validation trials							
V1	2.00	16	30	0	300	800	30, 120
V2	0.50	16	90	0	300	800	30, 120
V3	0.50	4	90	0	500	800	30, 120
V4	2.00	16	60	0	500	700	30, 120
V5	2.00	16	30	0.50	400	600	30, 120
V6	1.25	10	90	1.00	500	600	30, 120

$$X_i = \frac{x_i - \frac{1}{2}\left(x_{i,high} + x_{i,low}\right)}{\frac{1}{2}\left(x_{i,high} - x_{i,low}\right)} \tag{1}$$

Empirical Model Building

Empirical leaching models of the form given in Eq. 2 were constructed from the extraction results by multiple linear least squares regression (mLLSR). The empirical model building procedure is fully described elsewhere [9, 22].

$$\begin{aligned}\hat{y}_a = \beta_0 &+ \beta_1X_1 + \beta_2X_2 + \beta_3X_3 + \beta_4X_4 + \beta_5X_5 + \beta_6X_6 + \beta_7X_7 + \tilde{\beta}_{12}X_1X_2 \\ &+ \tilde{\beta}_{13}X_1X_3 + \tilde{\beta}_{14}X_1X_4 + \tilde{\beta}_{15}X_1X_5 + \tilde{\beta}_{16}X_1X_6 + \tilde{\beta}_{24}X_2X_4 + \tilde{\beta}_{26}X_2X_6 \\ &+ \tilde{\beta}_{17}X_1X_7 + \tilde{\beta}_{27}X_2X_7 + \tilde{\beta}_{37}X_3X_7 + \tilde{\beta}_{47}X_4X_7 + \tilde{\beta}_{57}X_1X_7 + \tilde{\beta}_{67}X_6X_7\end{aligned}$$

where

$$\begin{aligned}&\tilde{\beta}_{12} = \beta_{12} + \beta_{35},\ \tilde{\beta}_{13} = \beta_{13} + \beta_{25},\ \tilde{\beta}_{14} = \beta_{14} + \beta_{56},\ \tilde{\beta}_{15} = \beta_{15} + \beta_{23} + \beta_{46} \\ &\tilde{\beta}_{16} = \beta_{16} + \beta_{45},\ \tilde{\beta}_{24} = \beta_{24} + \beta_{36},\ \tilde{\beta}_{26} = \beta_{26} + \beta_{34}\end{aligned} \tag{2}$$

Symbol definitions: $\hat{y}_a$ = model-predicted extraction for element a, β_i = model parameter for experimental factor i, β_{ij} = model parameter for the second-order interaction between experimental factors i and j, $\tilde{\beta}_{ij}$ = model parameter for the second-order interaction that is aliased with another second-order interaction.

Results and Discussion

Characterization of Laterite Ore

The elemental and mineralogical characteristics of the starting laterite ore were characterized by ICP-OES, XRD, and SEM-EDS (Fig. 2). This ore was primarily composed of iron in the form of $Fe(OH)_3$; the other major component was Si, which was primarily present in the form of insoluble SiO_2. A small amount of Al was also present as $Al(OH)_3$. The primary elements of interest were present at concentrations of Sc = 42.3 mg/kg and Fe = 41.9 wt%.

(a) Elemental composition

Element	Composition (wt%)	
Sc	**42.3 mg/kg**	**± 0.8**
Co	782 mg/kg	± 12
Ni	1.2	± 0.01
Fe	**41.9**	**± 0.8**
Al	2.0	± 0.1
Mg	1.5	± 0.1
Mn	0.70	± 0.01
Ti	0.03	± 0.001
Na	0.01	± 0.001
Ca	0.01	± 0.003
Si*	0.21	± 0.13

(b) XRD Spectrum (c) SEM

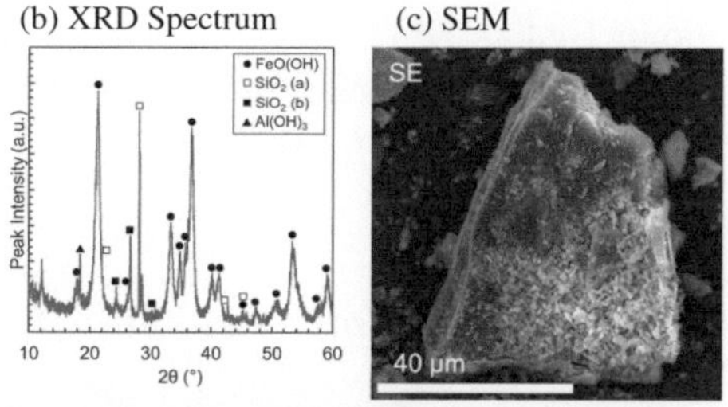

(d) SEM-EDS elemental mapping

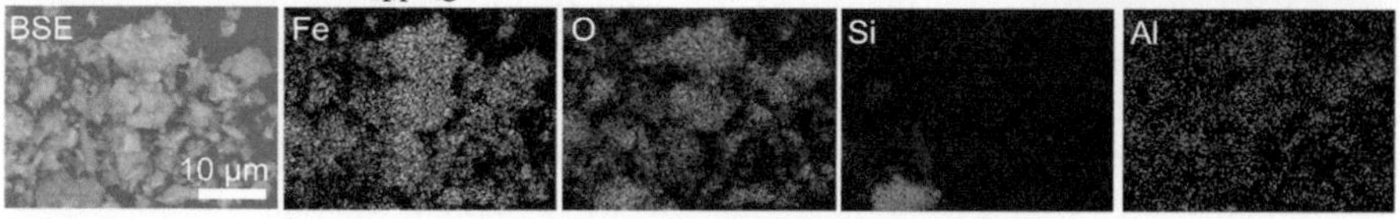

*Soluble Si

Fig. 2 Characterization of the starting laterite ore. **a** elemental composition measured by ICP-OES (average of four digestions), **b** XRD spectrum of the ore with peak identification, **c** SEM secondary electron image, **d** SEM-EDS elemental mapping (including a backscattered electron image)

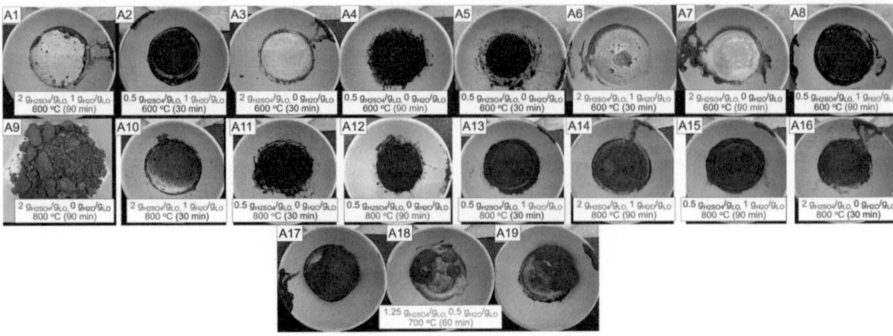

Fig. 3 Physical appearances of the acid-roasted ore samples. The acid roasting conditions are indicated for each sample

Physical Changes During Acid Roasting

The physical appearances of the acid-roasted ore samples were dependent on the acid roasting parameters, namely, the acid-to-ore ratio, the water-to-ore ratio, the roasting temperature, and the roasting time (Fig. 3). These physical appearance differences are indicative of different physicochemical outcomes of the acid roasting process, which can contribute to more, or less, extraction in the leaching stage. Notably, the samples roasted at 600 °C had a larger variation in color, with the ones having 2 $g_{H_2SO_4}/g_{LO}$ turning green-white, and the ones having 0.5 h $g_{H_2SO_4}/g_{LO}$ becoming dark brown, and the samples roasted at 800 °C exhibiting a more uniform light brown color. This change in color variation suggests that the conversion from white $Fe_2(SO_4)_3$ to red brown Fe_2O_3 (Reaction 3) occurs at different temperatures, depending on the concentration of H_2SO_4 in the system. In bauxite residue, this conversion has been shown to occur in the 650–700 °C range at an acid-to-residue ratio of 1 $g_{H_2SO_4}/g$ [12]. Additionally, in the samples which had water added prior to roasting, the resulting solids were more foam-like and porous, likely due to water evaporation during the roasting process.

Acid Roasting–Water Leaching Extraction Models

The ARWL extraction was measured for each of the trials indicated in Table 1. Linear empirical models of the form given by Eq. 2 were fit to these results, and the factor effects are presented in Fig. 4. At 95% confidence, these models were determined to be sufficient to explain the variation in the data (F-test); however, these models were found to be inadequate to fully explain the data (R-test), likely indicating the presence of non-linear effects in the true behavior of the system. However, these factor effects can be compared amongst each other to assess the relative importance of the different operating parameters on the extraction of the various elements of interest.

In both elements presented, the factor effects that had the largest positive significant impacts were $X_1X_6 + X_4X_5$ (acid ratio × roasting temperature + roasting water addition × agitation rate), X_6 (roasting temperature), X_7 (leaching duration), and X_1X_7 (acid ratio × leaching duration). The common factors having the largest negative significant impacts were X_3 (roasting duration), X_6X_7 (roasting temperature × leaching duration), and $X_1X_3 + X_2X_5$ (acid ratio × roasting time + leaching water ratio × agitation rate) for Fe.

The objective of this process is to maximize Sc extraction while minimizing Fe extraction, thus the differences between the Sc and Fe models must be studied to selectively extract Sc. The major difference between the Fe and Sc models was that increasing X_1 (acid ratio) had a strong effect on increasing Fe extraction, while this primary factor was insignificant for Sc extraction. This difference is likely caused by the difference in thermal stability of $Sc_2(SO_4)_3$ (onset of thermal decomposition: 700 °C) and $Fe_2(SO_4)_3$ (onset of thermal decomposition: 545 °C) [12]—during roasting, the $Fe_2(SO_4)_3$ decomposes more readily, meaning that in the samples with more sulfation (high X_1), more unreacted $Fe_2(SO_4)_3$ will likely be available for leaching after roasting; however, since $Sc_2(SO_4)_3$ is more thermally stable, it is less likely to decompose, and thus less sensitive to the amount of H_2SO_4 added to the system. Similarly, this phenomenon can also explain the large effect of the X_1X_6 interaction (as both X_4 and X_5 primary factors were not statistically significant, we can assume the X_4X_5 interaction is also not significant); as higher temperature increases the rate of thermal decomposition, [23] more $[SO_4]^{2-}$ is consumed, requiring larger H_2SO_4 additions to achieve high extraction. Additionally, the rate of thermal decomposition can explain the negative effect of roasting duration (X_3), as longer time at higher temperature will result in more conversion to insoluble species.

An important consideration in factor effects analysis is the identification of factors which do not have a significant impact on the response. In both cases, despite having a noticeable impact on the physical appearance of the samples, X_4 (the addition of water prior to the sulfation step) did not have any significant impact. In previous studies, this step was added to aid the mixing between the acid and solids, with the added water evaporating in the sulfation step; [12, 20] however, sufficient mixing appears to have been achieved in this system without requiring addition of water. Additionally, the agitation rate (X_5) had no significant impact in either case, suggesting that diffusion through the liquid-phase boundary layer is not the rate-limiting step in the leaching of the acid-roasted particles [24]. In addition, several of the second-order interactions (X_2X_7, X_3X_7, X_4X_7, X_5X_7, $X_1X_4 + X_5X_6$, $X_1X_2 + X_3X_5$, $X_2X_6 + X_3X_4$) had no or nearly no significant impact, indicating a lack of complex synergistic or antagonistic interactions.

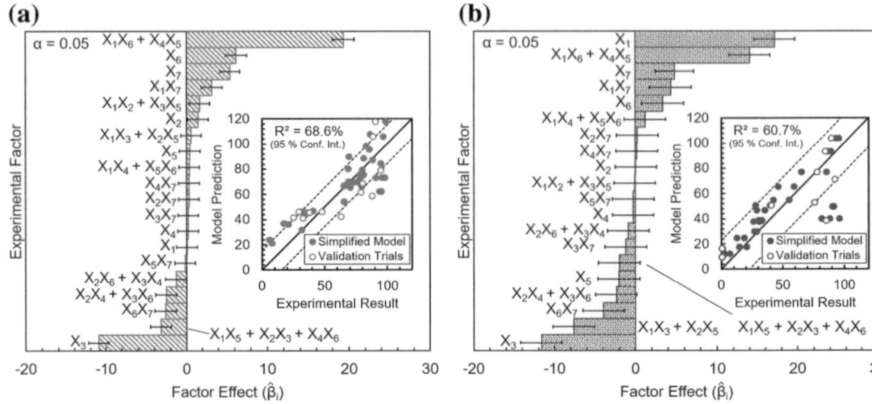

Fig. 4 Factor effect coefficients for the empirical extraction models. The error bars represent the 95% confidence interval for each of the factors (t distribution, using the standard deviation of the three replicate trials as an estimate of the system variance). The inset panels indicate the correlation between the empirical models and the experimental data. **a** Scandium, **b** Iron

Analysis of Sc and Fe Extraction Correlations

To assess the results in terms of maximizing Sc extraction while minimizing Fe extraction, the correlation between Sc and Fe extractions was assessed (Fig. 5). In high H_2SO_4 addition trials ($X_1 = 1$), the extraction of Sc was approximately linear with Fe extraction, allowing high extractions of Sc, at the cost of high Fe extraction; however, at lower H_2SO_4 addition ($X_1 = -1$), the extractions of Sc appeared to be uncoupled from Fe extraction, allowing selective extraction of Sc with minimal Fe co-extraction. The trials in which the acid ratio was 0.5 $g_{H_2SO_4}/g_{LO}$ ($X_1 = -1$), the roasting duration was 90 min ($X_3 = 1$), and the roasting temperature was 800 °C, ($X_6 = 1$) had nearly zero iron extraction ($\leq 1\%$—group a), but Sc extraction ranged from 17 to 38%. In these trials, an added benefit was high cobalt extraction ranging from 75 to 100%. A second subset of trials, conducted at [$X_1 = -1$, $X_3 = -1$, $X_6 = 1$] (group b), showed high Sc (65–80%) extraction with only moderate Fe extraction (13–20%). The trials conducted at [$X_1 = -1$, $X_6 = -1$] had slightly higher Sc extraction (group c); however, this extraction came at the cost of higher Fe extraction. As such, we can conclude that depending on process goals, the optimum ARWL conditions likely lie in one of the first two regions. As mentioned previously, cobalt can also be extracted from LO. Co extraction was typically equal to or higher than Sc extraction (Fig. 5), particularly in regions a and b, and thus the optimal conditions for selective Sc extraction also correspond to regions of optimal Co extraction. The extraction of Ni followed a similar pattern to Co extraction.

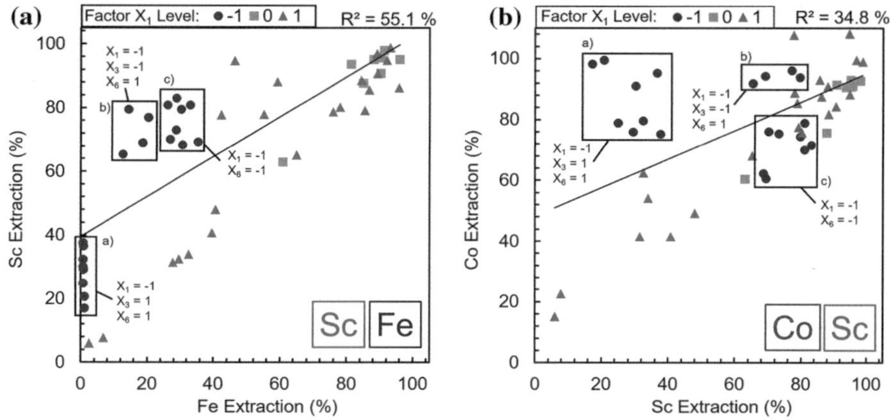

Fig. 5 Extraction correlations between Fe and Sc. The extraction of Sc compared with Fe extraction and Co compared with Sc for all trials. The factor level of X_1 (acid/LO mass ratio) is highlighted for each of the trials. The lines represent the linear correlation between all points. Trial groupings of particular interest (a, b, c) are highlighted, including the associated values of X_1, X_3 (roasting duration), and X_6 (roasting temperature)

The primary benefit of using the empirical modelling approach is that the results can be used to predict arbitrary extraction conditions, and optimal model parameters can be determined. Optimal conditions for three objectives were determined as (1) maximum Sc extraction, (2) minimum Fe extraction, and (3) maximum Sc to Fe extraction ratio. These settings correspond to test conditions V1, V2, and V3 and are presented in Table 2. As mentioned above, the models are inadequate for fully describing the extraction behavior, which explains the difference between the model-predicted and experimental results; however, further refinement of the model was deemed to be outside the scope of this study as the model was able to predict the general behavior of the system. The conditions giving maximum Sc/Fe ratio (V3) were chosen for further solvent extraction trials because the low extraction of Fe would increase the effectiveness of this step; however, depending on the process goals, other conditions may be more appropriate, such as the A13 conditions, which allow for 79.5% extraction of Sc, with only 14.4% co-extraction of Fe—this set of conditions is also beneficial in terms of operating parameters as it corresponds to low acid consumption, low water consumption, and short roasting duration.

Table 2 Summary of model-predicted and observed optimum trials. The trial parameters correspond to the conditions given in Table 1. Trials V1–V3 were predicted by the empirical model and were not used to produce the empirical model

Optimization objective	Trial ID	Model-predicted extractions (% Sc, Fe)	Experimental extractions (% Sc, Fe)
Maximum Sc extraction	V1	30 min: 105.7, 93.4 120 min: 117.8, 103.8	30 min: 88.5, 84.5 120 min: 90.7, 89.3
Minimum Fe extraction	V2	30 min: 41.8, 16.5 120 min: 41.1, 9.6	30 min: 24.9, 0.70 120 min: 32.4, 0.70
Maximum Sc/Fe extraction ratio	V3	30 min: 46.1, 16.5 120 min: 45.5, 9.6	30 min: 29.4, 0.70 120 min: 37.6, 0.70
High Sc extraction with low Fe extraction	A13	30 min: 66.7, 24.5 120 min: 66.0, 17.6	30 min: 65.3, 12.6 120 min: 79.5, 14.4

Selective Extraction Mechanism Investigation

To validate the proposed extraction mechanism, samples of acid-roasted ore were characterized by SEM-EDS (Fig. 6). In a sample which had high extraction of both Sc and Fe (A16), Fe and S were strongly co-localized, suggesting that Fe was highly sulfated (rxn 1), and de-sulfation (rxn 3) did not occur extensively, leading to high but unselective extraction. In contrast, sample V3 which had moderate Sc extraction with nearly no Fe extraction had considerably less S in the Fe particles, indicating extensive de-sulfation, while leaving some soluble sulfates (such as $Sc_2(SO_4)_3$). Trial A13, which had high Sc extraction, with low-moderate Fe extraction was between the two extremes, showing less Fe–S co-localization than A16 but more than V3, suggesting some de-sulfation, but it was not complete. These observations are consistent with the proposed extraction mechanism described above.

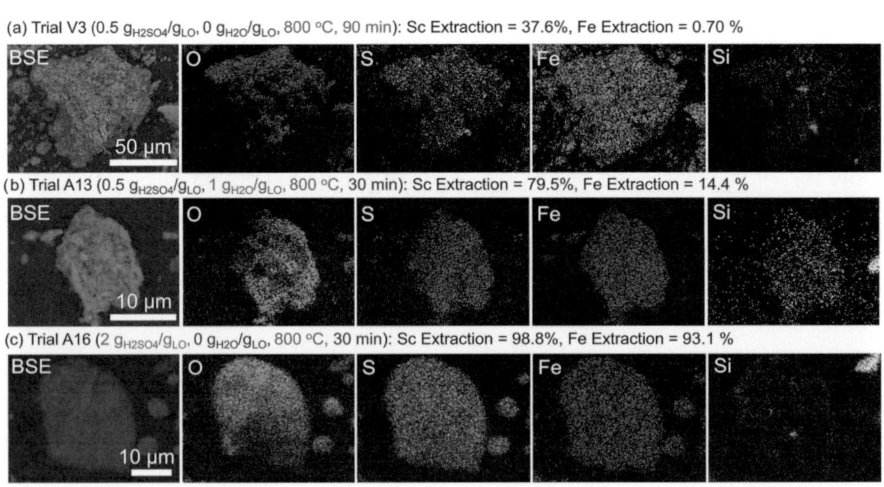

Fig. 6 SEM-EDS investigation of the selective extraction mechanism. BSE and Fe, O, S, Si elemental mapping are presented for acid-roasted ore samples for **a** V3, **b** A13, and **c** A16. Sc could not be detected as it was below the EDS detection limit

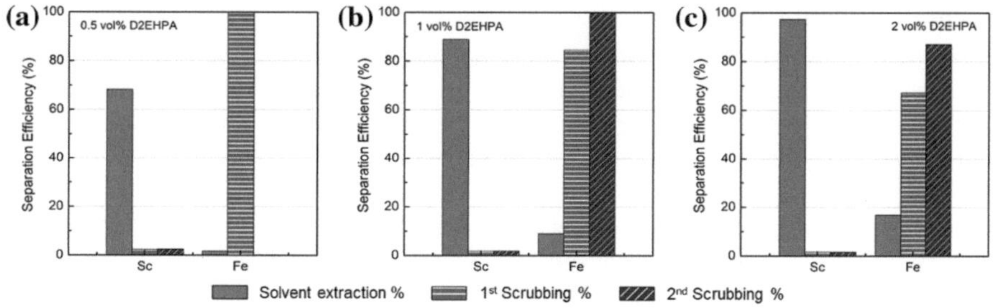

Fig. 7 Separation efficiency of Sc and Fe at different extractant concentrations. **a** 0.5 vol.% D2EHPA; **b** 1 vol.% D2EHPA; **c** 2 vol.% D2EHPA

Solvent Extraction

The leachate solution obtained from the V3 ARWL conditions (37.6% Sc extraction and 0.7% Fe co-extraction) was subjected to solvent extraction (SX) to separate Sc from Fe. The Sc and Fe separation efficiencies for each step in the SX process, using either 0.5, 1, or 2 vol.% of D2EHPA, are presented in Fig. 7. With an increase in D2EHPA concentration, the solvent extraction percentage of both Sc and Fe increased considerably, while the scrubbing percentage decreased. This is because there are more extractant molecules available for extracting and detaining Sc and Fe at high concentration. The overall results indicated that Sc could be readily separated from Fe by solvent extraction process, showing high separation factor (Eq. 3) over 80 in all cases. Investigation into the stripping of Sc from the D2EHPA was outside the scope of this study; however, previous studies have shown that efficient stripping can be achieved with strong basic solutions (such as 5 M NaOH) or fluoride salts [25].

$$SF = \frac{[M_1]_{org}/[M_1]_{aq}}{[M_2]_{org}/[M_2]_{aq}} = \frac{D_1}{D_2} \tag{3}$$

Conclusions

In summary, we have investigated the use of acid roasting–water leaching to selectively extract scandium from a nickel laterite ore. This process uses concentrated sulfuric acid to convert scandium-bearing minerals into water-soluble sulfates, and then uses a roasting step to convert the undesired iron sulfates to insoluble Fe_2O_3. Preliminary factor effects analysis and optimization demonstrated that high extraction of Sc can be achieved with minimal Fe co-extraction, which simplifies downstream separation by solvent extraction using D2EHPA. This work demonstrates that the ARWL of nickel laterite ore can be used to address the growing demand for scandium in a sustainable and economical way.

Author Contributions G.A. conceived and supervised the research. J.A. designed the ARWL experiments, performed material characterizations, analyzed/ interpreted the results, performed XRD analysis, and drafted the manuscript. R.Y. performed the ARWL experiments. J.K. designed and performed the solvent extraction trials.

Acknowledgements We thank Mr. Jiakai Zhang for his help with SEM-EDS. Access to the electron microscopy facility in Ontario Centre for the Characterization of Advanced Materials (OCCAM) and the Walter Curlook Materials Characterization and Processing Laboratory is acknowledged.

References

1. Røyset J, Ryum N (2005) Int Mater Rev 50:19
2. Ahmad Z (2003) JOM 55:35
3. U.S. Geological Survey, Mineral Commodity Summaries (2019)
4. Samson IM, Chassé M (2016) Encyclopedia of Geochemistry. Springer
5. Shaoquan X, Suqing L (1996) Hydrometallurgy 42:337
6. Wang W, Pranolo Y, Cheng CY (2011) Hydrometallurgy 108:100
7. Bohlsen M (2019) Investor Intel
8. Kaya Ş, Topkaya YA (2016) Rare Earths Ind., pp 171–182. Elsevier
9. Anawati J, Azimi G (2019) Waste Manag 95:549
10. Binnemans K, Jones PT, Blanpain B, Van Gerven T, Pontikes Y (2015) J Clean Prod 99:17
11. Borra CR, Pontikes Y, Binnemans K, Van Gerven T (2015) Miner Eng 76:20
12. Borra CR, Mermans J, Blanpain B, Pontikes Y, Binnemans K, Van Gerven T (2016) Miner Eng 92:151
13. Önal MAR, Borra CR, Guo M, Blanpain B, Van Gerven T (2015) J Sustain Metall 1:199
14. Liu Z, Zong Y, Li H, Jia D, Zhao Z (2017) J Rare Earths 35:896
15. Narayanan RP, Kazantzis NK, Emmert MH, Sustain ACS (2018) Chem Eng 6:1478
16. You Z, Li G, Zhang Y, Peng Z, Jiang T (2015) Hydrometallurgy 156:225
17. Yu D, Utigard TA, Barati M (2014) Metall Mater Trans B 45:662
18. Frankiewicz TC (1978) 4110106
19. Bainbridge DW (1973) Metall Trans 4:1655
20. Guo X, Li D, Park KH, Tian Q, Wu Z (2009) Hydrometallurgy 99:144
21. Swamy Y, Kar B, Mohanty J (2003) Hydrometallurgy 69:89
22. Zhang J, Anawati J, Yao Y, Azimi G (2018) ACS Sustain Chem Eng
23. Tagawa H, Saijo H (1985) Thermochim Acta 91:67
24. Reid S, Tam J, Yang M, Azimi G (2017) Sci Rep 7:1
25. Wang W, Cheng CY (2011) J Chem Technol Biotechnol 86:1237

Part II
Rare Earth Metals

Supercritical Fluid Extraction of Rare Earth Elements from Waste Fluorescent Lamp

Jiakai Zhang and Gisele Azimi

Abstract With promoted use of fluorescent lamps and the increasing amount of fluorescent lamp waste, its phosphors have become an ideal secondary source for critical materials, such as yttrium, europium, and terbium. Conventional recycling processes based on hydrometallurgy rely on a large volume of acids and organic solvents and generate large volumes of hazardous waste. Here, a novel environmentally friendly process based on supercritical fluid extraction was developed to recycle the critical rare-earth elements from waste fluorescent lamps. The proposed process uses supercritical CO_2 as the solvent, which is inert, safe, and abundant, along with tributyl-phosphate nitric acid (TBP–HNO_3) chelating agent. Rare-earth element extraction efficiencies of 50% was achieved without sample pretreatment. Pretreating samples with mechanical activation (ball milling) for 1 h resulted in a 20% improvement in extraction efficiency. High-resolution transmittance electron microscopy showed that during mechanical activation the sample becomes polycrystalline with nano-sized crystallite size, resulting in enhanced leaching efficiency.

Keywords Supercritical fluid · Mechanical activation · Rare-earth element · Fluorescent lamp · Phosphors · TEM

Introduction

Rare-earth elements (REEs) are among the most important strategic materials due to their optical, electrical, and magnetic properties [1]. REEs are important functional materials for realizing designed functionalities in electronic information, metallurgical machinery, petrochemical industry, energy and environment, and national

J. Zhang · G. Azimi (✉)
University of Toronto, 200 College Street, Toronto, ON M5S3E5, Canada
e-mail: g.azimi@utoronto.ca

J. Zhang
e-mail: kevinjk.zhang@mail.utoronto.ca

G. Azimi et al. (eds.), *Rare Metal Technology 2020*, The Minerals, Metals & Materials Series, https://doi.org/10.1007/978-3-030-36758-9_9

defense fields. The global demand for REEs has been growing considerably with annual growth rate ranging from 3 to 8% in recent years [2, 3]. The global consumption of REEs was 105,000 tonnes in 2011, and it is expected to be 200,000–240,000 tonnes by 2020 [2]. According to the United States Geological Survey, China has accounted for the vast majority of the world's rare-earth production since the mid-1990s, and China's share of the global market was about 86% in 2017 [4]. However, because of the serious environmental pollution and ecological destruction to local areas caused by rare-earth mining and smelting, China has begun a campaign to limit rare-earth mining, smelting, and exporting.

For tackling the sustainability challenges associated with REE supply, green mining, and efficient use of resources, new strategies have been initiated to mine these elements from secondary sources. An important application of REEs is the production of luminescent materials, which are phosphors used in cathode ray tubes and fluorescent lamps (FLs). Nearly 15% of REEs are used for FLs manufacturing [5]. The quantity of waste fluorescent lamps is expected to increase. With increasing awareness of environmental protection, FLs take the major share of the lighting market due to its lower energy consumption in comparison to that of traditional incandescent lamps, which helps with the reduction of greenhouse gas emission and promotion of energy conservation [6, 7]. For example, with the goal of achieving 20% energy efficiency improvement, 11 million households are expected to replace inefficient light bulbs by 2020 within the European Union [8]. Fluorescent lamps are ideal secondary sources for REEs, including terbium (Tb), europium (Eu), and yttrium (Y). Fluorescent lamps contain a high concentration of REEs, around 230,000 g/t, which is about 15 times higher than that in the primary REE ores [5]. Environmentally sustainable recycling and management of waste phosphors facilitates reduction of landfilled wastes, mitigation of the environment and health concerns due to the mercury contained, and efficient use of valuable materials.

Conventional recycling processes utilize hydrometallurgy to recycle REEs from waste phosphors. A summary of previous studies focused on recycling of REEs from the spent FLs via hydrometallurgy is presented in Table 1. The hydrometallurgical method can be adapted to different resources scales with relatively low energy consumptions. Hence, it is often the preferred waste valorization option. However, as can be seen in Table 1, the hydrometallurgical process is a reagent-intensive process with the consumption of large volumes of strong acids, which produces potentially secondary waste streams. Moreover, the previous studies mostly focused on the extraction efficiencies of Y and Eu from red phosphors and high extraction reported. Valuable materials like Tb was discarded as residue in many reported processes since blue phosphor that contains Tb is resistant to acid attack [7, 9]. Considering the drawbacks of hydrometallurgical processes, an alternative process for recycling waste phosphor with low cost, low environmental impact, and low energy consumption but high recovery efficiency for REEs is desired.

A novel process using supercritical fluid has shown to be effective for REEs recovery from secondary sources. The process utilizes supercritical CO_2 as the

Table 1 Overview of previous studies focused on recycling of REEs from spent FLs via hydrometallurgy

References	Reagent(s)	Temp (°C)	Time (min)	L:S ratio (mL/g)	Leaching efficiency
Rabah [10]	H_2SO_4/HNO_3 mixture, 4 M	125	12 h	N.A.	Eu: 92.8%; Y: 96.4%
De Michelis et al. [11]	2.0 mol/L H_2SO_4	90	240	5	Y: 85.0%
Yang et al. [12]	HNO_3 or HCl or H_2SO_4, 5 M	100	6 h	20	Eu: 95.5%, Y: 99.5%
Innocenzi et al. [13]	2.0 mol/L H_2SO_4 + 10 vol.% H_2O_2	70	3 h	5	Y: 100.0%
Tan et al. [6]	6 mol/L HCl	60	15	60	Tb: 89.4%, Eu: 93.1%, Y: 94.6%
Song et al. [14]	2.0 mol/L H_2SO_4	70	60	80	Eu: 91.1%, Y: 96.3%
Tan et al. [15]	6.0 mol/L HCl	60	15	60	Tb: 90.0%, Eu: 93.1%, Y: 94.6%

solvent, which is safe, inert, and abundant, in addition to tributyl-phosphate-nitric acid (TBP–HNO_3) chelating agent to extract REEs from solid matrices. Yao et al. [16] achieved over 85% of rare-earth elements (including La, Ce, Pr, and Nd) from the anode material of an end-of-life nickel–metal hydride (NiMH) battery from a Toyota Prius [16]. Zhang et al. [17] achieved around 95% extraction efficiencies of rare-earth elements, (including Nd, Dy, and Pr) from an end-of-life NdFeB magnet, and Shimizu et al. [18] achieved 99.7% Y and Eu but less than 7% Tb, Ce, and La from waste FLs [17, 18].

In this study, an SCFE process is developed to recycle REEs from waste FLs. For improvement of REEs extraction efficiency, the sample was mechanically activated with an oscillation mill. The effect of mechanical activation to sample crystal structure was investigated with high-resolution transmission electron microscopy (HR-TEM). Utilizing the design of experiment (DOE) methodology, the effects of multiple process parameters to extraction was evaluated, and the optimum operating conditions were determined by an empirical model constructed. The results show the feasibility of SCFE for REEs recovery from secondary sources and provide a scientific foundation for the realization of this alternative sustainable REEs recovery process.

Method and Material

Material

Waste lamp phosphors were acquired from Ontario Lamp Recyclers Inc. Large particles (such as glass and aluminum foils) were removed from the as-received phosphor sample by sieving, and −200 mesh powder sample was used in the experiments. Chemicals used are following: tri-n-butyl phosphate (TBP, $\geq$98%, VWR), concentrated nitric acid (15.7 M, 70 wt%, VWR), and carbon dioxide (CO_2, grade 4.0, Linde Canada).

TBP–HNO_3 Complex Preparation

The TBP–HNO_3 chelating agents complex was prepared by mixing TBP with 70 wt% HNO_3. The mixture was manually agitated for 5 min followed by 5 min settling time. The immiscible liquid phases were separated, and the top solution corresponds to the TBP–HNO_3 chelating agent.

Mechanical Activation

The phosphor sample was activated with an oscillation ball mill (Retsch MM400) in air. Phosphor samples (2 g) were placed in a tungsten carbide grinding jar (25 ml inner volume) with two tungsten carbide grinding balls (15 mm diameter). The sample was oscillated at 30 Hz for 60 min.

Design of Experiment

The experimental parameter settings for the SCFE runs were constructed using fractional factorial experimental design methodology [19]. The following operating parameters were investigated: temperature (X_1), pressure (X_2), residence time (X_3), sample: chelating agent (S:CA) ratio (X_4), and agitation rate (X_5). The parameter settings (X_i) were normalized and coded between −1 (low level) and +1 (high level) to allow for direct comparison of the relative impact of each parameter on the system response by direct comparison of the magnitude of the model coefficients.

Empirical Model Building and Statistical Methods

The results of the extraction experiments were utilized to build a set of empirical models to describe the extraction ($\hat{y_i}$) of the tested elements as a function of the five operating parameters (X_1–X_5, Eqs. 1, 2), where third-order and higher order interactions are considered as negligible.

$$\widehat{y}_i = \beta_0 + \tilde{\beta}_1 X_1 + \tilde{\beta}_2 X_2 + \tilde{\beta}_3 X_3 + \tilde{\beta}_4 X_4 + \tilde{\beta}_5 X_5 + \tilde{\beta}_{23} X_2 X_3 + \tilde{\beta}_{34} X_3 X_4 \quad (1)$$

where

$$\begin{array}{llll} \tilde{\beta}_1 = \beta_1 + \beta_{24} & + \tilde{\beta}_2 = \beta_2 + \beta_{14} & \tilde{\beta}_3 = \beta_3 + \beta_{15} & \tilde{\beta}_4 = \beta_4 + \beta_{12} \\ \tilde{\beta}_5 = \beta_5 + \beta_{13} & \tilde{\beta}_{23} = \beta_{23} + \beta_{45} & \tilde{\beta}_{34} = \beta_{34} + \beta_{25} & \end{array} \quad (2)$$

The experimental data was fit to this empirical model by multiple Linear Least Squares Regression (mLLSR) according to Eq. (3). $\widehat{\beta}$ is the model parameter vector, containing each of the model parameters $\left(\widehat{\beta}_0, \widehat{\beta}_1, \widehat{\beta}_2, \ldots\right)$, X is the experimental design matrix (Table 2), and Y is the response vector, containing each of the measured experimental extraction efficiencies. More details regarding the Empirical Model building can be found in the previous work [17].

$$\widehat{\beta} = \left(X^T X\right)^{-1} \left(X^T Y_i\right) \quad (3)$$

Supercritical Fluid Extraction Process

SCFE experiments were carried out in a 100 ml high-pressure reactor system manufactured by Supercritical Fluid Technology Inc., USA. A schematic diagram of the extraction system can be found in previous work [17]. Lamp phosphor

Table 2 The experimental design matrix (2_{III}^{5-2} fractional factorial design)

Run number	X_0	X_1	X_2	X_3	$X_4 = X_1X_2$	$X_5 = X_1X_3$
1	+	−	−	−	+	+
2	+	+	−	−	−	−
3	+	−	+	−	−	+
4	+	+	+	−	+	−
5	+	−	−	+	+	−
6	+	+	−	+	−	+
7	+	−	+	+	−	−
8	+	+	+	+	+	+

powder along with TBP–HNO_3 was loaded into the reactor, then CO_2 was injected, and pressure and temperature were increased to reach supercritical conditions. For all tests, the mass of the fluorescent lamp sample was fixed at 1 g (for example, 2 ml of the chelating agent was utilized for a test with a S:CA of 0.5 g/ml.). After completion of the reaction, the system was depressurized and the solution containing rare-earth complex was collected in EPA vials.

Characterization of the Phosphor Sample

Compositional analysis of the phosphor samples before and after SCFE was conducted using acid digestion in concentrated aqua regia (3 HCl:1 HNO_3, approximately 20 mL per 0.2 g sample) at 200 °C (9 °C/min ramp up, 20 min dwell at 200, 5 °C/min ramp down—in a MARS6 Xpress microwave digestion system), followed by Inductively Coupled Plasma Optical Emission Spectrometry (ICP-OES, Perkin Elmer Optima 7300 DV) using the following wavelengths: Ce 413.764 nm, Y 361.104 nm, Eu 381.967 nm, Tb 350.917 nm, La 398.852 nm, Sr 407.771 nm, and Sb 206.836 nm. The crystal morphology of the unprocessed sample and the mechanically activated sample was performed using high-resolution transmission electron microscopes (HRTEM, FEI Titan 80–300 LB) and transmission electron microscope (TEM, Jeol 2010F).

Results and Discussion

Characterization of the Phosphor Sample

The composition of the phosphor sample was analyzed with ICP-OES, the results are summarized in Table 3. As expected, the sample contains a high concentration of yttrium, and high-value rare-earth materials including europium and terbium.

Table 3 Composition analysis of the phosphor sample

Element	Y	La	Ce	Eu	Tb
Concentration (mg/kg)	32,756 (±773)	5,494 (±456)	4,349 (±493)	2,244 (±92)	1,562 (±156)
Element	Sr	Sb	Ca	Al	Ba
Concentration (mg/kg)	1,837 (±154)	116 (±13)	11,943 (±835)	2,351 (±198)	464 (±73)

Extraction Results

For SCFE of non-mechanical activated samples and mechanical-activated samples, the operating parameters and testing ranges were selected based on preliminary tests and previous studies conducted by Yao et al. [20] and Zhang et al. [17]. The relative effect of each operating parameter was assessed in a 2^{5-2}_{III} fractional factorial design, and three replicated center points trials. The 2^{5-2}_{III} fractional factorial design matrix and the extraction results for both non-mechanical-activated sample and mechanical-activated sample are summarized in Table 4. The average extraction efficiencies for non-mechanical-activated sample are 26.8% Ce, 36.2% Y, 33.6% Eu, 15.6% Tb, 25.7% La, 55.6% Al, 75.9% Ca, 43.5% Sr, and 51.9 Sb%. In comparison, the extraction efficiencies of REEs from the mechanical-activated samples are about 20–30% more than that of the non-mechanical-activated samples. The experimental results were used to build empirical extraction models to enable a quantitative comparison of relative effect of each tested operating parameter. In addition to providing an estimate of the between-run variance, the center point runs allowed a qualitative assessment of the quadratic effect of tested parameters.

Empirical Extraction Modelling

The main effect of primary test parameters and the second-order interactions were estimated by the construction of an empirical extraction model by multiple Linear Least Squares Regression (mLLSR). These models are not to provide mechanistic insight into the extraction process, but system response to operation parameters are investigated. Model parameters are evaluated at a 95% confidence interval, and statistically significant model parameters are illustrated in Table 5. More details about the empirical model building can be found in previous work [17].

As can be seen in Table 5, the REEs demonstrated similar extraction trend, which is expected as they share a similar chemical property. Among all operating parameters, increasing solid to chelating agent ratio (decreasing chelating agent), X4, had the largest negative effect on REEs extraction and base metal extraction. A high solid to chelating agent ratio hinders extraction because there are fewer acid molecules to attack the REEs containing phosphor sample, thus fewer REEs are liberated from the solid matrix. In addition, there are fewer TBP molecules available for supercritical CO_2 soluble REE–NO_3–TBP complex formation. With the fractional factorial experimental results and empirical extraction model constructed, the relative effect of the operating parameters on the system and extraction efficiency of REEs can be evaluated and predicted. As the temperature is not statistically significant at a 95% confidence level, a low-temperature level was chosen in the following test for a lower energy cost. For maximization of REE extraction, the model suggested following operation parameters: temperature = 40 °C, pressure = 4500 psi, time = 3 h, agitation rate = 750 rpm, solid to chelating agent

Table 4 Overview of the experimental design matrix with corresponding processing parameters and extraction efficiencies for Ce, Y, Eu, Tb, La, Sr, and Sb

Run Number (standard order)	Temperature X_1 (°C)	Pressure X_2 (psi)	Time X_3 (h)	S:CA X_4 (g/mL)	Agitation X_5 (rpm)	Non-mechanical-activated Extraction Efficiency (%) Ce, Y, Eu, Tb, La, Al, Ca, Sr, Sb	Mechanical-activated Extraction Efficiency (%) Ce, Y, Eu, Tb, La, Al, Ca, Sr, Sb
1	40	3000	1	1.8	1500	19.8, 28.4, 25.6, 9.2, 18.7, 51.8, 71.3, 40.3, 46.0	20.7, 40.6, 37.0, 7.4, 23.9, 17.9, 72.0, 36.2, 46.7
2	60	3000	1	0.2	750	23.5, 31.5, 27.2, 12.3, 23.4, 51.7, 72.1, 41.0, 45.7	53.0, 70.8, 66.8, 50.9, 58.0, 47.6, 84.5, 63.2, 71.3
3	40	4500	1	0.2	1500	35.9, 44.3, 42.5, 26.4, 35.2, 62.8, 78.7, 53.2, 54.4	46.4, 62.3, 58.7, 39.0, 46.2, 33.6, 81.7, 55.0, 59.8
4	60	4500	1	1.8	750	23.1, 34.2, 31.8, 12.8, 22.2, 55.5, 75.0, 40.9, 46.9	21.2, 40.3, 15.7, 16.6, 17.0, 1.5, 76.0, 48.2, 58.8
5	40	3000	3	1.8	750	20.1, 30.1, 29.0, 8.8, 19.8, 51.2, 83.2, 35.6, 49.6	11.2, 44.3, 31.2, 3.0, 31.0, 17.3, 76.4, 48.2, 54.1
6	60	3000	3	0.2	1500	36.5, 43.4, 41.6, 25.7, 34.8, 62.2, 75.7, 48.3, 52.9	32.3, 64.0, 53.0, 32.2, 39.3, 23.7, 78.0, 45.3, 56.0
7	40	4500	3	0.2	750	34.5, 43.6, 40.0, 23.1, 32.1, 58.2, 76.8, 50.2, 56.6	57.8, 71.7, 70.3, 55.8, 61.4, 47.2, 84.7, 61.7, 68.5
8	60	4500	3	1.8	1500	28.6, 37.2, 34.5, 16.9, 26.7, 59.2, 73.7, 43.9, 46.6	26.1, 53.3, 49.1, 24.8, 39.0, 22.9, 77.2, 50.2, 51.6
9	50	3750	2	1.0	1125	16.2, 28.5, 25.5, 3.0, 16.1, 47.3, 75.3, 34.9, 51.1	31.8, 50.1, 44.2, 21.7, 37.7, 33.6, 77.0, 48.1, 53.8
10	50	3750	2	1.0	1125	35.5, 44.1, 41.6, 24.4, 33.0, 61.0, 76.9, 49.7, 58.3	25.9, 51.0, 44.7, 19.5, 39.9, 27.5, 79.0, 52.3, 61.8
11	50	3750	2	1.0	1125	20.4, 32.3, 29.4, 8.6, 20.3, 50.7, 76.6, 39.5, 62.0	38.9, 51.9, 50.9, 26.4, 43.8, 34.7, 79.6, 53.5, 52.1

Table 5 Factor effect coefficients for the empirical extraction models of mechanical-activated sample

	Ce	Y	Eu	Tb	La	Al	Ca	Sr	Sb
β_0	33.7	55.7	47.8	24.6	39.8	26.5	78.9	51.1	58.4
$\hat{\beta}_1$	–	–	–	–	–	–	–	–	1.1
$\hat{\beta}_2$	4.3	–	–	–	–	–	1.1	2.8	1.3
$\hat{\beta}_3$	–	2.4	3.2	4.4	3.2	–	–	–	–
$\hat{\beta}_4$	−13.8	−11.3	−14.5	−19.9	−11.8	−11.6	−3.4	−5.3	−5.6
$\hat{\beta}_5$	–	–	–	–	–	–	−1.6	−4.3	−4.8
$\hat{\beta}_{23}$	5.8	3.2	8.1	10.2	6.1	7.4	–	–	1.2
$\hat{\beta}_{34}$	–	1.8	3.7	4.9	4.1	–	1.1	3.2	–

*Note: "–" denotes for non-statistically significant factors

ratio = 0.2. Under these optimized parameters, maximum extraction efficiencies of valuable metals were obtained (56.1% Ce, 73.2% Y, 71.2 Eu, 51.3% Tb, 55.4% La, 49.2% Al, 86.4% Ca, 58.2% Sr, 73.1% Sb).

Effect of Mechanical Activation on Physicochemical Changes of Phosphor Powder

Crystal lattice structure modification and chemical composition transformation can contribute to reactivity increment of phosphor solids via mechanical activation (ball milling). Conventional ball milling reduces the size of particles or disperses samples into a given media. High-energy ball mills are usually made of steel, tungsten carbide, or zirconia, which are used to increase the impact of energy and reduce contamination of the product. The high impact force, shear force, and collision energy introduced by mechanical activation result in crystal structure change of phosphor particles [14, 15]. Previous studies have shown that the XRD pattern of mechanically activated sample is significantly different from that of unprocessed samples [14, 15]. The XRD spectrum shows peak broadening and peak intensity reduction that is correlated with milling time and rotation speed of ball milling. With the increasing milling time and rotation speed, peak intensity decreases, and peak width increases. This is because of crystallite size change and the introduction of strain within the crystal lattice from elongated milling time [21, 22]. Mechanical forces create crystal defects, such as point defects and dislocations. The XRD peak intensity reduction and peak broadening observed suggest amorphicity increment within the sample, which is a contributing factor for extraction improvement [6, 14, 15, 21, 22]. The mechanical activation process also lowers apparent activation energy for leaching experiment. Loy et al. used shrinking core model for leaching kinetics of REEs from FLs modeling, and they found the unprocessed sample had a higher apparent activation energy (the apparent activation energies for the unprocessed sample and mechanical-activated sample were 68 kJ mol^{-1} and 1.4 kJ mol^{-1}, respectively) [22].

HRTEM/STEM imaging was conducted to visualize crystal structure, defects formation, and amorphous content increment of samples mechanically activated for 0 and 60 min. Figure 1a–c is TEM/STEM, selected area diffraction (SAD), and HRTEM/STEM images of the sample with 0 min mechanical activation (unprocessed sample). As can be seen in the SAD Figure, the diffraction pattern has shown a well-ordered structure, which suggests the unprocessed sample had a high level of crystallinity. The HRTEM/STEM image also provides evidence that the unprocessed sample has a high level of crystallinity, since the atoms are organized in a perfect lattice structure. The low extraction of REEs from the unprocessed sample could be because of this close-packed network and the chemical bonding. The closed packed network and chemical bonding hinder breaking of existing chemical bonds and formation of new chemical bonds during the extraction process [21]. Therefore, the REEs cannot be fully extracted from the solid matrix to form a complex with the chelating agent and dissolve in the supercritical CO_2, which ultimately results in low extraction efficiency.

Figure 1d–f is TEM/STEM, selected area diffraction (SAD), and HRTEM/STEM images of the sample with 60 min mechanical activation. As can be seen in the SAD Figure, the diffraction pattern has shown a beaded ring, which suggests the

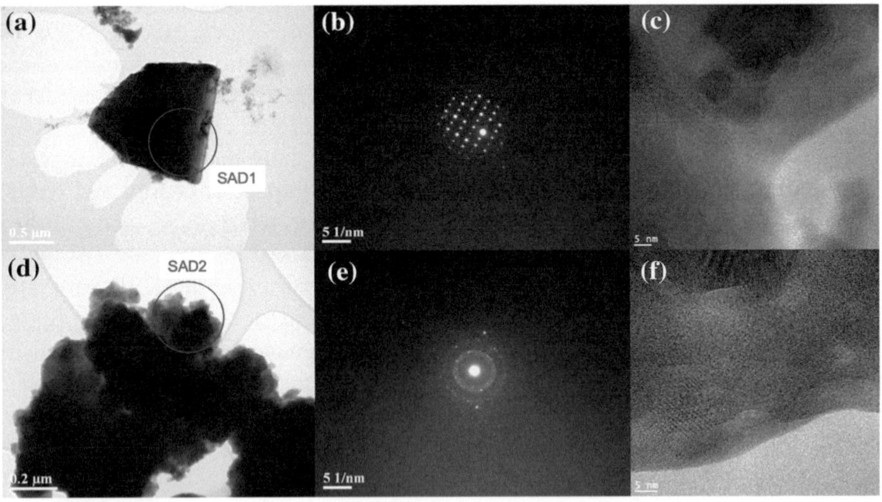

Fig. 1 **a** TEM/STEM image of a particle from the sample without mechanical activation; **b** corresponding SAD pattern of the sample without mechanical activation; **c** HRTEM/STEM image of the grain boundary of a crystalline particle without mechanical activation; **d** TEM/STEM image of a particle from the sample with 60 min mechanical activation; **e** corresponding SAD pattern of the sample with 60 min mechanical activation; **f** HRTEM/STEM image of the grain boundary of a crystalline particle without mechanical activation

sample with 60 min mechanical activation has gained a high degree of polycrystallinity. In the HRTEM/STEM image, some small degree of amorphization is observed at the edge of the particle. The image also shows that many grain boundaries and nanocrystallites are formed during the mechanical activation process. The majority of particles are nanocrystallites and a small degree of amorphousness is observed. The improved supercritical fluid extraction with mechanical activation can be because of irregular packing of the solid material. In addition, the high number of defects including dislocations and unidimensional defects provide diffusion pathways, which enhance diffusion and promote chemical reactions. Therefore, the REEs can be extracted more efficiently from the solid matrix to forming complexes with the chelating agent and dissolve in supercritical CO_2.

As mentioned above, mechanical activation could also cause chemical composition changes using high-energy ball milling, which is referred to as mechanosynthesis, mechanochemical synthesis, reactive grinding, and reaction millings. A typical example of such a process is milling a mixture of simple oxides, such as mechanochemical synthesis of $LaMnO_3$ from La_2–O_3 and Mn_2O_3 powders [23]. Since the waste fluorescent lamp phosphor powders contain different types of metal oxides, it is plausible to hypothesis a similar phenomenon occurred during the mechanical activation process. Tan et al. conducted an XPS analysis on both samples with and without mechanical activation [15]. They found the mechanical activation process caused $LaPO_4$ decomposition and generation of dissolvable rare-earth oxides. Therefore, the leaching performance was improved because of crystal structure change and chemical decomposition.

Conclusions

A novel green and environment-friendly process was developed by utilizing sc–CO_2 as a solvent along with TBP–HNO_3 chelating agent for REEs extraction from waste fluorescent lamp phosphors. This process minimized the usage of chemicals, including acids and organic solvents, which significantly reduced waste generated from the process in comparison with traditional hydrometallurgical processes. Fractional factorial design of experiment methodology was utilized to investigate the effect of five operating parameters, achieving over 70% extraction of REEs from the waste fluorescent lamp phosphor sample.

This study and previous studies on recycling of NdFeB magnet and NiMH battery [17, 20] demonstrate aeriometallurgy, known as supercritical fluid extraction, which is potentially a viable alternative to hydrometallurgy for the recovery of valuable materials from secondary sources. The future investigation could focus on the REEs purification and chelating agent recycling process. In addition, the examination of other contaminants present in the fluorescent lamp phosphor sample is required, such as mercury residue. The ultimate goal of this work is to develop a sustainable and economical process for the recycling of post-consumer waste streams.

References

1. Wu Y, Yin X, Zhang Q, Wang W, Mu X (2014) The recycling of rare earths from waste tricolor phosphors in fluorescent lamps: a review of processes and technologies. Resour Conserv Recycl 88(100):21–31
2. Tan Q, Li J, Zeng X (2015) Rare earth elements recovery from waste fluorescent lamps: a review. Crit Rev Environ Sci Technol 45(7):749–776
3. Roskill (2019) Global industry, market & outlook 2018—rare earths. https://roskill.com/market-report/rare-earths/
4. Roskill (2019) Global industry, market & outlook 2018—rare earths
5. Swain N, Mishra S (2019) A review on the recovery and separation of rare earths and transition metals from secondary resources. J Clean Prod 220:884–898
6. Tan Q, Deng C, Li J (2016) Innovative application of mechanical activation for rare earth elements recovering: process optimization and mechanism exploration. Sci Rep 6:1–10
7. Wu Y, Wang B, Zhang Q, Li R, Yu J (2014) A novel process for high efficiency recovery of rare peroxide system. RSC Adv 4:7927–7932
8. Machacek E, Richter JL, Habib K, Klossek P (2015) Recycling of rare earths from fluorescent lamps: value analysis of closing-the-loop under demand and supply uncertainties. Resour Conserv Recycl 104(2015):76–93
9. Gijsemans L, Forte F, Onghena B, Binnemans K (2018) Recovery of rare earths from the green lamp phosphor $LaPO_4$: Ce^{3+}, Tb^{3+} (LAP) by dissolution in concentrated methanesulphonic acid. RSC Adv 8(46):26349–26355
10. Rabah MA (2008) Recyclables recovery of europium and yttrium metals and some salts from spent fluorescent lamps. Waste Manag 28(2):318–325
11. De Michelis I, Ferella F, Varelli EF, Vegliò F (2011) Treatment of exhaust fluorescent lamps to recover yttrium: Experimental and process analyses. Waste Manag 31(12):2559–2568
12. Yang F, Kubota F, Baba Y, Kamiya N, Goto M (2013) Selective extraction and recovery of rare earth metals from phosphor powders in waste fluorescent lamps using an ionic liquid system. J Hazard Mater 254–255(1):79–88
13. Innocenzi V, De Michelis I, Ferella F, Vegliò F (2013) Recovery of yttrium from cathode ray tubes and lamps' fluorescent powders : experimental results and economic simulation. RSC Adv **33**:2390–2396
14. Song G et al (2017) Improvement in rare earth element recovery from waste trichromatic phosphors by mechanical activation. J Clean Prod 151(2017):361–370
15. Tan Q, Deng C, Li J (2017) Enhanced recovery of rare earth elements from waste phosphors by mechanical activation. J Clean Prod 142:2187–2191
16. Yuxiang Yao GA, Farac NF (2017) Supercritical fluid extraction of rare earth elements from nickel metal hydride battery. ACS Sustain Chem Eng 6:1417–1426
17. Zhang J, Anawati J, Yao Y, Azimi G (2018) Aeriometallurgical extraction of rare earth elements from a NdFeB magnet utilizing supercritical fluids. ACS Sustain Chem Eng 6 (12):16713–16725
18. Shimizu R, Sawada K, Enokida Y, Yamamoto I (2005) Supercritical fluid extraction of rare earth elements from luminescent material in waste fluorescent lamps. J Supercrit Fluid 33 (3):235–241
19. Box GEP, Hunter JS, Hunter WG (2005) Statistics for experimenters: design, innovation, and discovery, 2nd edn. Wiley
20. Yao Y, Farac NF, Azimi G (2017) Supercritical fluid extraction of rare earth elements from Nickel metal hydride battery. ACS Sustain Chem Eng 6(1):1417–1426
21. Van Loy S, Binnemans K, Van Gerven T (2018) Mechanochemical-assisted leaching of lamp phosphors: a green engineering approach for rare-earth recovery. Engineering 4(3):398–405

22. Van Loy S, Binnemans K, Van Gerven T (2017) Recycling of rare earths from lamp phosphor waste: Enhanced dissolution of $LaPO_4$: Ce^{3+}, Tb^{3+} by mechanical activation. J Clean Prod 156:226–234
23. Zhang Q, Saito F (2000) Mechanochemical synthesis of $LaMnO_3$ from La_2O_3 and Mn_2O_3 powders. J Alloys Compd 297(1–2):99–103

Supercritical Fluid Extraction of Rare-Earth Elements from a Canadian Ore

Jiakai Zhang, Kimberly Watada, Maziar E. Sauber and Gisele Azimi

Abstract Rare-earth elements (REEs) are essential in many critical green technologies. Conventional REE processing methods rely on large volumes of acids and organic solvents and generate significant volumes of hazardous waste. In recent years, supercritical fluid extraction (SCFE) has emerged as an alternative green extraction process for the recovery of REEs from primary and secondary resources. In this study, the SCFE process was used for the extraction of REEs from a Canadian ore. The process utilizes supercritical CO_2 as the solvent along with the tributyl-phosphate–nitric acid chelating agent. A fractional factorial design of the experiment was used to investigate the effect of various operating factors including temperature, pressure, solid to chelating agent ratio, residence time, and agitation rate on the extraction efficiency. The results of this work demonstrate the viability of utilizing SCFE as a cleaner and more sustainable option to extract REEs from primary resources.

Keywords Rare-earth concentrate · Supercritical fluid · Extraction · Tributyl phosphate

Introduction

In recent years, a steady, reliable, and secure supply of critical metals has become increasingly important to major industrialized countries that seek to sustain their industrial base and to develop advanced technologies, such as clean energy. In light

J. Zhang · K. Watada · G. Azimi (✉)
University of Toronto, 200 College Street, Toronto, ON M5S3E5, Canada
e-mail: g.azimi@utoronto.ca

J. Zhang
e-mail: kevinjk.zhang@mail.utoronto.ca

M. E. Sauber
CanmetMINING, Natural Resources Canada, 555 Booth Street,
Ottawa, ON K1A 0G1, Canada

G. Azimi et al. (eds.), *Rare Metal Technology 2020*, The Minerals, Metals & Materials Series, https://doi.org/10.1007/978-3-030-36758-9_10

of this, Canada, with its significant critical metal reserves, has an opportunity to supply some of the global demand for critical metals. However, the metallurgy for Canadian ores containing REEs involves a complex sequence of concentration, leaching, impurity removal, individual separation, refinement, alloying, and forming stages before they can be used in the production of permanent magnets, consumer electronics, and other high value-added high-tech products.

Pyrometallurgy and hydrometallurgy are the traditional approaches for rare-earth element (REE) extraction. Pyrometallurgy is often used in the processing of high-grade ores because this method is not always suitable for low-grade ores due to its high energy requirement, the difficulty of treating a small amount of material, and additional burdens on environmental treatment [1]. Pyrometallurgy relies upon calcination of the feed with a mixture of reducing and fluxing agents. It has drawbacks of high energy consumption and large volumes of greenhouse gas emissions. In addition, during calcination, REEs may partition into slag phases, and an additional hydrometallurgical step would then be required to recover these elements [1]. Hydrometallurgy is a preferred alternative for REEs extraction from rare-earth-containing ores and concentrates because of its adaptability to resources scale and relatively low energy consumption. Although it is a promising process, the technique has the drawback of being reagent intensive, often requiring high consumption of strong acids, organic solvents, production of secondary streams of potentially hazardous wastes, and complex process flowsheets [1]. Considering the drawbacks of pyrometallurgy and hydrometallurgy, a process with low energy consumption and small waste generation that yields a high extraction of REEs is highly desirable for the extraction of rare-earth-containing ores and concentrates.

Supercritical fluid extraction (SCFE) is an emerging green technology that recently has been used for the recovery of various metals and is gaining interest in the extraction and separation of lanthanides and actinides [2]. This is a fast-growing field due to the desirable properties of supercritical fluids (SCFs) as solvents. SCFE process has been developed in the late 1970s to extract or produce high-value products and then expanded to less specialized applications. Major commercial SCFE processes have been focused on the energy industry, including direct liquefaction of coal [3] and enhanced oil recovery from petroleum reservoirs [2], as well as the chemical industry, including food and pharmaceuticals. Other examples include coffee decaffeination [4], vegetable oil production [5], extraction of organics like alcohols from aqueous solutions [6], regeneration of catalysts [7], and activated carbon that is used for adsorbing organic contaminants in industrial plants [8].

Supercritical fluid is defined as any substance that has been heated and compressed to above its critical temperature and pressure, where distinct liquid and gas phases do not exist. SCFs have high diffusivity and low viscosity (gas-like property) and high solvability (liquid-like property). These properties enable them to penetrate and transport solutes from different matrices at a higher rate and more efficiently compared with that in a liquid phase [9]. Close to the critical point, small changes in pressure or temperature result in large changes in density, allowing many properties of a supercritical fluid to be fine-tuned. Carbon dioxide and water are the most commonly used supercritical fluids, being used for decaffeination and power generation, respectively [9]. Carbon dioxide has moderate critical properties. Furthermore, it is inert, safe, and abundant and easy to recycle; thus, it does not generate secondary waste. Moreover, the final extraction products from this process are easily recoverable [2]. Hence, it is commonly used as a solvent in research and industry.

To dissolve a metal species in supercritical CO_2, three conditions must be met [10]:

1. The metal ion charge must be neutralized via bonding or association with a negatively charged species.
2. Metal coordination must be satisfied through bonding.
3. The product of the metal–ligand complex must be nonpolar, so a homogenous supercritical mixture can be obtained (CO_2 is nonpolar).

Hence, the application of direct SCFE of metal ions and organometallic compounds with supercritical CO_2 was demonstrated impossible because of the requirement of charge neutralization and the weak solute–solvent interactions. The addition of chelating agents is necessary to form a metal chelated complex, which makes the metal ion and organometallic compounds soluble in supercritical fluid [11, 12]. The efficiency of SCFE is mainly affected by the solubility of chelating agents and corresponding metal chelates in supercritical CO_2 [13]. Many chelating agents have been tested for SCFE of metal ions and organometallic compounds. Among those chelating agents, dithiocarbamates, β-diketones, organophosphorus agent, and macrocyclic compounds have been studied in more depth [13]. Tributyl-phosphate–nitric acid chelating agent, TBP–HNO_3, has demonstrated desirable properties for REEs extraction [2, 9, 13]; thus, it was chosen in this study.

In the early 1990s, a few researchers investigated the use of supercritical CO_2 as an extraction medium for REE extraction [14–16]. The early work of SCFE of REEs was mainly focused on REE extraction from synesthetic mixtures of rare-earth oxides, a summary of which is provided by Yao et al. [9]. Previous studies have developed SCFE of REEs from secondary resources, such as nickel–metal hydride battery [9] and neodymium iron boron magnet [17]. To date, only a few studies have investigated the recovery of REEs from primary resources using SCFE, a summary of which is provided in Table 1.

In this study, an SCFE process was developed to recover REEs from a Canadian ore (Nechalacho ore deposit located in Northwest Territories). With a systematic experimental approach, this work was also devoted to finding the optimum SCFE process. A fractional factorial design of experiment methodology was utilized to design the experimental matrix to investigate the effect of various operating factors including temperature, pressure, solid to chelating agent ratio, residence time, and agitation rate on the extraction efficiency. The findings of this study are expected to help enable the sustainable extraction of REEs from primary resources using *aeriometallurgy*—i.e., metallurgy with air or gas—as a green alternative to conventional hydrometallurgy, pyrometallurgy, and solvometallurgy routes.

Table 1 A summary of previous studies on SCFE of REEs from primary resources

References	Feed	Solvent/ chelating agent	Temperature (°C)	Pressure (MPa)	Time (h)	Extraction efficiency (%)
Samsonov et al. [18]	Monazite concentrate	CO_2/HDEHP–HNO_3	45	20	1	La: 8; Ce: 4; Pr: 8 Nd: 11; Sm: 6; Gd: 6; Tb:11; Dy: 48; Ho: 56;
Samsonov et al. [19]	Monazite	CO_2/D2EHPA–HNO_3	NA	NA	NA	La: 61; Ce: 29; Pr: 61; Nd: 62; Sm: 58;
	Phosphogypsum	CO_2/TBP–HNO_3	NA	NA	NA	La: 92; Ce: 94; Pr: 94; Nd: 95
		CO_2/D2EHPA–HNO_3	NA	NA	NA	La: 73; Ce: 92; Pr: 93; Nd: 95
Sinclair et al. [10]	Roasted bastnäsite	CO_2/TBP–HNO_3	65	34	1.5	La: 3; Ce: 100; Pr: 99; Nd:100

Materials and Method

The ore samples used in this study were first pretreated to prepare a high-grade REE concentrate: Drill core samples received from Avalon's Nechalacho basal zone deposit located in Northwest Territories were stage crushed, blended, and a composite sample was made. The composite sample was then finely ground to approximately 80% passing 25 μm in a ball mill. The ground feed was subjected to a locked cycle flotation test followed by a magnetic separation to mimic the flowsheet designed for this deposit by Avalon Advanced Materials Inc. The details of this pretreatment are beyond the scope of this paper. Tributyl phosphate (TBP, $\geq$98%) and concentrated nitric acid (70 wt%) were obtained from VWR and carbon dioxide (CO_2, grade 4.8) was obtained from Linde Canada.

Feed Material Characterization

The particle size of the sample was analyzed with a laser particle size analyzer (Malvern Mastersizer 2000). For composition determination, the sample was digested with alkali fusion followed by acid dissolution. For alkali fusion, a sample of 49.75 wt% $Li_2B_4O_7$: 49.75 wt% $LiBO_2$: 0.50 wt% LiBr flux was utilized at a ratio of 7.2 g_{flux}/g_{sample}. Fusion was performed at 1050 °C for 9 min, followed by dissolution in 100 mL of 4 wt% HNO_3. A clear solution was obtained after stirring for 5 min which ensured the complete dissolution of particles in the solution. The elemental composition of the solution was characterized using inductively coupled plasma mass spectroscopy (ICP-MS Thermo Scientific iCAP Q ICP-MS). The composition obtained from this method was used as the baseline for extraction efficiency calculations.

$TBP–HNO_3$ Complex Preparation

The $TBP–HNO_3$ chelating agent complex was prepared by mixing TBP with 70 wt% HNO_3. The mixed solution was manually mixed in a separation funnel for 5 min, then allowed to settle. The immiscible liquid phases were separated, and the top organic phase ($TBP–HNO_3$ chelating agent) was collected.

Experimental Design

Supercritical fluid extraction was performed on the concentrated sample. A fractional factorial design of experiment methodology was used to design the experimental matrix. It enables testing of a large number of process parameters while maximizing workflow efficiency by reducing the total number of trials. The impact of the interaction of multiple variables can also be studied. In this work, operating parameters were temperature, pressure, residence time, agitation rate, and sample to chelating agent ratio. A resolution III fractional factorial design was utilized, so there were eight experimental trials, $2^{5-2} = 8$. In addition, three central point runs were conducted to evaluate the reproducibility of the experimental results statistically. The testing parameters conditions and the experimental matrix are presented in Tables 2 and 3, respectively.

Table 2 SCFE testing parameters and conditions

Factor	Level
X_1: Temperature	−1: 40 °C; +1: 60 °C
X_2: Pressure	−1: 3000 psi; +1: 4500 psi
X_3: Time	−1: 1 h; +1: 2 h
X_4: Agitation rate	−1: 750 rpm; +1: 1500 rpm
X_5: Sample to chelating agent ratio	−1: 1 g/mL; +1: 0.2 g/mL

Table 3 Design of experiment matrix

Run number	X_0	X_1	X_2	X_3	$X_4 = X_1X_2$	$X_5 = X_1X_3$
1	+	−	−	−	+	+
2	+	+	−	−	−	−
3	+	−	+	−	−	+
4	+	+	+	−	+	−
5	+	−	−	+	+	−
6	+	+	−	+	−	+
7	+	−	+	+	−	−
8	+	+	+	+	+	+

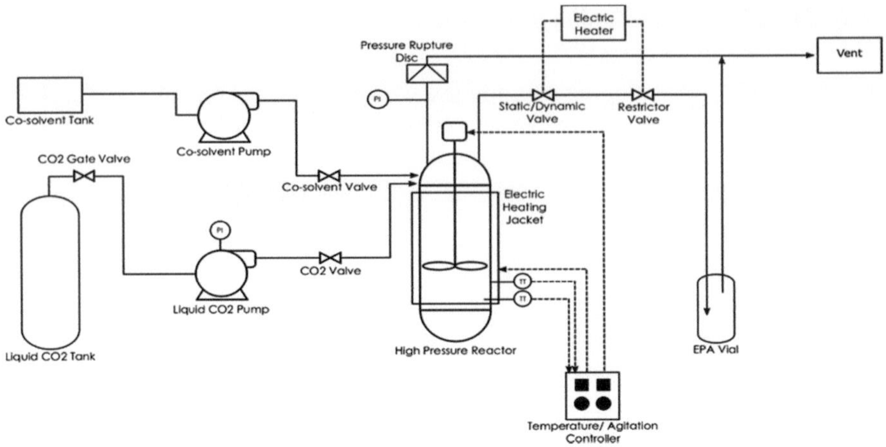

Fig. 1 Schematic diagram of the SCFE process

Supercritical Fluid Extraction Process

Supercritical fluid extraction (SCFE) was conducted in a 100 mL 316L stainless steel reactor manufactured by Supercritical Fluid Technology Inc., USA. The concentrated sample was loaded into the reactor with TBP–HNO_3, followed by CO_2 injection until the design pressure and temperature were reached. For all tests, the mass of the ore sample was fixed at 1 g. After extraction, the system was depressurized, and the solution of rare-earth complex was collected in EPA vials. The schematic diagram of the process is presented in Fig. 1.

Results and Discussion

Sample Characterization

The concentrated sample used as feed for the SCFE step was analyzed and showed a P_{80} (80% passing size) equal to 24 μm. The particle size distribution is shown in Fig. 2. The sample composition was determined by ICP-MS, and the results are summarized in Table 4.

Supercritical Fluid Extraction Results

After SCFE, the solid residues were characterized with alkali fusion followed by ICP-MS to calculate extraction efficiencies. The extraction results are presented in

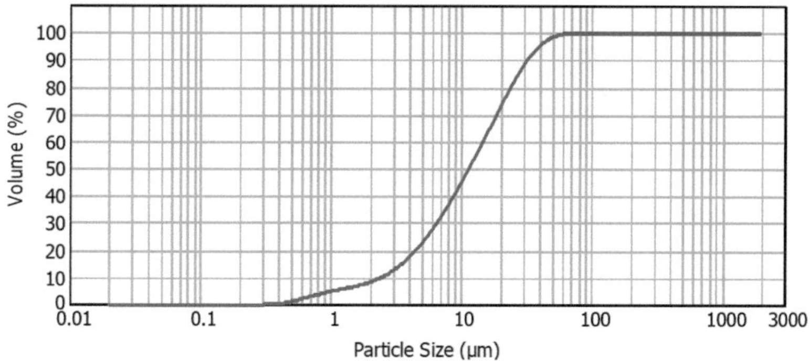

Fig. 2 Particle size distribution

Table 4 Compositional analysis of the ore using alkali fusion followed by ICP-MS

Element (mg/kg)	Y 12,001 (±42)	La 13,578 (±9)	Ce 28,886 (±162)	Pr 3,849 (±5)	Nd 14,224 (±7)	Sm 3,550 (±6)	Gd 3,743 (±9)
Element (mg/kg)	Tb 609 (±1)	Dy 3,231 (±9)	Ho 556 (±1)	Er 1,359 (±3)	Yb 989 (±1)		
Element (mg/kg)	Al 21,181 (±140)	Fe 62,777 (±344)	Ca 10,898 (±24)	Na 4,859 (±42)	K 10,922 (±103)	Zr 134,706 (±3498)	Ta 1,485 (±88)

Note The numbers in parentheses are standard deviations for each element in the sample. The standard deviation was calculated based on three replicated composition analysis for the sample

Table 5. The results of the extraction experiments were utilized to build a set of empirical models to describe the extraction ($\hat{y_i}$) of the tested elements as a function of the five operating parameters (X_1 to X_5), where third-order and higher order interactions are considered as negligible. An empirical model in the following format was constructed for each element.

$$\hat{y}_i = \beta_0 + \tilde{\beta}_1 X_1 + \tilde{\beta}_2 X_2 + \tilde{\beta}_3 X_3 + \tilde{\beta}_4 X_4 + \tilde{\beta}_5 X_5 + \tilde{\beta}_{23} X_2 X_3 + \tilde{\beta}_{34} X_3 X_4 \quad (1)$$

where

$$\begin{aligned} &\tilde{\beta}_1 = \beta_1 + \beta_{24} \quad \tilde{\beta}_2 = \beta_2 + \beta_{14} \quad \tilde{\beta}_3 = \beta_3 + \beta_{15} \quad \tilde{\beta}_4 = \beta_4 + \beta_{12} \\ &\tilde{\beta}_5 = \beta_5 + \beta_{13} \quad \tilde{\beta}_{23} = \beta_{23} + \beta_{45} \quad \tilde{\beta}_{34} = \beta_{34} + \beta_{25} \end{aligned} \quad (2)$$

The experimental data was fit to this empirical model by multiple Linear Least Squares Regression (mLLSR) according to, Eq. 3. $\widehat{\beta}$ is the model parameter vector, containing each of the model parameters $\left(\widehat{\beta}_0, \widehat{\beta}_1, \widehat{\beta}_2, \ldots\right)$, X is the experimental

Table 5 SCFE extraction results for the ore sample

Run ID	Temperature (°C)	Pressure (psi)	Time (h)	Agitation rate (rpm)	S:CA (g:mL)	Extraction Efficiency (%) Na, Al, K, Ca, Fe, Y, Zr, La, Ce, **Pr, Nd**, Sm, Gd, **Tb, Dy**, Ho, Er, Yb
1	40	3000	1	1500	0.2	32.8, 30.5, 39.6, 52.0, 41.9, 42.3, 82.4, 46.2, 48.1, **49.1, 49.7**, 51.2, 46.7, **46.8, 41.1**, 43.9, 44.4, 40.4
2	60	3000	1	750	1	39.5, 32.2, 43.7, 58.8, 43.8, 41.7, 84.9, 53.8, 53.1, **54.8, 54.5**, 53.9, 49.8, **36.3, 42.3**, 44.6, 45.9, 40.5
3	40	4500	1	750	0.2	38.3, 34.9, 42.7, 57.7, 43.5, 46.9, 86.7, 52.8, 51.9, **54.2, 54.9**, 54.6, 52.5, **50.6, 43.7**, 47.8, 48.3, 44.3,
4	60	4500	1	1500	1	19.8, 18.8, 35.6, 44.1, 24.5, 30.4, 74.7, 30.7, 34.8, **37.5, 38.2**, 39.1, 32.2, **34.8, 30.2**, 28.9, 33.6, 29.4
5	40	3000	2	1500	1	34.7, 29.5, 35.9, 49.1, 36.6, 42.5, 81.5, 44.7, 44.5, **45.2, 45.1**, 46.5, 44.8, **46.5, 40.4**, 45, 44.6, 41.6
6	60	3000	2	750	0.2	41.7, 35.6, 42.2, 63.9, 45.4, 49.4, 91.2, 58.1, 57.3, **57.4, 56.8**, 55.9, 54.5, **52.5, 47**, 49.3, 49.0, 44.7
7	40	4500	2	750	1	24.1, 23.6, 24.9, 52.4, 39.1, 45.5, 84.7, 46.6, 44.1, **45.5, 46.5**, 49.6, 49.1, **49.8, 44.2**, 48.5, 48.9, 47.9
8	60	4500	2	1500	0.2	28.3, 20.7, 28.1, 56.6, 36.7, 38.9, 71.8, 48.5, 50.5, **50.8, 51.3**, 50.1, 42.4, **42.6, 35**, 38.6, 38.9, 34.2
9	50	3750	1.5	1125	0.6	31.6, 26.6, 30.9, 58.3, 42.4, 49.8, 86.3, 53.8, 53.1, **53.8, 53.2**, 54.0, 53.6, **51.9, 46.4**, 49.7, 49.7, 47.2
10	50	3750	1.5	1125	0.6	24.9, 23.8, 27.9, 54.0, 37, 44.5, 87.4, 51.8, 52.5, **51.4, 51.6**, 51.6, 50.5, **47.6, 43.6**, 47.2, 47.3, 45.2
11	50	3750	1.5	1125	0.6	33.9, 26.7, 33.7, 55.2, 36.8, 46.1, 83.9, 53.4, 47.5, **53.2, 52.3**, 52.5, 51.1, **44.1, 44.1**, 48.9, 47.1, 39.9

design matrix, and Y is the response vector containing each of the measured experimental extraction efficiencies.

$$\hat{\beta} = \left(X^T X\right)^{-1}\left(X^T Y_i\right) \quad (3)$$

All calculations were performed in RStudio, and the factor effects are presented in Fig. 3. SCFE parameters are as following: X_1 = Temperature, X_2 = Pressure, X_3 = Time, X_4 = Agitation rate, X_5 = S to CA ratio.

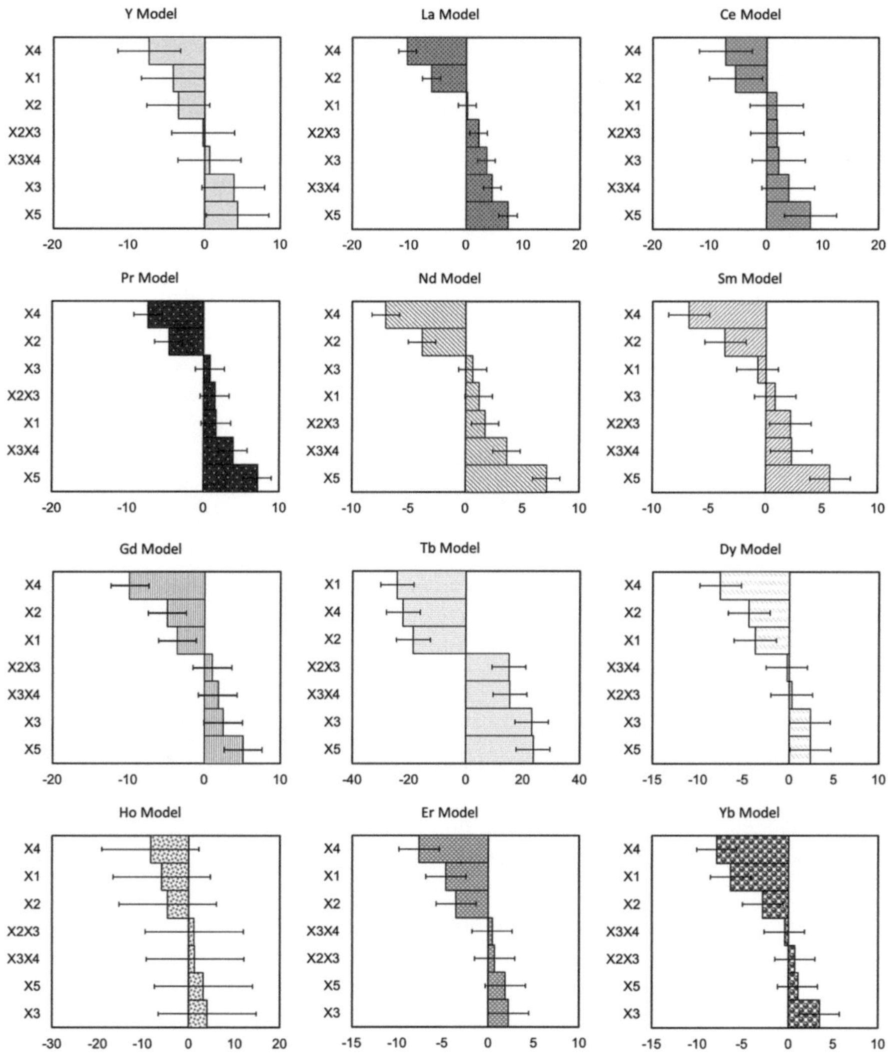

Fig. 3 Factor effect coefficient plot for rare-earth elements

The experimental design for this project had been purposely designed with three mid-point runs for variance estimation and the presence of data curvature. The tests were carried out in a randomized order (generated by Python random number generator). The standard deviations of the three mid-point runs are Na 4.7%, Al 1.7%, K 2.9%, Ca 2.2%, Fe 3.2%, Y 2.7%, Zr 2.8%, La 1.0%, Ce 3.1%, Pr 1.2%, Nd 0.8%, Sm 1.2%, Gd 1.7%, Tb 3.9%, Dy 1.5%, Ho 7.0%, Er 1.5%, and Yb 3.8%. The statistical analysis was carried out with estimated variance to determine statistical significance. The error bars represent the 95% confidence interval for the estimated model parameter coefficients. Factors with error bar not crossing zero are considered significant. The experiment matrix results in a resolution III design, which means, assuming that only first-order factors (main effect) are significant. First-order factors may be confounded with two-factor interactions. Second-order interactions are aliased with each other. As can be seen in Fig. 3, X_2 (pressure) and X_4 (agitation rate) have a significant negative effect on the extraction of REEs, and X_5 (solid to chelating agent ratio) has a significant positive effect on the extraction of REEs.

The negative effect of pressure (X_2) can be explained by the fact that at a lower pressure, the supercritical fluid has gas-like properties, which has a higher diffusion coefficient. Thus, it can penetrate and transport solutes from different matrices at a higher rate. The negative effect of agitation rate (X_4) can be explained by the fact that high agitation creates a high shear environment, which could cause the bonding between metal and chelating agent unstable. The positive effect of solid to chelating agent ratio (X_5) can be explained by the fact that increasing chelating agent to sample ratio enhances extraction because at higher CA content, more chelating agent is available which results in greater complexation degree.

As can be seen in Fig. 3, temperature (X_1) is not significant to extraction behavior of light rare-earth elements (LREEs), except for Gd. However, increasing temperature resulted in lower extraction efficiency of heavy rare-earth elements (HREEs). The relationship between the solubility of a solute in SCFE and the density is described by Chrastil model (Eq. 4):

$$\ln S = k \ln \rho + \frac{A}{T} + B \tag{4}$$

where S (g/L) is the solubility of a solute in supercritical fluid; ρ (g/L) is the density of the supercritical fluid; k is the association number that describes the number of solvent molecules associated with the complex; T is the temperature in K; A and B are empirical parameters. Because increasing temperature at constant pressure decreases the density ρ and A/T term in Eq. 4, it decreases the solubility of the solute, thus decreasing the extraction efficiency.

Residence time (X_3) had a positive effect on the extraction efficiency of REEs because longer duration ensures the dissolution process goes to completion, and efficiency is extraction per unit time, thus it also increases.

Conclusions

In this work, an SCFE process was developed to recover REEs from a Canadian ore. A fractional factorial design of experiment methodology was utilized to design a resolution III experimental matrix for SCFE experiments. Five operating parameters, i.e., temperature, pressure, residence time, agitation rate and solid to chelating agent ratio were tested. The highest extraction of 55% was observed in run #6 with the following operating conditions: temperature was 60 °C, pressure was 3000 psi, time was 2 h, agitation rate was 750 rpm, and solid to chelating agent ratio was 0.2. This study demonstrates that supercritical fluid extraction (aeriometallurgy) is potentially a viable alternative approach for REE recovery from primary resources. Future work will focus on extraction efficiency improvement, process parameter optimization, and extraction mechanism exploration.

Acknowledgements The authors would like to acknowledge the financial support provided by Nation Resources Canada (NRCan, Project No. 3000686549).

References

1. Tunsu C, Petranikova M, Gergorić M, Ekberg C, Retegan T (2015) Reclaiming rare earth elements from end-of-life products: a review of the perspectives for urban mining using hydrometallurgical unit operations. Hydrometallurgy 156:239–258
2. Zhang J, Anawati J, Yao Y, Azimi G (2018) Aeriometallurgical extraction of rare earth elements from a NdFeB magnet utilizing supercritical fluids. ACS Sustain Chem Eng 6:16713–16725
3. Kershaw JR, Jezko J (1982) Supercritical gas extraction of south african coals. Sep Sci Technol 17(1):151–166
4. J. Montagu, M. Configurations, B. Deflection, and S. Technologies, United States Patent (19), No. 19 (1993)
5. Bulley NR, Fattori M, Meisen A, Moyls L (1984) Supercritical fluid extraction of vegetable oil seeds. J Am Oil Chem Soc 61(8):1362–1365
6. Laitinen A, Kaunisto J (1999) Supercritical fluid extraction of 1-butanol from aqueous solutions. J Supercrit Fluids 15(3):245–252
7. Gallyamov RF, Sagdeev KA, Sagdeev AA, Gumerov FM, Galimova AT (2014) Regeneration of the catalysts by supercritical fluid extraction. Int J Anal Mass Spectrom Chromatogr 02 (01):1–14
8. Robertson AM, Lester JN (2006) Recovery of s-triazine herbicides and associated breakdown products from granular activated carbon using supercritical fluid extraction. Water Environ Res 67(6):899–905
9. Yao Y, Farac NF, Azimi G (2018) Supercritical fluid extraction of rare earth elements from nickel metal hydride battery. ACS Sustain Chem Eng 6(1):1417–1426
10. Sinclair LK, Baek DL, Thompson J, Tester JW, Fox RV (2017) Rare earth element extraction from pretreated bastnäsite in supercritical carbon dioxide. J Supercrit Fluids 124:20–29
11. Wai CM, Waller B (2000) Dissolution of metal species in supercritical fluids principles and applications. Ind Eng Chem Res 39(12):4837–4841

12. Lin Y, Liu C, Wu H, Yak HK, Wai CM (2003) Supercritical fluid extraction of toxic heavy metals and uranium from acidic solutions with sulfur-containing organophosphorus reagents. Ind Eng Chem Res 42(7):1400–1405
13. Ding X, Liu Q, Hou X, Fang T (2016) Supercritical Fluid extraction of metal chelate: a review. Critical Rev Anal Chem 1–20
14. Laintz KE, Wai CM, Yonker CR, Smith RD (1991) Solubility of fluorinated metal diethyldithiocarbamates in Supercritical carbon dioxide. J Supercrit Fluids 4(3):194–198
15. Wai CM, Lin Y, Brauer R, Wang S, Beckert WF (1993) Supercritical fluid extraction of organic and inorganic mercury from solid materials. Talanta 40(9):1325–1330
16. Lin Y, Wai CM (1994) Supercritical fluid extraction of lanthanides with fluorinated β-Diketones and tributyl phosphate. Anal Chem 66(13):1971–1975
17. Zhang G, Anawati J, Yao J, Azimi Y (2018) Aerio-metallurgical extraction of rare earth elements from NdFeB magnets utilizing supercritical fluids, ACS Sustain Chem Eng 6:6713–16725
18. Batorshin GS et al (2015) Recovery of rare earth elements, uranium, and thorium from monazite concentrate by supercritical fluid extraction. Radiochemistry 57(4):343–347
19. Kulyako YM, Samsonov MD, Myasoedov BF, Trofimov TI, Malikov DA (2016) Supercritical fluid extraction of rare earth elements, thorium and uranium from monazite concentrate and phosphogypsum using carbon dioxide containing tributyl phosphate and di-(2-ethylhexyl)phosphoric acid. Russ J Phys Chem B 10(7):1078–1084

Optimizing Zr and REE Recovery from Zircon Through a Better Understanding of the Mechanisms Governing Its Decomposition in Alkali Media

Yves Thibault, Joanne Gamage McEvoy and Dominique Duguay

Abstract Zircon can accommodate HREEs at concentrations where their recovery can be considered. However, one drawback is the aggressive conditions required to break down the zircon structure that typically involves high-temperature treatment (>500 °C) in alkali media. In this context, the study focuses on investigating the mechanisms governing alkali–zircon interaction with the objective of defining optimal Zr/REE recovery pathways. Here we report the progress of alkali fusion experiments coupled with characterization down to the nanometer scale. Our results reveal that zircon is preferentially attacked by specific alkali species infiltrating the crystal structure along a network of defects that are progressively annealed with increasing temperature. Considering that annealing kinetics is faster than the decomposition rate, we demonstrate that, for certain REE-rich zircons, a process can be designed where >95% Zr and REE recovery is achieved below 200 °C, while treatment at higher temperatures is detrimental, rendering the crystals less reactive.

Keywords Zirconium · Rare-earth elements · Zircon crystallinity · Alkali-based decomposition

Y. Thibault (✉) · J. Gamage McEvoy · D. Duguay
CanmetMINING, Natural Resources Canada, 555 Booth Street,
Ottawa, ON K1A 0G1, Canada
e-mail: Yves.Thibault2@canada.ca

G. Azimi et al. (eds.), *Rare Metal Technology 2020*, The Minerals, Metals & Materials Series, https://doi.org/10.1007/978-3-030-36758-9_11

Introduction

In addition to being a major source of Zr, zircon ($ZrSiO_4$) can accommodate rare-earth elements (REEs), and in particular the heavy REEs (HREE) at concentrations where their recovery can also be considered. Structurally, zircon is a tetragonal orthosilicate consisting of chains of SiO_4 tetrahedra sharing edges with ZrO_8 dodecahedra [1, 2]. With such a stable crystalline structure, zircon displays very high thermal and chemical resistance. Consequently, Zr recovery flowsheets typically include aggressive high-temperature treatment (>500 °C) with melted alkali carbonates or hydroxides to decompose zircon [3–5]. Nevertheless, as zircon contains low, yet significant amounts of U and Th (<5000 ppm), natural occurrences of zircon of great geological age (e.g. >500 million years) that have been exposed to enough actinide-induced radiation dose to progressively transform the crystalline structure to the amorphous (metamict) state are common [6, 7]. Interestingly, although enhanced reactivity of radiation-damaged zircon in natural surficial processes has been reported [8], the impact of zircon crystallinity on its decomposition during alkali fusion processes have not been well documented. In this context, the objective of the present study is to investigate if optimal low-crystallinity zircon decomposition pathways could be identified. The approach taken was to perform controlled alkali fusion experiments coupled with detailed characterization of the zircon crystalline nature and of the mechanisms governing its interaction with the melted alkali media.

Materials and Methods

Characterization

Powder X-ray diffraction (XRD) analyses were conducted to provide the mineralogical composition of the feed, residues, and precipitates, as well as to monitor the crystallinity of zircon as a function of temperature. The X-ray diffractograms were collected with a Rigaku D/MAX 2500 rotating anode diffractometer using monochromatic Cu Kα radiation (λ = 0.15406 nm) at 40 kV and 200 mA. The structural properties of zircon grains were also investigated through laser Raman spectroscopy. The spectra were acquired from 126 to 2030 cm^{-1} using a Renishaw InVia Reflex Raman spectrometer fitted with a green 514 nm excitation laser focused to $\approx$2 μm.

To obtain the chemical composition of the pristine zircon grains as well as information on the textural nature and elemental distribution within reacted crystals after the alkali-based experiments, backscattered electron (BSE) imaging, and X-ray microanalysis by wavelength-dispersive spectrometry (WDS) were performed on polished sections using a JEOL JXA 8230 electron probe X-ray microanalyzer (EPMA). As the experimental products are highly water soluble, preservation of the reactive phases was achieved by polishing the resin-impregnated material using a diamond film with isopropanol as a lubricant.

Materials

The feed used is an ore concentrate from an old geological formation (≈2 Ga), which, based on X-ray diffraction analysis, consists mainly of zircon, quartz, feldspars, and biotite. Mineralogical characterization revealed that zircon is highly liberated, representing approximately 65% of the mass with an average grain size of about 20–25 μm. The zircon chemical composition, as determined by EPMA, is highly variable, but the average composition indicates that it contains a significant amount of REEs (Table 1). Split fractions of the pristine feed, fired for 2 h at specific temperatures up to 800 °C, were also characterized and used as starting materials.

Alkali-Based Decomposition Experiments

The alkali-based zircon decomposition experiments were performed along the Na–K joint within the carbonate and the hydroxide systems. In order to minimize the formation of water-insoluble alkali zirconosilicate (e.g. Na_2ZrSiO_5), the (Na,K):($ZrSiO_4$) molar ratio was kept in excess of 6:1 (e.g. [5]).

For the carbonate system, ACS-grade Na_2CO_3 and K_2CO_3 were first mixed in the required amount and fired for 12 h at 650 °C. The resulting carbonate solid solution ($[Na_{0.7}K_{0.3}]_2CO_3$) was finely ground, and thoroughly mixed with the feed. The material was then loaded in a covered 99.8% pure alumina crucible and heated in a temperature-controlled muffle furnace. The temperature was increased at a rate of 400 °C h^{-1} followed by a 2.5-hour dwell at a peak temperature of 950 °C after which the furnace was shut off.

For the hydroxide system, the feed and required amounts of 50% w/w NaOH and 50% w/v KOH aqueous solutions to produce an equimolar composition (1NaOH:1KOH) were introduced in a PTFE-lined Alloy 400 vessel surrounded by a heating mantle. The temperature, controlled using a PFA-coated K-thermocouple inserted in the load, was increased to a peak temperature of 195 °C in 45 min followed by a dwell of 30–90 min. From the onset of boiling of the aqueous

Table 1 Average selected REE contents of zircon (expressed as wt% oxides, N = 40) in the pristine feed as measured by EPMA

Y_2O_3	La_2O_3	Ce_2O_3	Nd_2O_3	Sm_2O_3	Gd_2O_3	Dy_2O_3	Er_2O_3
1.12	0.19	0.60	0.65	0.27	0.36	0.23	0.16

solution (≈125 °C), up to peak temperature, the water is lost through evaporation resulting in a concentrated [$Na_{0.5}K_{0.5}$]OH melt. After the experiment, the material was either preserved for characterization or leached to estimate recovery.

Results and Discussion

Zircon Crystallinity

Considering its modal abundance in the concentrate (≈65%), the XRD peaks assigned to zircon have anomalously low intensity compared to those from the other minerals present in the concentrate suggesting a low degree of crystallinity. This is further supported by the fact that upon progressive heating up to 800 °C, a steady increase in zircon diffraction peaks intensity, initiated at a temperature as low as 250 °C, and accompanied by a shift to higher 2θ values (Fig. 1a), indicative of a contraction of the unit cell, all reflect an atomic rearrangement to a more crystalline structure [6].

Additional structural information can be gained by investigating the zircon at the crystal scale. Raman spectra, obtained using a 514 nm excitation laser focused to 2 μm, were collected at specific areas of numerous grains. The acquired spectra show variable characteristics but the vast majority have no clear vibrational features, once again, suggesting poor crystallinity for most of the zircons (Fig. 1b; [7]). On the other hand, after heating up to 800 °C, all grains show rich spectra characteristic of well-crystallized zircon. Interestingly for the pristine feed, using the U and Th concentrations measured by EPMA at the same grain locations where the Raman spectra were obtained, as well as the estimated age of the zircon, we can

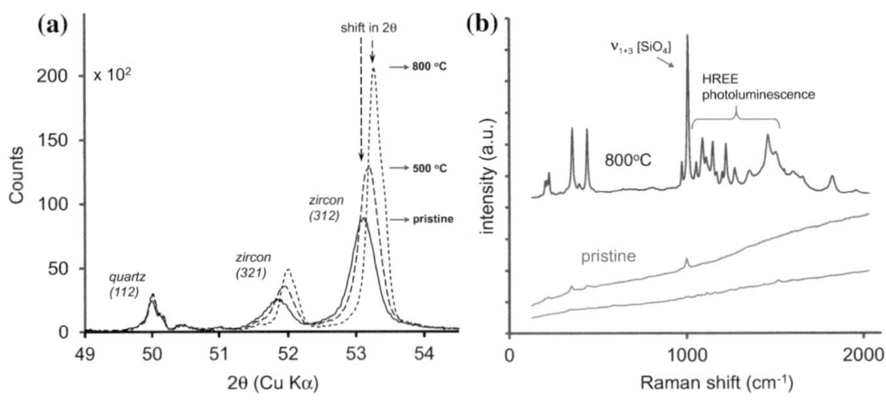

Fig. 1 **a** XRD patterns showing the progressive increase in intensity and shift in 2θ of the (321) and (312) zircon diffraction peaks; **b** typical Raman spectra obtained for zircon in the pristine feed and after heating to 800 °C

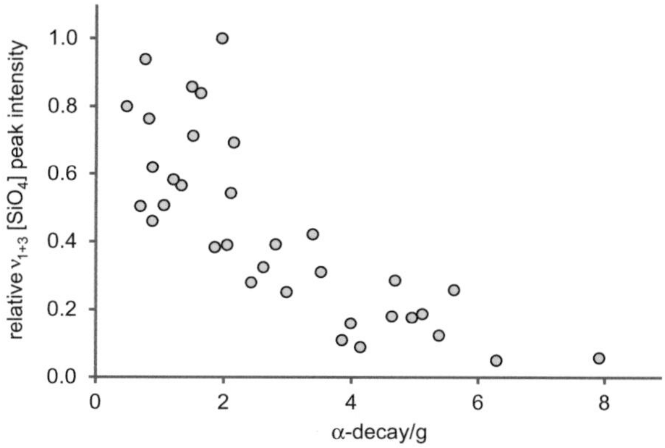

Fig. 2 Correlation between the radiation dose and the ν_{1+3} [SiO_4] Raman peak intensity for various zircon grains in the pristine feed

calculate the radiation exposure, expressed as α-decays/g [6]. As illustrated in Fig. 2, there is a clear negative correlation between the calculated radiation dose and the net peak intensity of the ν_{1+3} [SiO_4] stretching vibration (e.g. Fig. 1b: Raman shift ≈1000 cm^{-1}) characteristic of crystalline zircon. This strongly supports that poor crystallinity is related to the propagation of actinide-induced defects (metamictization).

Considering that reactive processes are preferentially initiated at defects, radiation-damaged zircon should present less resistance to decomposition, leading to potential opportunities for efficient recovery of Zr and REEs. However, the observed increase in crystallinity with temperature will progressively anneal defects and should render the zircon more refractory. Consequently, as the rate of recrystallization can be faster than the decomposition rate, the reaction temperature for an optimal alkali fusion process should be minimized.

Zircon Decomposition in a [$Na_{0.7}K_{0.3}$]$_2$$CO_3$ Melt

As can be seen on the Na–K carbonate phase diagram shown in Fig. 3a, the freedom to minimize temperature and keep the carbonate phase fully liquid is limited to 709 °C, a temperature where, as discussed above, we can expect significant recrystallization of the zircon. Nevertheless, a zircon decomposition experiment using the pristine feed and a [$Na_{0.7}K_{0.3}$]$_2$$CO_3$ composition was performed within the carbonate liquid field at 750 °C for a duration of 2.5 h. From the BSE image obtained on a polished section of the preserved experimental product, it is clear that the decomposition is quite limited and consists of veinlets infiltrating the grains, possibly along regions initially rich in defects (Fig. 4). The distribution of the alkalis, as determined by EPMA WDS X-ray mapping, indicates a clear enrichment of potassium compared to sodium in the reacted areas of the grains suggesting that it preferentially interacts with the zircon phase (Fig. 4).

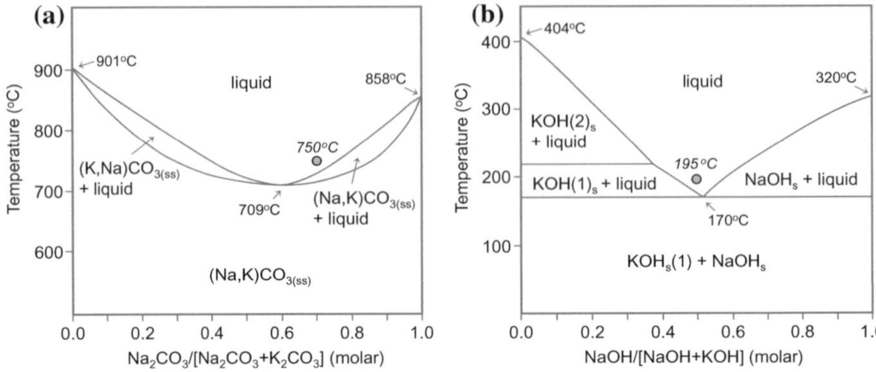

Fig. 3 **a** Na_2CO_3–K_2CO_3 and **b** NaOH–KOH phase diagrams as generated by FactStage. The conditions (alkali proportion, temperature) for the zircon decomposition experiments performed in the carbonate and hydroxide systems are shown as filled circles. Some minor fields were omitted for clarity. s: solid; ss: solid solution

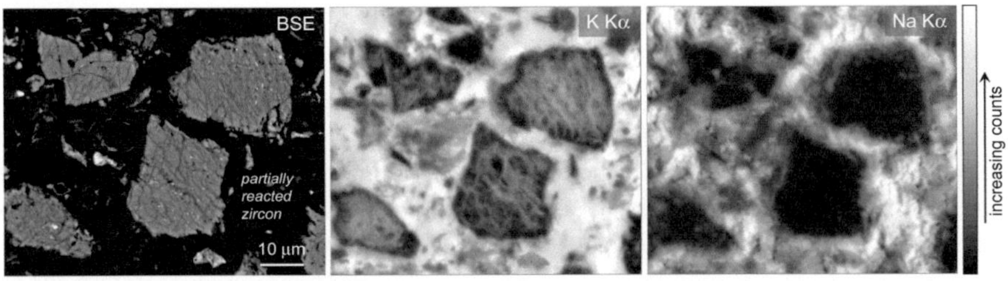

Fig. 4 BSE image and WDS K Kα and Na Kα X-ray maps obtained on a polished section across the preserved product of a 750 °C experiment using the pristine feed and a $[Na_{0.7}K_{0.3}]_2CO_3$ composition

Zircon Decomposition in a [Na$_{0.5}$K$_{0.5}$]OH Melt

For the hydroxide system, as shown on the NaOH–KOH phase diagram (Fig. 3b), a fully liquid concentrated hydroxide phase can be maintained at a temperature as low as 170 °C with a mixture close to equimolar sodium and potassium endmembers ($[Na_{0.5}K_{0.5}]OH$; [3]). In the context of a feed rich in radiation-damaged zircon, such a composition seems beneficial, as in addition to minimizing temperature to prevent recrystallization, potassium, which appears more reactive, is kept as a reagent component. Consequently, decomposition experiments were performed in a $[Na_{0.5}K_{0.5}]OH$ melt at a temperature of 195 °C and durations from 30 to 90 min. Interestingly, even after only 30 min, as can be seen on the BSE image in Fig. 5, the extent of decomposition for the pristine feed is very advanced, although distinct degrees of reactivity for different zircon crystals are observed likely reflecting

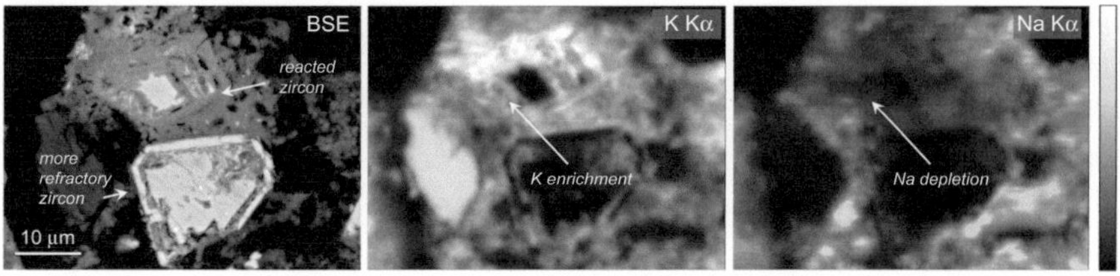

Fig. 5 BSE image and WDS Na Kα and K Kα X-ray maps obtained on a polished section across the preserved product of a 195 °C experiment using the pristine feed and a $[Na_{0.7}K_{0.5}]OH$ composition for a duration of 30 min

distinct defect densities from grain to grain (Fig. 2). Once again, the alkali-bearing phases replacing zircon are highly enriched in potassium compared to sodium (Fig. 5; cf. [3])

To evaluate the impact of the progressive temperature-induced annealing of defects on zircon decomposition, similar experiments at 195 °C and a duration of 90 min were conducted on the pristine feed as well as the feed fired at 500 and 800 °C. To compare recovery, the products were leached following a method proposed by Liu et al. [5]. First three 30-min water leach cycles were performed at 50 °C to dissolve the alkali silicates and hydrolize the alkali zirconate. The filtered residual solid was then leached in diluted HCl at 65 °C to transform the water-insoluble hydrated zirconia and REE hydroxide in the form of soluble oxy-chlorides and chlorides. After filtering the acidic solution, the residue was recuperated and the filtrate was left to evaporate to obtain a precipitate. Finally, both the residue and precipitate were dried, weighed and characterized by XRD and EPMA. The only phase present in the residue is zircon, while, as expected, zirconium oxychloride hydrate and REE chloride hydrate was identified in the precipitate (Fig. 6a) Zirconium recovery was also calculated based on the compositions and the relative weight of the residue and precipitate. The results indicate that, with the pristine feed, 98% zirconium can be recovered when treated at 195 °C for 90 min. However, for the feed fired at 500 and 800 °C, the recovery dropped to 88% and 53% respectively, emphasizing the strong dependence of defects on decomposition. Comparison of the REE contents relative to Zr in the precipitate is very consistent with those of the average initial zircon composition indicating comparable recovery values (Fig. 6b). Finally, the alkali and silica contents of the precipitate are very low ($Na_2O + K_2O < 0.4$ wt%; $SiO_2 < 2.5$ wt%) suggesting that no significant amount of intermediate water-insoluble alkali zirconosilicate remained after the decomposition experiments. This is important for efficient reagent regeneration as it indicates that nearly all the alkalis were dissolved during the initial water leach.

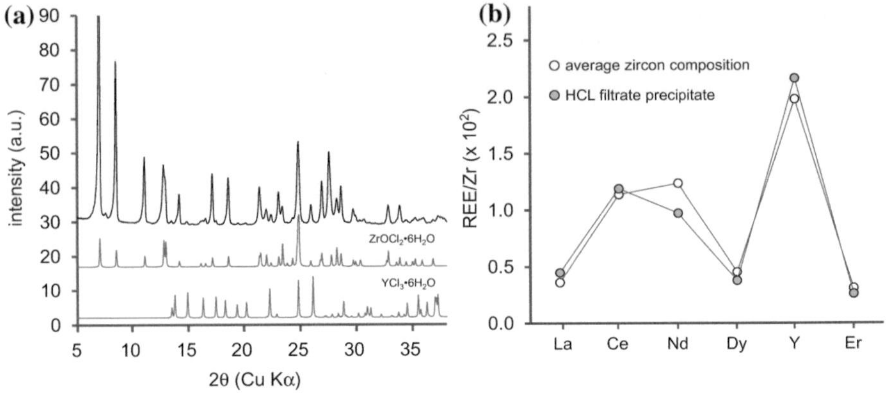

Fig. 6 Characterization of the precipitate recovered from the filtrate of the HCl leach for a decomposition experiment on the pristine feed at 195 °C for 90 min (see details in text): **a** XRD pattern showing the phases identified; **b** comparison of the REE:Zr ratios in the precipitate to those calculated from the feed average zircon composition

Conclusions

This study emphasizes the importance of detailed characterization of zircon crystallinity to optimize its decomposition for Zr and REE recovery, in particular as natural occurrences of zircon of great age that have been damaged by actinide-induced radiation are fairly common [6, 7]. For such radiation-damaged zircon, minimization of decomposition temperature is critical. Our results demonstrate that, not only can more than 95% recovery be achieved below 200 °C in the presence of $[Na_{0.5}K_{0.5}]OH$, but that treatment at higher temperatures is detrimental, as recrystallization will render the crystals less reactive. In addition to significant energy reduction compared to a conventional alkali fusion, the significant decrease in decomposition temperature can also address engineering challenges related to process containment in caustic media (e.g. corrosion, fracturing, clogging), as it allows the use of materials (e.g. PTFE) that are not adequate for higher temperature treatment.

Acknowledgements Financial support for this project was provided by Natural Resources Canada through a special fund for the REE and Chromite R&D Initiative. The authors would like to thank Derek Smith (CanmetMINING) for support with the XRD characterization and Mary Jane Walzak (Surface Science Western) for performing the Raman spectroscopy analyses. Rolando Lastra provided helpful suggestions throughout the course of this study.

References

1. Deer WA, Howie RA, Zussman J (1997) Rock-forming minerals Volume 1a: Orthosilicates. The Geological Society, London
2. Finch RJ, Hanchar JM (2003) Structure and chemistry of zircon and zircon-group minerals. Rev Mineral Geochem 53:1–25. In: Hanchar JM, Hoskin HO (eds) Zircon. Mineralogical Society of America, Washington, DC
3. Abdelkader AM, Daher A, El-Kashef E (2008) Novel decomposition method for zircon. J Alloys Compd 460:577–580
4. Wang Z, Xu Q, Xu M, Wang S, You J (2015) In situ spectroscopic studies of decomposition of $ZrSiO_4$ during alkali fusion process using various hydroxides. RCS Adv 5:11658–11666
5. Liu J, Song J, Qi T, Zhang C, Qu J (2016) Controlling the formation of Na_2ZrSiO_5 in alkali fusion process for zirconium oxychloride production. Adv Powder Technol 27:1–8
6. Ewing RC, Meldrum A, Wang LM, Weber WJ, Corrales R (2003) Radiation effects in zircon. Rev Mineral Geochem 53(1):387–425. In: Hanchar JM, Hoskin HO (eds) Zircon. Mineralogical Society of America, Washington, DC
7. Zhang M, Salje EKH, Franan I, Graeme-Barber A, Daniel P, Ewing RC, Clark AM, Leroux H (2000) Metamictization of zircon: Raman spectroscopic study. J Phys Condens Matter 12:1915–1925
8. Hay DC, Dempster TJ (2009) Zircon alteration, formation and preservation in sandstones. Sedimentology 56:2175–2191

An Innovative Process for Extracting Scandium from Nickeliferous Laterite Ore

Jihye Kim and Gisele Azimi

Abstract Laterite ores contain significant amounts of scandium, a strategic material with versatile applications. In this study, a two-stage process was developed to concentrate and recover scandium from nickeliferous laterite ore. In the first step, carbothermic smelting was performed at 1400–1600 °C using lignite as a reductant and calcia and/or silica as a flux. This process resulted in a slag phase concentrated in Sc and a metallic iron phase enriched with nickel and cobalt. Under the optimum conditions, scandium was successfully concentrated in the slag phase more than 14 times than that in the starting material. In the second step, the slag phase was leached using NaOH cracking to recover Sc. A fractional factorial design methodology was utilized to investigate the effect of various operating parameters during the smelting and the leaching processes and to optimize the processes. After optimization, 88% of scandium was recovered during the NaOH cracking process.

Keywords Carbothermic reduction · NaOH cracking · Scandium · Laterite ore · Ferronickel · Smelting · Cobalt

Introduction

Laterite ore is an important deposit as it contains appreciable amounts of nickel and cobalt. It is mainly composed of iron, silicon, magnesium, nickel, cobalt, and in some cases scandium. Scandium shows a close association with ferrous iron in Ni–Co laterite deposits; thus, laterite ore is considered to be one of the most promising Sc resources in many regions, particularly in Australia and Europe [1].

Scandium is an expensive and rare element because of its wide applications in the aerospace industry, metallurgy, nuclear technology, and fuel cell. Sc is not rare

J. Kim · G. Azimi (✉)
University of Toronto, 200 College Street, Toronto, ON, Canada M5S3E5
e-mail: g.azimi@utoronto.ca

J. Kim
e-mail: jhye.kim@mail.utoronto.ca

G. Azimi et al. (eds.), *Rare Metal Technology 2020*, The Minerals, Metals & Materials Series, https://doi.org/10.1007/978-3-030-36758-9_12

as named; it is known that approximately 22 mg/kg of Sc is present in the Earth's crust [2, 3]. However, the challenge associated with Sc supply is that it is widely distributed and not concentrated in particular minerals, which result in the absence of primary mine for Sc production. Sc is only obtained as a by-product of rare-earth elements (REEs), iron–uranium, tungsten, and titanium mining [4, 5]. It is supplied for only 5–12 tonnes in the form of scandium oxide (Sc_2O_3) annually worldwide and consequentially its high price hinders wider industrial applications of Sc [6, 7]. As a result, Sc is rapidly becoming a critical material and recovery of Sc from primary and secondary sources is an emerging research area. However, recovering Sc is problematic because of the low concentration of Sc and the presence of impurities, such as Fe, Al, Ca, Mn, and Mg. Accordingly, there is an urgent need for concentrating Sc in a specific phase and recovering Sc with high extraction efficiency.

In this study, a two-stage innovative process was developed with an aim to extract scandium from a laterite ore. The first stage involves carbothermic reduction in which the iron is reduced at 1400–1600 °C and separated as a metallic phase, while scandium is enriched in a slag phase. In the second stage of the process, a two-step NaOH cracking process was used to recover Sc from the slag.

This work is devoted to developing an innovative process for extracting scandium from nickeliferous laterite ore. A systematic study was conducted to find the optimum operating conditions of carbothermic reduction and alkaline cracking, and to elucidate the mechanisms behind each process. The developed process provides several advantages of concentrating Sc in the slag phase, achieving high final Sc extraction efficiency (>88%), and producing ferronickel alloy as the by-product.

Experimental

Materials

A ground nickeliferous laterite ore was obtained from New Caledonia. Coal, sulfuric acid, and nitric acid were supplied by VWR (USA). Calcium oxide (99.5% purity, −325 mesh) was supplied by Materion (USA). Silicon dioxide (99.5% purity, −10 μ) and discandium trioxide (>99.9% purity) were obtained from Alfa Aesar (USA). Sodium hydroxide was purchased from Caledon Laboratories Ltd. (Canada).

Experimental Procedure

The dried laterite ore samples were mixed with lignite and flux (CaO and/or SiO_2). The mixture was then pressed into a pellet by a pneumatic/hydraulic press

(250 MPa, 3 min; Pro Point Equipment Ltd., UK). The pellets were placed in a graphite crucible and subjected to smelting in a vertical tube furnace (Carbolite Gero Limited, UK) at different smelting temperatures for 1.5 h under N_2 atmosphere. Next, the pellets were cooled to room temperature and the metallic Fe and slag phases were manually separated. The ground slag sample was mixed with 50 wt% NaOH and the mixture was baked in a muffle furnace (Gesswein Canada, Canada). After baking, the sample was leached in acid using a magnetic stirrer (Corning PC-420C, USA).

Factorial Design of Experiments and Empirical Model Building

A systematic study was carried out to investigate the effect of operating parameters on the composition of a slag phase and extraction efficiency of Sc for the first and second processes, respectively. Table 1 shows the detailed information on operating parameters for each process.

A fractional factorial design methodology was used to design two 2_{IV}^{4-1} experimental matrices and to build empirical models. The experimental data was then fitted to the empirical model (Eq. 1) using Multiple Linear Least Squares Regression (mLLSR, Eq. 2).

$$\hat{y}_i = \hat{\beta}_0 + \hat{\beta}_1 X_1 + \hat{\beta}_2 X_2 + \hat{\beta}_3 X_3 + \hat{\beta}_4 X_4 + \hat{\beta}_{12} X_1 X_2 + \hat{\beta}_{13} X_1 X_3 + \hat{\beta}_{14} X_1 X_4 \quad (1)$$

where $\hat{\beta}$ is the vector including each of the model parameters, β_0 corresponds to the baseline bias, β_1 corresponds to X_1, β_2 corresponds to X_2, β_3 corresponds to X_3, β_4 corresponds to X_4, and β_{ij} corresponds to the second-order interaction of the two different factors.

$$\hat{\beta} = \left(X^T X\right)^{-1} \left(X^T Y_i\right) \quad (2)$$

where X_{i} is the experimental design matrix, and Y_{i} is the response matrix including the actual response.

Table 1 Summary of operating parameters for the carbothermic reduction and NaOH cracking processes

Factor	X_1	X_2	X_3	X_4
Carbothermic reduction	Smelting temperature	Carbon-to-ore mass ratio	Flux-to-ore mass ratio	Flux type
NaOH cracking	Baking temperature	NaOH-to-slag ratio	Baking time	Liquid-to-slag ratio

Morphological, Mineralogical, and Compositional Characterization

The morphological characteristics of the samples were investigated using scanning electron microscopy–energy dispersive spectroscopy (SEM-EDS, Hitachi SU 3500). The mineralogical characterizations were obtained with X-ray diffraction (XRD, Rigaku MiniFlex 600). Electron probe microanalyzer (EPMA, JEOL JXA8230 5-WDS electron microprobe) was used for the phase mapping. For compositional analysis, microwave digestion was performed at 220 °C (MARS 6 Xpress system, CEM Corporation). The concentration of Sc and base metals (Fe, Cr, Ni, Co, Mn, Al, Mg, Si, and Ca) in the leachate solutions was then measured by inductively coupled plasma–mass spectrometry (ICP-MS, Thermo Scientific iCAP Q) and inductively coupled plasma–optical emission spectrometry (ICP-OES, PerkinElmer Optima 8000), respectively. For the silica content measurement, an X-ray fluorescence spectrometer (XRF, Bruker S2 Ranger) was utilized.

Results and Discussion

Characterization of Laterite Ore

A thorough characterization of feed material was performed using ICP-OES, ICP-MS, XRD, and SEM-EDS. As shown in Table 2, the main constituent was determined to be Fe (41.9 wt%); the laterite ore also contains Al (2.2 wt%), Mg (1.8 wt%), Ni (1.2 wt%), Mn (0.9 wt%), and Co (0.1 wt%). Based on XRF analysis, the Si content was 3.7 wt%. Along with base metals, this ore contains 35.9 mg/kg of Sc and trace amounts of other REEs. Given the economic interest and concentration, this study only focuses on Sc recovery.

On the basis of XRD results, laterite ore is composed of goethite ($FeO(OH)$), talc ($Mg_3Si_4O_{10}(OH)_2$), birnessite ($H_{4.344}MnO_{4.172}$), and bayerite (α-$Al(OH)_3$). SEM image showed that the laterite ore particle has a small particle size with a rough surface (Fig. 1). Elemental mapping results clearly showed that there are two types

Table 2 Elemental composition of feed laterite ore. (a) Bulk elements; (b) Rare-earth elements

(a) Bulk elements							
	Fe	Al	Mg	Ni	Mn	Ti	Co
(b) Rare-earth elements							
	Sc	Y	Nd	Gd	Tb	Dy	Lu
Composition (wt%)	41.89	2.16	1.75	1.22	0.85	0.18	0.08
Composition (mg/kg)	35.91	1.15	3.37	4.11	1.11	0.85	8.85

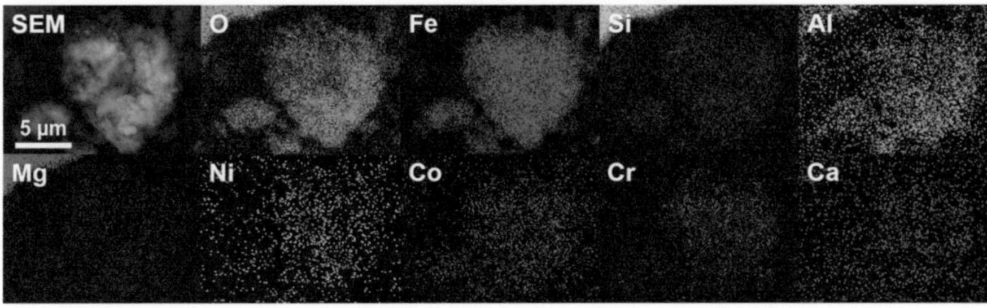

Fig. 1 Secondary electron image and EDS elemental mapping results of laterite ore

of particles in this system: iron oxide, and silicate. The former is enriched with Al, Ca, Ni, and Co, while the latter has a close association with Mg, Al, and Ca. These are in good agreement with XRD and ICP results.

Carbothermic Reduction

Carbothermic reduction experiments were conducted based on the 2_{IV}^{4-1} experimental matrix designed using a fractional factorial design methodology. The results showed that Sc reports into the slag phase, while Fe, Ni, and Co report into the metallic phase. Figure 2a presents the ordered factor effect coefficients with enough significance ($\alpha = 0.05$) on the concentration of Sc in the slag phase.

As can be seen in Fig. 2a, the factor with the most significant positive impact on the Sc concentration in the slag phase is smelting temperature (X_1). This is due to the effect of X_1 on the viscosity of the slag phase. Viscosity is in inverse proportion to fluidity. Therefore, a low viscosity is preferred to clearly separate the slag phase

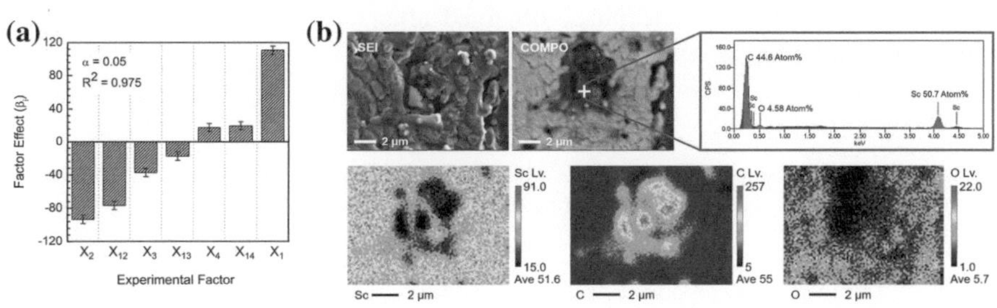

Fig. 2 **a** Ordered chart of factor effect coefficients for the empirical model of Sc concentration in a slag phase; **b** SEM, EDS, and EPMA mapping results of the product obtained after the smelting of scandium oxide at a carbon mass ratio of 0.5

from the metallic phase. The relationship between viscosity and temperature is shown in Eq. 3 [8, 9].

$$\eta = \eta_O \times \exp\left(E_a/\mathrm{RT}\right) \quad (3)$$

where η is the viscosity of slag, R is the universal gas constant, E_a is the activation energy, and T is the temperature (K).

Flux type (X_4) also has a positive effect on the Sc concentration in the slag phase (calcia is preferred). In nature, Sc is highly associated with basic or ultrabasic rocks, whereas it has a low association with acidic rocks [10, 11]. Therefore, adding calcia is preferable to increase the Sc concentration in the slag phase as it increases the basicity of the slag.

The concentration of Sc in the slag phase is negatively affected by the flux-to-ore mass ratio (X_3). This is because of the fact that increasing the flux amount leads to an increased amount of slag; thus, the Sc concentration in the slag phase is diluted. Carbon-to-ore mass ratio (X_2) also has a negative impact. This observation can be explained by the scandium carbide formation. At a high carbon-to-ore mass ratio, there is a higher possibility to form scandium carbide. The formation of scandium carbide from scandium oxide under the same smelting conditions was confirmed by EPMA as presented in Fig. 2b, verifying the negative effect of carbon ratio.

On the basis of factor effect coefficients, an empirical model of Sc concentration in a slag phase was prepared for process optimization as follows:

$$\begin{aligned}\hat{y}_i = {} & 143.5 + 110.5X_1 - 93.3X_2 - 37.2X_3 + 17.4X_4 - 76.7X_1X_2 \\ & - 17.6X_1X_3 + 19.6X_1X_4\end{aligned} \quad (4)$$

The optimum conditions for maximizing Sc concentration in a slag phase were determined to be high levels of smelting temperature (X_1) and flux type (X_4), and low levels of carbon-to-ore mass ratio (X_2) and flux-to-ore mass ratio (X_3). Under these conditions, the slag phase contained 530 mg/kg of Sc, which is more than 14 times higher concentration than that in the starting material (36 mg/kg) (1.7% absolute average relative deviation). The elemental composition of feed laterite ore and that of slag and metallic phase obtained after carbothermic reduction under optimized conditions is tabulated in Table 3.

Table 3 Elemental composition of feed laterite ore and that of slag and metallic phase obtained after the carbothermic reduction under optimized conditions

Composition (wt%)	Fe	Cr	Al	Mg	Ni	Co	Mn	Ti	Sc (mg/kg)
Laterite ore	41.9	3.5	2.2	1.8	1.2	0.1	0.9	0.2	36
Slag phase	0.9	0.05	34.7	20.2	0.003	0.005	8.8	0.02	530
Metallic phase	88.6	7.6	0.005	0.002	2.1	0.1	1.6	0.04	<0.02

NaOH Cracking

Similar to the first process, a fractional factorial design methodology was used to design the 2_{IV}^{4-1} experimental matrix for the NaOH cracking process. Figure 3 shows the ordered factor effect coefficients with enough significance ($\alpha = 0.05$) for the extraction efficiency of Sc.

As shown in Fig. 3, Sc extraction from slag is positively influenced by baking temperature (X_1), NaOH-to-slag ratio (X_2), and baking time (X_3), while negatively affected by the liquid-to-slag ratio (X_4). The positive impact of baking temperature (X_1) is because molten sodium hydroxide has strong corrosivity. NaOH melts at around 320 °C, which is in the range of our process (200–400 °C) [12, 13]. Molten sodium hydroxide is known to be highly corrosive and easily dissolve silicates. Furthermore, the increase of baking temperature expedites the kinetics of the silicate decomposition reaction [14]. The positive impact of the NaOH-to-slag ratio indicates that NaOH is the limiting reagent in the digestion reaction. Therefore, when more NaOH molecules are added to the system, more scandium molecules convert to the water-soluble hydroxide form. The positive effect of baking time is due to the slow kinetics of silicate digestion reaction. At a low level of baking time (30 min), the reaction could not be completed [14, 15]. The synergistic effect of baking temperature and time was observed as the reaction kinetics becomes faster at higher baking temperature.

The negative effect of the liquid-to-slag ratio is due to the gelation of silicic acid (H_4SiO_4), as follows:

$$SiO_{2(s)} + 2H_2O_{(l)} = H_4SiO_{4(aq)} \tag{5}$$

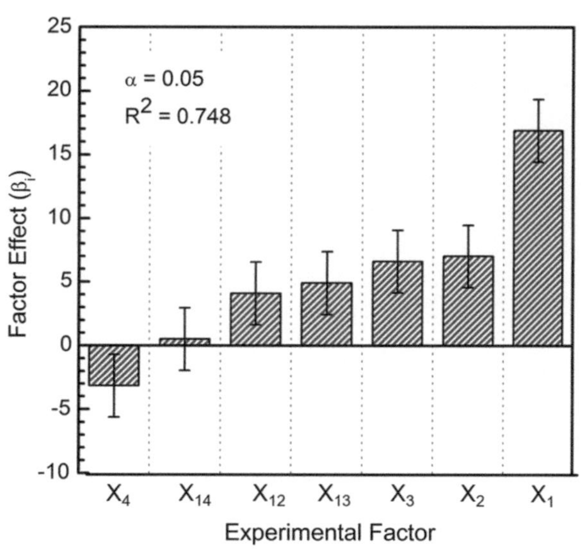

Fig. 3 Ordered chart of factor effect coefficients for the empirical model for Sc extraction during NaOH cracking process

Under acidic conditions, silica easily transforms into silicic acid. Silicic acid tends to polymerize and gelate, entrapping the slag particles and consequentially hampering Sc extraction [16].

Based on the factor effect coefficients, an empirical Sc extraction model was built and presented in Eq. 6.

$$\hat{y}_i = 42.9 + 16.9X_1 + 7.0X_2 + 6.6X_3 - 3.2X_4 + 4.1X_1X_2 + 4.9X_1X_3 \quad (6)$$

Using this equation, the optimized conditions for maximizing Sc extraction efficiency from slag were determined to be high levels of baking temperature (X_1, 400 °C), NaOH-to-slag ratio (X_2, 5 mL/g), and baking time (X_3, 90 min), and low level of liquid-to-slag ratio (X_4, 4 mL/g) and the corresponding predicted extraction efficiency of Sc was 85.5%. Under these optimum conditions, experiments were performed to assess the predictability of the model and 88.4% of Sc was recovered, verifying high accuracy of the empirical Sc extraction model (4.0% absolute average relative deviation).

Conclusions

Here a two-stage process was developed with an aim to concentrate and extract a valuable material, i.e., scandium, from nickeliferous laterite ore, while producing a metallic iron enriched in Ni and Co as a by-product. The carbothermic reduction was conducted at 1400–1600 °C using lignite as a reductant and calcia and/or silica as a fluxing reagent. A systematic study was conducted utilizing a fractional factorial design methodology to investigate the quantitative effect of four variables, namely, smelting temperature, carbon-to-ore mass ratio, flux-to-ore mass ratio, and flux type, on the concentration of Sc in a slag phase. On the basis of the results, an empirical model was built to optimize the process for maximizing Sc concentration in a slag phase and the optimum conditions were determined to be high smelting temperature and low carbon ratio without flux. Under these conditions, Sc was concentrated in the slag phase more than 14 times compared with that in the feed material.

After carbothermic reduction, the NaOH cracking process was carried out to extract Sc from the slag phase. For this process, four operating parameters, namely, baking temperature, NaOH-to-slag ratio, baking time, and liquid-to-slag ratio, were investigated. After the process optimization, the optimized conditions were determined to be high baking temperature, high NaOH ratio, long baking time, and low liquid ratio, and 88.4% of Sc was extracted under these conditions, indicating the high accuracy of the empirical extraction model for Sc.

References

1. Jaireth S, Hoatson DM, Miezitis Y (2014) Geological setting and resources of the major rare-earth-element deposits in Australia. Ore Geol Rev 62:72–128. https://doi.org/10.1016/j.oregeorev.2014.02.008
2. Greenwood NN, Earnshaw A (1997) Chemistry of the elements, 2nd edn. Butterworth-Heinemann
3. Lide DR (2005) CRC handbook of chemistry and physics, 85th edn. CRC Press, Boca Raton
4. USGS (2018) Mineral commodity summaries
5. Wang W, Cheng CY (2011) Separation and purification of scandium by solvent extraction and related technologies: a review. J Chem Technol Biotechnol 86:1237–1246. https://doi.org/10.1002/jctb.2655
6. Riggall S (2015) Australian scandium supply—a paradigm shift for a strategic metal. In: California ASM international, AeroMat'15, Long Beach
7. Kaya Ş, Yagmurlu B, Dittrich C, Friedrich B (2017) Potential of scandium supply from ores and industrial products in the near future. In: 2nd European rare earth resources conference, ERES2017, Santorini, Greece
8. Mudersbach D, Drissen PM, Kühn M, Geiseler J (2001) Viscosity of slags. Steel Res 72:86–90. https://doi.org/10.1016/j.fuel.2012.11.014
9. Seetharaman S, Mukai K, Sichen D (2005) Viscosities of slags—an overview. Steel Res Int 76:267–278. https://doi.org/10.1002/srin.200506008
10. Horovitz CT (2012) Scandium: its occurrence chemistry physics, metallurgy, biology and technology. Elsevier, Oxford
11. Williams-Jones AE, Vasyukova OV (2018) The economic geology of scandium, the runt of the rare earth element litter. Econ Geol 113:973–988. https://doi.org/10.5382/econgeo.2018.4579
12. Douglas TB, Dever JL (1954) Anhydrous sodium hydroxide: the heat content from 0 degrees to 700 degrees C, the transition temperature, and the melting point. J Res Natl Bur Stand 53:81–90. https://doi.org/10.6028/jres.053.010
13. Perry DL, Phillips SL (1995) Handbook of inorganic compounds. Taylor & Francis
14. Han YF, Sun TC, Li J et al (2011) Removing of Si in the NaOH molten salt reaction of titanium slag to produce TiO_2. Adv Mater Res 418–420:387–392. https://doi.org/10.4028/www.scientific.net/AMR.418-420.387
15. Mu WN, Zhai YC (2010) Desiliconization kinetics of nickeliferous laterite ores in molten sodium hydroxide system. Trans Nonferrous Met Soc China (English Ed) 20:330–335. https://doi.org/10.1016/S1003-6326(09)60142-4
16. Alkan G, Yagmurlu B, Cakmakoglu S et al (2018) Novel approach for enhanced scandium and titanium leaching efficiency from bauxite residue with suppressed silica gel formation. Sci Rep 8:1–11. https://doi.org/10.1038/s41598-018-24077-9

Recovery of Strategic Materials from Canadian Bauxite Residue by Smelting Followed by Acid Baking–Water Leaching

John Anawati and Gisele Azimi

Abstract Sustainable sourcing of raw materials is becoming an increasingly important factor to consider in the modern economic and technological landscape. The valorization of bauxite residue, a by-product of the Bayer process for alumina production, offers an opportunity to use a material commonly considered to be a waste stream as an abundant and readily available resource. In this study, a two-step process was developed to extract valuable materials from bauxite residue, employing carbothermic smelting, producing crude metallic iron and a slag phase which concentrates scandium, and other elements of interest, which are then extracted by acid baking–water leaching. Preliminary process tests were carried out, and fundamental investigation and characterizations were used to gain an understanding of the underlying physicochemical mechanisms. This waste valorization process is intended to be integrated into a larger near-zero-waste process to sustainably recover the valuable components of bauxite residue to help build the circular economy.

Keywords Bauxite residue · Carbothermic smelting · Acid baking–water leaching · Waste valorization

G.A. conceived and supervised the research. J.A. designed and performed the experiments, analyzed/interpreted the results, and drafted the manuscript.

J. Anawati · G. Azimi (✉)
Department of Chemical Engineering and Applied Chemistry, Laboratory for Strategic Materials, 200 College Street, Toronto, ON M5S 3E5, Canada
e-mail: g.azimi@utoronto.ca

G. Azimi
Department of Materials Science and Engineering, University of Toronto, 184 College Street, Toronto, ON M5S 3E4, Canada

G. Azimi et al. (eds.), *Rare Metal Technology 2020*, The Minerals, Metals & Materials Series, https://doi.org/10.1007/978-3-030-36758-9_13

Introduction

Aluminum is the second-most used metal in the world, with over 60 million tonnes produced globally every year [1]. The most common method for primary aluminum production is the combination of the Bayer process, in which Al_2O_3 is extracted from bauxite ore via high-temperature/pressure NaOH leaching and recovered by crystallization, and the Hall–Héroult process, where the Al_2O_3 is electrolytically reduced to metallic aluminum. A major side-product of this process is bauxite residue (BR), the alkaline solid waste resulting from the Bayer process, produced at a rate of 150 million tonnes per year [2]. Bauxite residue, which is primarily composed of Fe_2O_3, FeO(OH), AlO(OH), $Al(OH)_3$, $CaTiOSiO_4$, TiO_2, SiO_2, and $Na_6(Al_6(Si_{5.5}(H_4)_{0.5})O_{24})(NaCl)_{0.4}(NaOH)_{0.7}$, [3, 4] is typically treated as a waste product due to its high alkalinity (pH 10–13) and complex mineralogy. Currently, less than 3% of BR produced worldwide is used productively, primarily as an additive to cement, with the rest landfilled in dry heaps or tailing ponds, leading to an estimated worldwide inventory of 3 billion tonnes [2, 4].

Bauxite residue stockpiles are present all over the world, and are rich in several important materials, such as iron (Fe), aluminum (Al), titanium (Ti), scandium (Sc), hafnium (Hf), rare-earth elements (REEs), and others. [3] As such, this "waste product" has the potential to be used as an abundant, economical, and sustainable resource for several valuable materials and a viable alternative to primary mining and extraction. One particularly valuable BR component is Sc, a rare element which typically exists geologically as a minor element in other materials and which has a market value of US $\$4600/kg_{Sc_2O_3}$ [5] due to its low global production, and important applications for producing lightweight Al–Sc alloys, lasers, high-intensity lamps, and solid-oxide fuel cells. [5, 6] Because of its scarcity and high value, any material with Sc concentrations above 20 $mg_{Sc/kg}$ is considered a potential Sc resource [7]. As such, with a Sc concentration of up to 120 $mg_{Sc/kg}$ [4] (31 $mg_{Sc/kg}$ in the sample studied in this work), bauxite residue represents a potential sustainable low-cost Sc resource.

Several previous studies have investigated the valorization of bauxite residue as resource for Sc and other valuable materials [8–10]; however, no single large-scale strategy has been adopted globally because the physical characteristics of BR have made efficient and economical extraction and purification of the target materials particularly challenging. Because residues from different geographic locations and alumina refineries have different compositions and characteristics, each residue necessitates a solution adapted for each BR feed.

Previous studies on BR valorization have focused primarily on direct acid leaching for the extraction of valuable materials, which is technically straightforward; however, this method generally has disadvantages of high reagent, water, and energy consumption, poor selectivity for the target elements, and high production of highly acidic waste [11]. Other more complex approaches to extraction have also been investigated, such as dry digestion, acid roasting, alkaline roasting, alkaline leaching, ionic liquid leaching, and bioleaching; [12–16] however, these methods have not yet been deployed at large scales. One common challenge for the efficient extraction and subsequent separation and purification of these valuable materials, in particular Sc, is the close association with Fe. Scandium often occurs as a substitution impurity in Fe minerals due to its similar ionic radius and charge (3+). Thus, efficient Sc extraction is associated with high co-extraction of Fe [17].

Recently, an alternative multi-step processing approach has been studied to address the iron co-extraction challenges and to increase the profitability of the overall extraction process. This process employs pyrometallurgical carbothermic smelting to extract the iron content of the BR as crude metallic iron, leaving the Sc and other valuable materials in an iron-depleted slag phase which allows extraction by conventional means [18, 19]. In this approach, similar to the blast furnace process for the production of crude iron from iron ore, the oxidized Fe^{3+} in the BR is converted to metallic Fe by reacting with a reducing agent, usually carbon, according to Reaction 1—this crude iron is immiscible with the slag produced by the Al, Si, and Ca content of the BR, allowing phase separation. Typically, to reduce the BR melting temperature and system viscosity, a fluxing agent is added. Different fluxes have been investigated, such as calcium, magnesium, and silicon oxides, nickel laterite ore, and industrial waste products, such as basic oxygen furnace slag and municipal solid waste incineration fly ash [20–24].

$$Fe_2O_3 + 1.5C = 2Fe + 1.5CO_2 \qquad \text{(Reaction1)}$$

The previous studies on the use of smelting to recover iron, scandium, and other materials from BR have typically employed direct acid leaching of the Fe-depleted slag [18, 25]; however, as mentioned above, direct acid leaching typically consumes large quantities of reagents, and produces large volumes of hazardous waste. It has previously been shown that the use of acid baking–water leaching (ABWL), a hybrid pyro-hydro-metallurgical process, can enhance the efficiency of direct acid leaching of BR by reducing acid and water consumption, increasing extraction efficiency, and reducing leaching times [3]. In ABWL, the solid feed is mixed with concentrated sulfuric acid (95–98%H_2SO_4), then baked at 200–400 °C to ensure conversion of the constituent minerals to readily-soluble sulfate materials. The acid-baked solid is then leached with water at ambient conditions (1 atm, 25 °C), allowing the rapid dissolution of the soluble sulfate phases and extraction of the target materials. ABWL has not previously been combined with carbothermic smelting to efficiently extract the valuable materials from BR slag.

In this work, BR was smelted under two contrasting sets of processing conditions: mild conditions (low temperature and low flux addition) and strong conditions (high temperature and high flux addition), and the resulting slags were subjected to ABWL to extract the valuable materials (primarily Sc and Ti). The slags and the acid-baked slags were characterized by inductively coupled plasma optical emission spectroscopy (ICP-OES), secondary electron microscopy–energy dispersive spectroscopy (SEM-EDS), and X-ray diffraction to gain insight into the physicochemical mechanisms behind these processes and to inform future investigations into these processes.

Materials and Methods

Experimental Process Steps

The "carbothermic smelting plus acid baking–water leaching" process is presented schematically in Fig. 1. Wet bauxite residue was first dried at 50 °C for 24 h and manually crushed to a coarse powder. The dried BR was then mixed with 10–30 wt% (relative to BR) CaO flux (99.5%, Materion), and 10 wt% carbon (crushed lignite coal, VWR). The powder mixture was pelletized at 250 MPa to ensure complete contact between the BR, flux, and carbon. The pellets were smelted at 1400–1600 °C for 1 h in graphite crucibles under 1.4 L/min 99.998% Argon flow in a box furnace (Carbolite). Two trials were conducted, at different smelting conditions (trial A and trial B), as given in Table 1.

The smelted pellets were removed from the graphite crucibles, and the slag phases were manually separated from the metallic phase, and crushed into a fine powder. Samples of the crushed slags were then digested by alkali fusion plus acid digestion (Claisse LeNeo) for elemental characterization by ICP-OES (PerkinElmer Optima 8000). The crushed BR slag was then used for acid baking–water leaching trials.

The crushed BR slag (SBR) was mixed with concentrated H_2SO_4 at $1.84\ g_{H_2SO_4}/g_{SBR}$ in a porcelain crucible and thoroughly mixed with a glass rod.

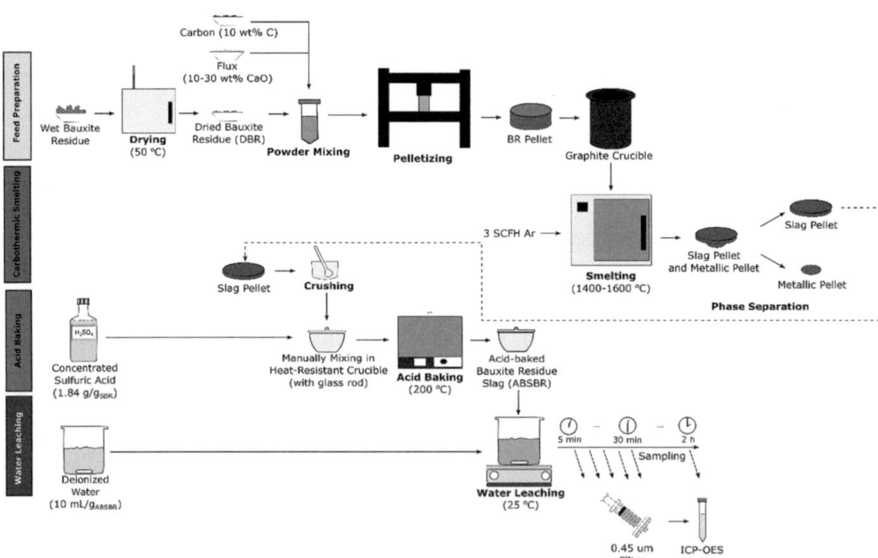

Fig. 1 Schematic diagram of the carbothermic smelting plus acid baking–water leaching process

Table 1 Elemental characterization of smelted BR slags and ABWL extraction results. The results for both Trial A and Trial B, alongside sample results for the direct ABWL of BR for comparison. Trial A and B were treated with the following ABWL conditions: 1.84 $g_{H_2SO_4}/g_{SBR}$, 200 °C (2 h), 10 mL_{H_2O}/g_{ABSBR}. The untreated BR (data from a previous study [3]) was treated with: 1.75 $g_{H_2SO_4}/g_{SBR}$, 200 °C (2 h), 9.5 mL_{H_2O}/g_{ABSBR}

Element	Slag composition	Recovery in Slag	ABWL extraction (%)		
			5 min	30 min	2 h
Trial A slag (Flux: 10 wt% CaO, Carbon: 10 wt% C, Temperature 1400 °C, 1 h)					
Sc	48.6 mg/kg	92.7%	15.6	47.4	58.0
Fe	3640 mg/kg	0.9%	15.3	59.5	105.7[b]
Al	20.5 wt%	103.3%[b]	16.5	55.8	67.0
Ti	6.06 wt%	94.7%	12.3	39.3	40.0
Ca	17.4 wt%	109%[a]	9.8	9.9	10.4
Na	7.60 wt%	86.7%	17.0	50.2	66.9
Trial B slag (Flux: 30 wt% CaO, Carbon: 10 wt% C, Temperature 1600 °C, 1 h)					
Sc	37.0 mg/kg	77.0%	60.7	69.6	80.8
Fe	9110 mg/kg	2.4%	6.0	10.1	20.3
Al	18.2 wt%	100.0%	56.9	63.8	65.0
Ti	4.43 wt%	75.5%	25.3	26.3	25.8
Ca	37.6 wt%	101.5%[a]	4.1	4.2	4.2
Na	5660 mg/kg	7.0%	1.3	1.1	2.2
Untreated BR (dry)					
Sc	31.1 mg/kg	N/A	46.9	55.6	57.9
Fe	24.3 wt%	N/A	45.8	52.9	55.4
Al	11.8 wt%	N/A	60.3	75.9	81.2
Ti	3.75 wt%	N/A	33.8	37.2	38.3
Ca	2.34 wt%	N/A	77.4	91.0	89.3
Na	5.21 wt%	N/A	72.7	100.9[b]	107.2[b]

[a]Ca is present from both the starting BR and the addition of CaO flux
[b]In some cases, calculated values slightly above 100% are reported. This can be attributed to potential heterogeneity in the composition of the flux samples, and experimental variability in the slag characterization procedure, leading to slight discrepancies in the extraction efficiency and recovery calculations. These results should be interpreted as indicating approximately 100% extraction/recovery

The acid–SBR mixture was baked at 200 °C for 2 h in air in a muffle furnace (Degussa–Ney). The acid-baked SBR (ABSBR) was leached in deionized water at room temperature (25 °C) in a glass reactor equipped with a magnetic stirrer (400 rpm). The leaching solution was sampled at 5, 30, and 120 min and filtered with a 0.45 μm nylon syringe filter (Fisher Scientific) for quantification of extracted materials by ICP-OES.

Morphological Characterization

The micro-scale morphologies of the BR slags and acid-baked BR slags were characterized by SEM (Hitachi SU3500) using secondary electron (SE) and back-scattered electron (BSE) detection. Elemental mapping was performed during SEM imaging using EDS. Prior to observation, the samples were carbon coated to ensure sample conductivity.

Mineralogical Characterization

The mineralogical phases of the BR slags and acid-baked BR slags were characterized by X-ray diffraction (XRD–Rigaku MiniFlex 600). A scan rate of 1.25°/min was employed. The peak intensities were normalized to have a uniform maximum peak intensity for visualization purposes. The peak identification was carried out by comparing the measured spectra to the ICDD spectral database.

Results and Discussion

Smelting—ABWL Characterizations and Extraction Results

Two smelting trials were conducted at different smelting conditions: Trial A, which employed a lower smelting temperature (1400 °C) and flux composition (10 wt% CaO, relative to mass of BR), and trial B, which employed a high smelting temperature (1600 °C) and flux composition (30 wt% CaO). The smelted slag composition, the estimated proportion recovered in the slag, and ABWL extraction results are given in Table 1. The recovery proportion estimation was based on the mass of slag recovered following the manual separation of the phases, and should be seen as an approximate value, as some slag may have remained adhered to the metal phase or the graphite crucible, affecting the recovery mass. Relative to untreated dried BR, the concentration of the target Sc, and most of the non-Fe matrix materials (Al, Ca, Ti) are higher in the slags compared with the feed BR because they strongly partition into the slag. Trial A had higher concentrations of metals because the dilution effect of the CaO flux was less important. Importantly, the concentration of Fe, the major matrix component in BR (24.3 $wt\%_{Fe}$) was reduced to 0.4–0.9 wt% in the slags. This Fe reduction indicates that the majority of the Fe content of the BR is recovered in the metallic phase, and that the ABWL feed is depleted in Fe, which leads to reduced Fe co-extraction in comparison with the direct ABWL of untreated BR. Furthermore, the Na content of slag A was increased, similar to the behavior of Al and Ti; whereas it was reduced to 0.6 wt% in slag B. This Na decrease suggests that the evaporation of Na [26] occurs between

1400 and 1600 °C. This evaporation is an important consideration at industrial scales due to Na corrosion, safety, and environmental challenges.

The smelting process was shown to enhance the ABWL process relative to untreated BR, as the extraction efficiencies of valuable materials were approximately equivalent, or superior to BR, alongside the fact that the starting materials had higher concentrations of valuable materials which can lead to larger absolute amounts of Sc extracted (62% overall Sc recovery after 2 h of leaching, compared with 58% for untreated BR). The relatively fast leaching kinetics of ABWL were conserved, with the majority of extraction occurring within 5–30 min. In both cases, Al extraction was high, suggesting that Al_2O_3 could be recovered from the process, increasing the Bayer process efficiency.

Although slag B has a lower Sc concentration, the extraction of Sc from that was considerably higher than that of trial A, suggesting that the more extreme smelting conditions lead to a higher degree of disrupting the Sc-bearing Al and Si phases in BR, [27, 28] making them more vulnerable to H_2SO_4 attack. Furthermore, while slag B had more residual iron following smelting, the extraction efficiency of Fe was low, with slow kinetics, compared to slag A, which had a higher Fe extraction (approximately 100% extraction after 2 h). Following ABWL, the trial A slag leachate had 1.03 mg_{Sc}/L and 141 mg_{Fe}/L, while trial B had 1.15 mg_{Sc}/L and 71 mg_{Fe}/L.

Morphological and Elemental Mapping of the Smelting and Acid-Baking Processes

The microstructure and elemental mapping of the slags before and after acid baking for both trials were observed by SEM-EDS (Fig. 2). In both trials, the smelting slags had a dense and smooth microstructure with a relatively uniform atomic composition (absence of bright spots in the BSE scans). In comparison, the acid-baked slags were formed by larger particle aggregates that had a non-uniform atomic composition and a rough surface, suggesting that these particles have a more active surface area available for rapid leaching.

The elemental mapping of Al, Si, Ca, and O were highly co-localized in the slags, suggesting a monolithic Al–Si–Ca–Oxide phase as the primary phase. These particles were comprised of closely associated phases with different compositions. After acid baking, the co-localization of Al, Si, and Ca was reduced, suggesting that the baking step effectively breaks down and transforms the crystal phases into new sulfur-bearing phases. Thus, Sc and other valuable materials that are trapped within the structures can be released and/or sulfated.

Two major differences are observed between the trial A and trial B during characterization. First, as described before, trial A had a relatively high Na content, which appeared to be co-localized primarily with Al. Additionally, Ti behavior appeared to be completely different in both trials. In slag A, a distinct Ti-rich

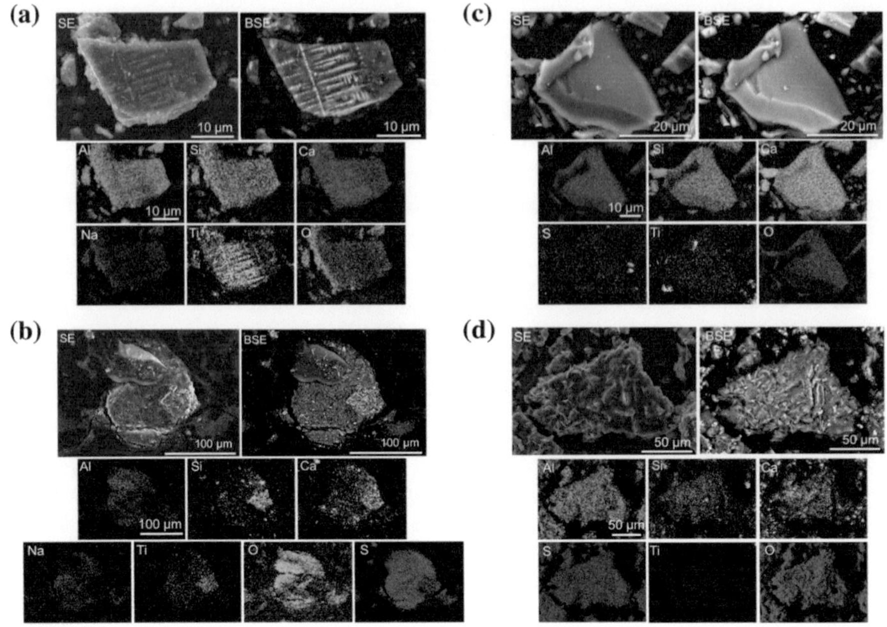

Fig. 2 SEM-EDS elemental mapping of the smelting slag and acid-baked slag for smelting Trials A and B. **a** Trial A—Slag, **b** Trial A—Acid-baked slag, **c** Trial B—Slag, **d** Trial B—Acid-baked slag

dendritic pattern formed on the surface of the particles. On the contrary, in slag B, Ti appeared to condense into small Ti-rich regions. The reason behind this observation will be discussed later in the paper.

Mineralogical Characterization of the Smelting and Acid-Baking Processes

To confirm that changes in the mineralogical phases of the slags were occurring during the acid baking step of the process, and to investigate the differences in extraction behaviors between the two trials, samples of Trial A and Trial B slags were characterized by XRD before and after acid baking (Fig. 3).

As expected, the XRD spectra of the slags were considerably different from that of untreated bauxite residue, because, during the smelting process, the BR melts and the liquid phase undergoes considerable chemical transformation as the iron content was reduced and separated as a distinct liquid phase, and recrystallized with a different chemical composition. Interestingly, the combination of smelting at a different temperature and with a different flux concentration led to considerable differences in mineralogical compositions both before and after acid baking.

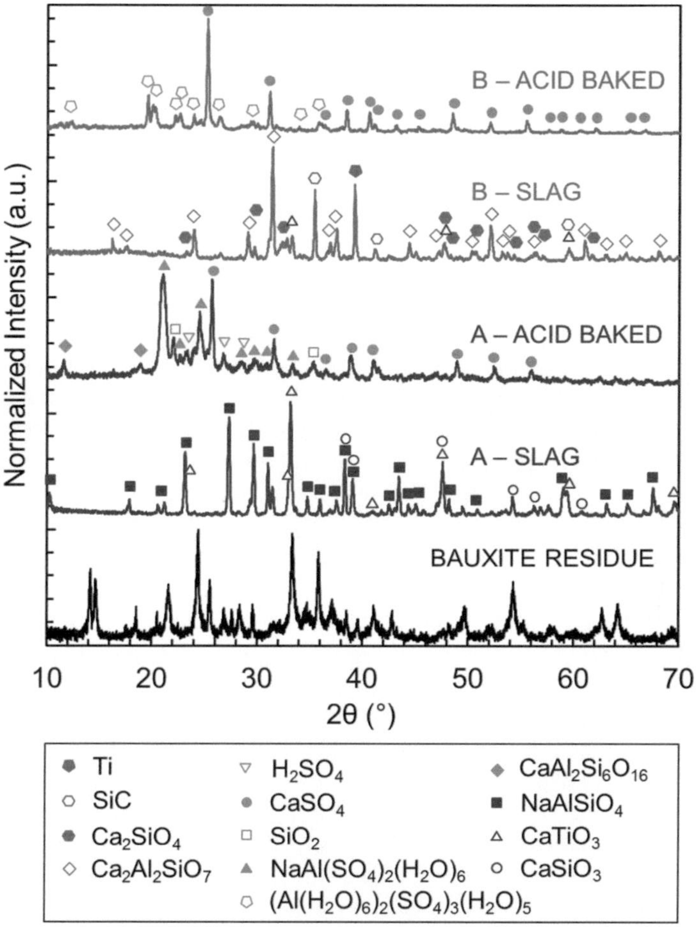

Fig. 3 Normalized X-ray diffraction (XRD) spectra and phase identification for the smelting slags and acid-baked slags for Trials A and B. The spectrum for untreated dry BR is also shown for comparison

In Trial B, in agreement with the SEM-EDS analysis, the primary detected phase was $Ca_2Al_2SiO_7$, with small amounts of Ca_2SiO_4 and $CaTiO_3$. However, on the contrary to the SEM-EDS observations, Ca, Si, and Al did not form a single monolithic Ca–Si–Al phase in Trial A, instead forming separate Na–Al–Si and Ca–Si phases which were closely associated.

As discussed above, Na was present in considerably higher concentrations in slag A compared with slag B due to the lower smelting temperature in Trial A, leading to lower Na evaporation from the system. As such, the primary detected phase in Trial A was $NaAlSiO_4$, while no Na-bearing phase was detected in Trial B. Furthermore, as Trial B had a higher Ca content in the starting material, the

Ca-bearing phases were consequently more prominent. The primary slag phases for trial A (nepheline—$NaAlSiO_4$) and B (gehlenite—$Ca_2Al_2SiO_7$) are known to react with acids; [29] thus, during the acid baking step, these phases are expected to break down, converting the constituent metals to soluble sulfates.

The Ti-rich dendrites which were observed in the SEM-EDS for trial A were shown to be $CaTiO_3$, which has previously been shown to form dendrite structures. [30, 31] In contrast, the dense Ti-rich spots observed in trial B were likely metallic Ti. It is possible that in the high temperature, highly reducing, oxygen-depleted environment during smelting, that the reduction of oxidized Ti species to metallic Ti becomes thermodynamically favoured. [32] Thermodynamically, reduced Ti is expected to alloy with Fe, and thus be absent from the slag [33]; however, this Fe–Ti alloying was not observed, as slag B contained 4.4 wt%Ti (corresponding to approximately 76% of the Ti in the starting BR). One possible explanation for this discrepancy is the possibility that the Ti forms within the slag phase after the slag–Fe separation occurs, and it does not have time to fully migrate to the iron phase and across the iron-slag interface before the system began to solidify.

In both acid-baked slags, anhydrous $CaSO_4$ was the most prominent phase, occurring as the breakdown product of the Ca-bearing phases (A: $CaTiO_3$, $CaSiO_3$, B: $Ca_2Al_2SiO_7$, $CaTiO_3$, Ca_2SiO_4). Because $CaSO_4$ was the primary phase in both slags, their extraction was essentially the same, reaching a plateau at 600–660 mg_{Ca}/L which is approximately equivalent to $CaSO_4$ solubility in complex aqueous systems [34]. In trial A, the main non-$CaSO_4$ phase present was tamarugite ($NaAl(SO_4)_2(H_2O)_6$) which is water soluble [35]; however, yugawaralite ($CaAl_2Si_6O_{16}$), which is not soluble, was also detected. This presence of an insoluble phase can explain the lower Sc extraction in trial A, as the Sc may have remained trapped in the insoluble $CaAl_2SiO_{16}$. Aside $CaSO_4$, the other primary phase in trial B was alunogen ($(Al(H_2O)_6)_2(SO_4)_3(H_2O)_5$). As Sc is known to associate with Al [27] this phase was likely the main Sc-bearing phase. Alunogen is known to be soluble, [35] which explains the high extraction of Sc in trial B, as the bulk Sc-bearing phases would quickly break down and dissolve away during water leaching, allowing any $Sc_2(SO_4)_3$ produced during acid baking to come into contact with water and dissolve.

Conclusions

This study validated the use of combined carbothermic smelting and acid baking–water leaching for the extraction of valuable materials such as scandium, titanium, and aluminum from bauxite residue with minimal iron co-extraction, and the recovery of iron as a crude metallic iron product. Advanced characterization techniques were used to study the physicochemical mechanisms behind the smelting and acid-baking steps of the process, and thus to gain a better understanding of how the different processing parameters affect the extraction outcomes. These tests offer insight into the process, particularly the behavior of sodium and titanium during smelting, and the behavior of calcium, aluminum, and silicon during acid baking–water leaching. Further work is currently underway to exhaustively and systematically study the effects of various smelting and ABWL process parameters on the extraction behaviors, to determine the optimum operating parameters for efficient performance, and to determine the most efficient and sustainable ways to integrate this process into a larger BR valorization process.

The ultimate goal of this work is to produce a fully integrated, economical, and sustainable, near-zero waste process in which the BR is fully utilized, thus answering the society's demand for critical materials, without producing additional negative environmental outcomes by reducing dependence on environmentally damaging primary extraction processes and decreasing the rate of bauxite residue accumulation in the environment.

Acknowledgements This work was supported by Rio Tinto (grant number 503532) and Natural Sciences and Engineering Research Council of Canada (NSERC) (grant number 503534). The authors gratefully acknowledge Rio Tinto for providing us with bauxite residue samples and for technical support. We thank Dr. Raiden Acosta for help with XRD, and Alkali Fusion, and Mr. Kok Long Ng for help with SEM-EDS. Access to the electron microscopy facility in Ontario Centre for the Characterization of Advanced Materials (OCCAM) and the Walter Curlook Materials Characterization & Processing Laboratory is acknowledged.

References

1. The International Aluminium Institute (2019) Primary aluminium production—global data for Jan 1973 to Jan 2019
2. Evans K (2016) J Sustain Metall 2:316
3. Anawati J, Azimi G (2019) Waste Manag 95:549
4. Liu Y, Naidu R (2014) Waste Manag 34:2662
5. U.S. Geological Survey (2019) Mineral commodity summaries
6. Røyset J, Ryum N (2005) Int Mater Rev 50:19
7. Shaoquan X, Suqing L (1996) Hydrometallurgy 42:337
8. Liu Z, Li H (2015) Hydrometallurgy 155:29
9. Borra CR, Blanpain B, Pontikes Y, Binnemans K, Van Gerven T (2016) J Sustain Metall 2:365
10. Binnemans K, Jones PT, Blanpain B, Van Gerven T, Pontikes Y (2015) J Clean Prod 99:17
11. Yagmurlu B, Dittrich C, Friedrich B (2017) J Sustain Metall 3:90
12. Rivera RM, Ulenaers B, Ounoughene G, Binnemans K, Van Gerven T (2018) Miner Eng 119:82
13. Qu Y, Lian B (2013) Bioresour Technol 136:16
14. Borra CR, Mermans J, Blanpain B, Pontikes Y, Binnemans K, Van Gerven T (2016) Miner Eng 92:151
15. Davris P, Balomenos E, Panias D, Paspaliaris I (2016) Hydrometallurgy 164:125
16. Yatsenko SP, Pyagai IN (2010) Theor Found Chem Eng 44:563
17. Borra CR, Pontikes Y, Binnemans K, Van Gerven T (2015) Miner Eng 76:20
18. Borra CR, Blanpain B, Pontikes Y, Binnemans K, Van Gerven T (2016) J Sustain Metall 2:28
19. Kaußen F, Friedrich B (2015) Chem-Ing-Tech 87:1535

20. Alkan G, Yagmurlu B, Ma Y, Xakalashe B, Stopic S, Dittrich C, Friedrich B (2018) In: Pontikes Y (ed) Proceedings of 2nd international bauxite residue valorisation and best practices conference, Athens, pp 215–222
21. Balomnenos E, Kastritis D, Panias D, Paspaliaris I, Boufounos D (2014) Light Met 2014:143
22. He A, Zeng J (2017) Mater Des 115:433
23. Lucas H, Alkan G, Xakalashe B, Friedrich B (2018) In: Proceedings of 2nd international bauxite residue valorisation and best practices conference, pp 263–270
24. Xakalashe B, Friedrich B (2018) In: Proceedings 2nd international bauxite residue valorisation and best practices conference, pp 233–240
25. Yagmurlu B, Alkan G, Xakalashe B, Friedrich B, Stopic S (2017) In: 35th International ICSOBA Conference Hamburg, Germany, 2–5 October, 2017, p 587
26. Brewer L, Margrave J (1955) J Phys Chem 59:421
27. Vind J, Malfliet A, Bonomi C, Paiste P, Sajó IE, Blanpain B, Tkaczyk AH, Vassiliadou V, Panias D (2018) Miner Eng 123:35
28. Liu Z, Zong Y, Li H, Zhao Z (2018) Miner Eng 119:263
29. Terry B (1983) Hydrometallurgy 10:135
30. Huang Y-J, Chiu H-T, Lee C-Y (2009) CrystEngComm 11:1904
31. Chen T, Zheng Y, Lu Z, Xu T, Liu Y, Meng X, Xu G, Han G (2019) Nanotechnology 1
32. Bose S (2007) High temperature coatings. Elsevier, pp 29–52
33. Murray JL (1981) Bull Alloy Phase Diagr 2:320
34. Azimi G, Papangelakis VG, Dutrizac JE (2007) Fluid Phase Equilib 260:300
35. Anthony JW, Bideaux RA, Bladh KW, Nichols MC (2005) Handbook of mineralogy, Mineralogical Society of America

Separation of Neodymium and Dysprosium by Molten Salt Electrolysis Using an Alloy Diaphragm

Tetsuo Oishi, Miki Yaguchi, Yumi Katasho and Toshiyuki Nohira

Abstract As a part of our ongoing study on a new rare-earth elements recycling process using molten salts and alloy diaphragms, selective permeation of Dy through an alloy diaphragm is investigated in LiCl–KCl eutectic melts containing $NdCl_3$ and $DyCl_3$ at 450 °C. In this process, the permeation of rare-earth elements is carried out via the following three steps: (a) cathodic reduction of rare-earth ions to form alloys on one side of the diaphragm, (b) diffusion of rare-earth atoms in the diaphragm, and (c) anodic oxidation of rare-earth atoms to ions on the other side of the diaphragm. A preliminary experiment reveals that Dy selectively permeates the diaphragm under an appropriate condition.

Keywords Rare-earth elements · Recycling · Molten salt electrolysis · Alloy diaphragm

Introduction

Rare-earth elements (REEs) are widely used in industries due to their unique characteristics. One of the most important products obtained from REEs is the Nd–Fe–B magnet, an ultra-strong permanent magnet indispensable for applications such as high-performance motors, wind power plants, and energy-saving consumer electronics. However, the stable supply of REEs is sometimes threatened because

T. Oishi (✉) · M. Yaguchi · Y. Katasho
National Institute of Advanced Industrial Science and Technology (AIST), 16-1 Onogawa, Tsukuba, Ibaraki 305-8569, Japan
e-mail: tetsuo.oishi@aist.go.jp

T. Nohira
Kyoto University, gokasho, Kyoto, Uji 611-0011, Japan

G. Azimi et al. (eds.), *Rare Metal Technology 2020*, The Minerals, Metals & Materials Series, https://doi.org/10.1007/978-3-030-36758-9_14

REEs resources are unevenly located in the earth's crust. In contrast to Nd, which is a relatively abundant element, heavy REEs including Dy are very scarce and unevenly distributed. Since Dy is added to the Nd–Fe–B magnets to improve their high-temperature stability and is indispensable for high-performance motors, the development of a new recovering process of REEs, especially Dy, from Nd–Fe–B magnets is of high research interest [1]. Thus far, a variety of processes have been proposed and investigated as new recycling processes of REEs [2–4]. These processes, however, have not been commercialized. This is attributable to the high process cost and the unstable price of REEs.

Based on this background, we have proposed and been investigating a very simple and economical REEs recycling process through which Dy or Nd metal can be recovered separately and directly from used Nd–Fe–B magnets [5–7]. A conceptual drawing of the process is shown in Fig. 1. In this process, REEs in the magnet are selectively dissolved into a molten salt by anodic oxidation. Then, the REE of interest is selectively permeated through an alloy diaphragm. Finally, the REE is recovered as metal or alloy by electrodeposition on a cathode. The key of the present process is the alloy diaphragm, which performs as a bipolar electrode and realizes the selective permeation. The reactions and processes occurring in this system can be summarized as follows: (a) RE ions in the anolyte are reduced at the surface of the diaphragm to form a RE alloy, (b) the RE atoms in the alloy diffuse to the other side of the diaphragm, and (c) the RE atoms that reach the other surface of the diaphragm are oxidized to RE ions in the catholyte. Since these three steps have a certain selectivity, which can be controlled by the electrolytic condition, the alloy diaphragm can effectively separate REEs in the molten salt. Future work involves the one-step recovery of Nd and Dy separately from wasted magnets.

Herein, we describe a permeation experiment using an alloy film for the selective permeation of Dy in LiCl–KCl eutectic melts containing $NdCl_3$ and $DyCl_3$ at 450 °C.

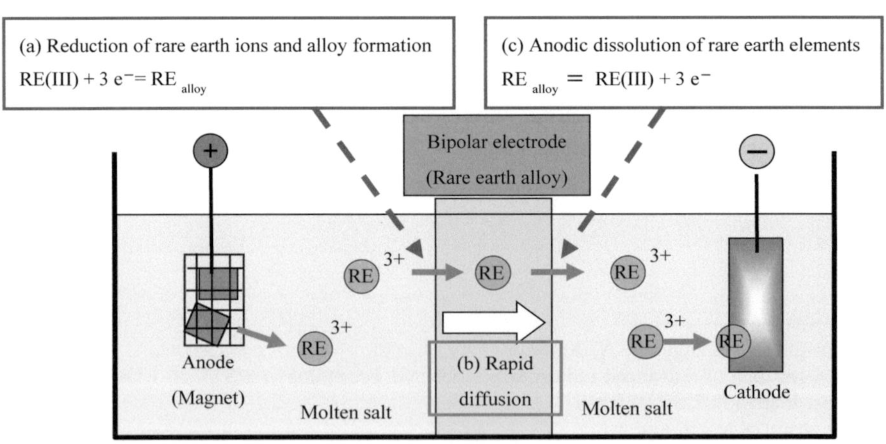

Fig. 1 A conceptual drawing of the new recycling process. [5] modified from the original form

Experimental

A schematic drawing of the electrolytic cell used in this work is depicted in Fig. 2. High purity alumina was used for the material of the cell. The diaphragm was set at the center of the system to separate anolyte from catholyte. A Ni–RE alloy film was used as the diaphragm because Ni easily forms RE alloys in a molten salt, and the Ni–RE alloying/dealloying phenomena have been extensively investigated [8, 9]. The preformed Ni–RE alloy film, however, is difficult to install into the electrolytic cell because it is a very fragile material. Therefore, a Ni film having a thickness of 0.1 mm and a surface area exposed to the melt of 4.9 cm^2 was employed as the starting material of the Ni–RE alloy film. The electrolyte was LiCl–KCl eutectic melts (44:56 wt%), which was first dried under vacuum at 180 °C for 4 h and then dissolved under 1 atm Ar atmosphere at 450 °C. After the temperature of the melt reached 450 °C, Ar gas was bubbled for 12 h. Then, a 0.5 mol% of $NdCl_3$ and $DyCl_3$ was added into the anolyte, which is hereafter called melt A. The catholyte melt, in turn, will be called melt B. The reference electrodes, which were Ag wires immersed in a LiCl–KCl eutectic melt containing 1 mol% of AgCl, were placed in a Pyrex glass tube having a thin bottom. The potential of these Ag^+/Ag electrodes was calibrated with reference to that of a Li^+/Li electrode, which was prepared by electrodeposition of Li metal on a Ni wire. All the potentials in this paper are given with reference to the Li^+/Li potential. The counter electrodes were a glassy carbon rod and an Al plate immersed in melt A and B, respectively. The permeation experiment was carried out by controlling the potential of the diaphragm using two potentiogalvanostats (PGS-A and PGS-B), as shown in Fig. 2. The diaphragm was controlled at +0.60 V by PGS-A. At this potential, both Nd and Dy form an alloy

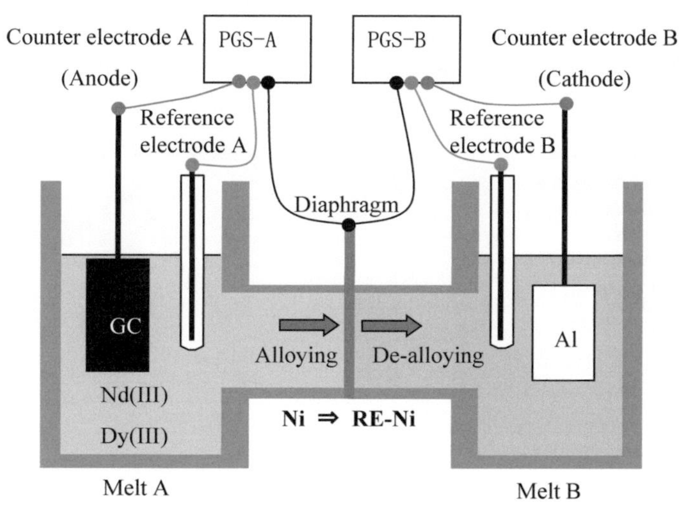

Fig. 2 A schematic drawing of the experimental setup

phase of $RENi_2$ (RE = Dy or Nd), but the formation rate of $DyNi_2$ is expected to be higher than that of $NdNi_2$. At the same time, the potential of the diaphragm was set at +1.0 V by PGS-B. This corresponds to the potential region at which RE–Ni alloys with low RE content such as $RENi_3$ or RE_2Ni_7 are more stable than $RENi_2$, thereby ensuring that Dy and Nd could be dissolved from the $RENi_2$ phase. After a certain current started to pass between the diaphragm and the Al plate immersed in melt B (measured by PGS-B), the potentiostatic electrolysis was terminated and the galvanostatic electrolysis started at a current of +20 mA, which corresponds to the current density of +4 mA/cm^2 on the diaphragm. The permeation experiment was occasionally stopped at predetermined time periods, and restarted after sampling a small part of melt B and changing the Al plate. The used Al plate and the sample of melt B were dissolved into hydrochloric acid solutions, and the amount of Nd and Dy permeated into melt B was evaluated by using inductively coupled plasma atomic emission spectrometry (ICP-AES).

Results and Discussion

Figure 3 shows the current passed between the diaphragm and the Al plate in melt B during the potentiostatic electrolysis measured by PGS-B. Since the starting material of the diaphragm was pure Ni, anodic currents were hardly observed at the beginning of the process. As the formation of the Ni–RE alloy proceeded from the melt A side, the anodic current gradually increased until reaching 10 mA/cm^2. These results indicate that almost all the diaphragm was converted from pure Ni to $RENi_2$, and that the anodic dissolution of Dy and/or Nd started at the surface of the melt B side of the diaphragm.

Figure 4 plots the amount of Dy and Nd permeated through the diaphragm against the amount of electricity measured by PGS-B, revealing that the amount of permeated Nd and Dy increased with the amount of passed electricity. Compared to

Fig. 3 Time dependence of the current density on the melt B side surface of the diaphragm during the potentiostatic electrolysis (first step of the permeation experiment)

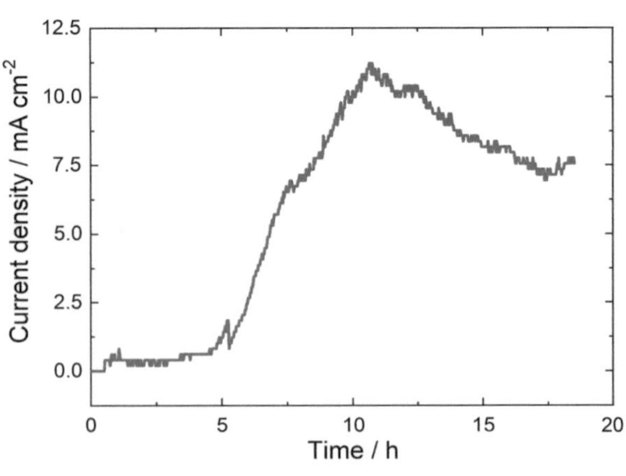

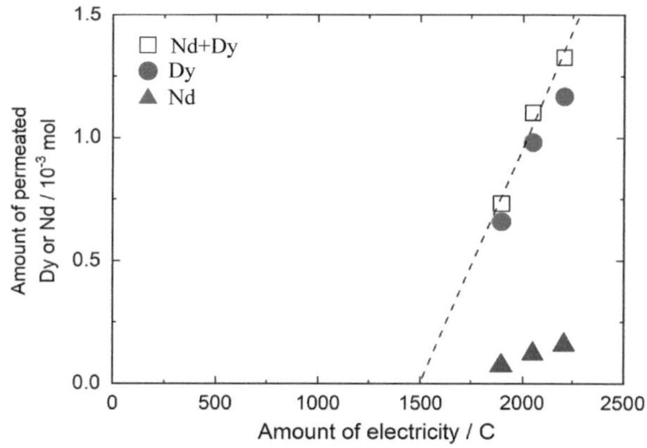

Fig. 4 The amount of permeated Nd and Dy through the diaphragm plotted against the amount of electricity measured by PGS-B

Nd, the permeated amount of Dy was about seven times larger, indicating that Dy was selectively permeated under the present condition. This selectivity is attributable to the difference in the alloying rate of Nd and Dy at the potential of +0.60 V, which was controlled by PGS-A. On the other hand, the amount of Nd and Dy in melt B increased after the amount of electricity reached 1,500 C, most likely due to the presence of moisture in the melt, which causes the RE ions in the LiCl–KCl melt to react spontaneously forming stable and insoluble chemicals of REOCl. In fact, precipitates were observed in the melt after the electrolysis, which was confirmed to be REOCl (mainly DyOCl) by XRD. Under this condition, the Dy and Nd ions that permeate to melt B during the early stage of the experiment react with the moisture and precipitate. This is the most plausible explanation for the low amount of Nd and Dy found in melt B at the lower electricity region according to Fig. 4. These results indicate that the experimental setup should be improved for more precise measurement.

Conclusions

A permeation experiment of Nd and Dy through an alloy diaphragm was conducted in LiCl–KCl eutectic melts containing $NdCl_3$ and $DyCl_3$ at 450 °C. The results indicated that Nd and Dy permeated through the diaphragm, and that selective permeation of Dy could be achieved under the appropriate condition. This is the first result that demonstrates the selective permeation of Dy through the alloy diaphragm in molten salts. This is also an important step to realize the new REEs recycling process through which Dy and Nd metal can be recovered separately and directly from used Nd–Fe–B magnets.

Acknowledgements This work was partly supported by the New Energy and Industrial Technology Development Organization (NEDO) of Japan.

References

1. Tanaka M, Oki T, Koyama K, Narita H, Oishi T (2013) Recycling of rare earths from scrap. Handb Phys Chem Rare Earths 43:159–211
2. Murase K, Machida K, Adachi G (1995) Recovery of rare metals from scrap of rare earth intermetallic material by chemical vapour transport. J Alloys Compd 217:218–225
3. Uda T, Jacob KT, Hirasawa M (2000) Technique for enhanced rare earth separation. Science 289:2326–2329
4. Takeda O, Okabe TH, Umetsu Y (2004) Phase equilibrium of the system Ag-Fe-Nd, and Nd extraction from magnet scraps using molten silver. J Alloys Compd 379:305–313
5. Oishi T, Konishi H, Nohira T, Tanaka M, Usui T (2010) Separation and recovery of rare earth metals by molten salt electrolysis using alloy diaphragm. Kagaku Kogaku Ronbunshu 36:299–303
6. Kobayashi S, Kobayashi K, Nohira T, Hagiwara R, Oishi T, Konishi H (2011) Electrochemical formation of Nd-Ni alloys in molten LiF–CaF_2–NdF_3. J Electrochem Soc 158:E142–E146
7. Konishi H, Ono H, Takeuchi E, Nohira T, Oishi T (2017) Electrochemical formation of Nd alloys using liquid metal electrodes in molten LiCl–KCl systems. Rare Met Technol 2017:93–102
8. Nohira T, Ito Y (2004) Formation of rare earth alloys by molten salt electrochemical process. Molten Salts 47:5–12
9. Konishi H, Nohira T, Ito Y (2001) Formation and phase control of Dy alloy films by electrochemical implantation and displantation. J Electrochem Soc 148:C506–511

Part III
PGM, Zn, V, Ti, U, Th, In, Ag, Fe

Electrodialysis in Hydrometallurgical Processes

P. Zimmermann, Ö. Tekinalp, L. Deng, K. Forsberg, Ø. Wilhelmsen and O. Burheim

Abstract Electrodialysis (ED) is an ionic exchange membrane process for separation of different components and species. In desalination, a large part of the energy is used to sustain a concentration difference between the solutions, but in the processes of exchanging in selective manners, the energy need is lower and more directed towards ohmic losses in the membranes. The latter has relevance to several hydrometallurgical industries, as they very often accumulate undesired species in their process streams and currently apply intensive (energy and chemical) routes to remove these species. Here, we describe the principle of ED and discuss opportunities for component and salt separation using ion-exchange membranes by providing a brief review of ED in the hydrometallurgical sector.

Keywords Electrodialysis · Metal separation · Hydrometallurgy · Review

---

P. Zimmermann (✉) · K. Forsberg · Ø. Wilhelmsen · O. Burheim
Department of Energy and Process Engineering, NTNU, Trondheim, Norway
e-mail: pauline.zimmermann@ntnu.no

Ø. Wilhelmsen
e-mail: oivind.wilhelmsen@ntnu.no

O. Burheim
e-mail: burheim@ntnu.no

Ö. Tekinalp · L. Deng
Department of Chemical Engineering, NTNU, Trondheim, Norway
e-mail: onder.tekinalp@ntnu.no

L. Deng
e-mail: liyuan.deng@ntnu.no

Ø. Wilhelmsen
SINTEF Energy Research, Trondheim, Norway

G. Azimi et al. (eds.), *Rare Metal Technology 2020*, The Minerals, Metals & Materials Series, https://doi.org/10.1007/978-3-030-36758-9_15

Introduction

Electrodialysis (ED) is a membrane separation process based on the movement of ionized species in solution under the influence of an electric field. ED can be applied to concentrate ionic solutions, deionize salt solutions, or separate ionic and non-ionic species [1]. The industrial application of ED has its roots in the demineralization of brackish water, as first reported in the 1950s [2, 3]. Due to constant improvement of process design and membrane materials, ED nowadays represents an interesting alternative for different applications in water and wastewater treatment, the food industry, and the chemical and hydro-metallurgical sector [4].

In metal production and processing, large amounts of aqueous streams containing metallic as well as non-metallic species are generated upon removing the wanted bulk metals. Some of the remaining diluted compounds can be of great value, e.g. rare-earth and noble metals, however, their concentration in the solution is often too low to allow for profitable recovery using established processes. ED with ion-selective membranes can be a cost-effective alternative to recover resources and thereby increase circular economy in the metal processing industry.

Another challenge in hydrometallurgy is the undesirable impurities in the raw materials. When leaching ores to extract the mineral of interest, unwanted compounds like halogens and metals accumulate in the aqueous streams. Their presence can make the electrolysis more energy demanding or possibly be harmful and their removal requires a lot of chemicals. A pre-treatment with ED can provide selective removal of problematic compounds and save the use of chemicals, energy, and money.

This paper outlines the principles of ED operation and a review of selected efforts of implementing ED into the hydrometallurgical or otherwise relevant processes.

Principle of Electrodialysis

ED in its simplest form is based on alternating membranes with selectivity for anions or cations and then to apply a current. The resulting outcome is desalination like, for instance, desalination of a NaCl solution as shown in Fig. 1, where sodium (Na^+) is illustrated as yellow ions and chloride (Cl^-) as red ions. An anion-exchange membrane (AEM) is a polymeric material containing fixed positive charges that reject other cations, and *vice versa* for cationic exchange membranes (CEM).

Desalination of a binary mixture like sodium chloride and water is a simple form of ED. However, in most realistic solutions one has a mixture of components and different concentration of the different components. Perhaps the most common example is sea water, which consists of not only water, Na^+, and Cl^-, but also ions like Mg^{2+}, SO^{2-}, K^+, Ca^{2+}, and more. In a simplified model for sea water solutions where one considers a lot of NaCl and much less $MgCl_2$, one can have

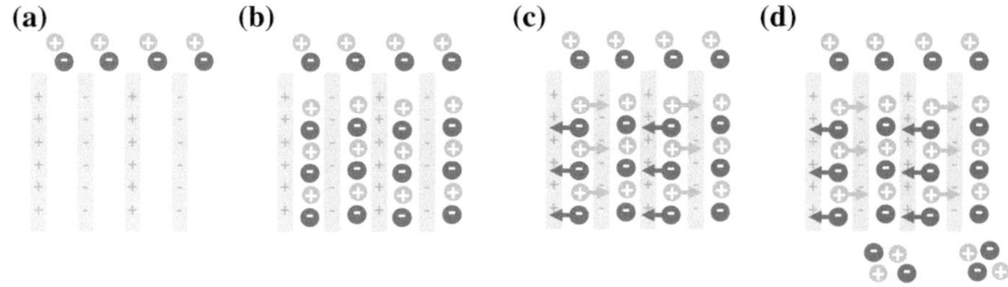

Fig. 1 Schematic description of ED with ion-selective membranes for the purpose of desalination. A solution containing salt (**a**) is fed into a membrane module (**b**), where a current is applied (**c**). In doing so, anions go through the AEM and the CEM, so that all ions gather and one ends up with a desalinated solution and a brine solution (**d**)

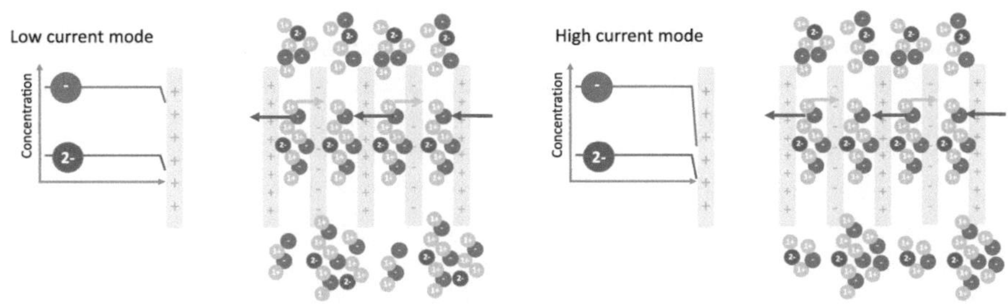

Fig. 2 Selective desalination by boundary layer concentration effects. Considering a simplified sea water solution consisting of NaCl and $MgCl_2$, one can have different removal rates of the different cationic species at different current rates

different relative removal of the different components at different rates, a phenomenon referred to as boundary layer separation. This is illustrated in Fig. 2.

In boundary layer separation, one can expect that at *low overall current*, one removes equally much of both components (here cations) so that after a short time the lower concentration specie (here Mg^{2+}) is removed almost completely. In this case, the degree of desalination is low, but the divalent magnesium ions are completely removed. In another mode where a *high overall current* is applied, the limiting current density of the lower concentration component (or specie) is reached immediately and one removes much more of the component (specie) of the higher concentration. In these cases, one can obtain a brine with little magnesium and thus lower the relative content of the low concentration component (specie). There are many examples of systems where boundary layer separation is of interest, and each of them requires careful examination as several different fluid dynamic phenomena can occur and care should be taken when measuring and modeling systems as such [5].

Role of the Membrane

So far we have mainly discussed the ED process in terms of simple principles for binary salt solutions and possibly more complex solutions of ternary and higher order. In these cases, the role of the membrane is to ensure that only cations go through the CEM and only anions go through AEM. However, ED systems offer many more solutions and also restrictions. A highly relevant example is selectivity of ions with different valences, as shown in Fig. 3. Here, one applies only CEM or AEM, but has alternated membranes that reject higher valency ions. These membranes typically have special coatings to obtain such properties [6]. This means that one can use ED for either sulphate or magnesium/calcium removal in the case of seawater and for many different applications in the hydrometallurgical industry by surface modification of existing ionic membranes.

When considering selectivity of membranes, the common figure of merit is the permselectivity of the membrane, α. The permselectivity is interpreted in different ways, but is ultimately an empirical coefficient that compares a measure value to a theoretical value, as described in detail in section “Energy Use in Electrodialysis”. The permselectivity is sometimes treated as an ability to discriminate a positive ion over a negative ion by the application of the liquid junction potential theory model [7]. Several membranes do, however, offer close to unity discrimination of one salt over the other and the liquid junction theory does not account for the contribution of water in the permselectivity. Therefore, and at higher concentrations, the permselectivity also describes the ability to not transport water alongside the membranes, a phenomena often referred to as osmotic drag [8, 9].

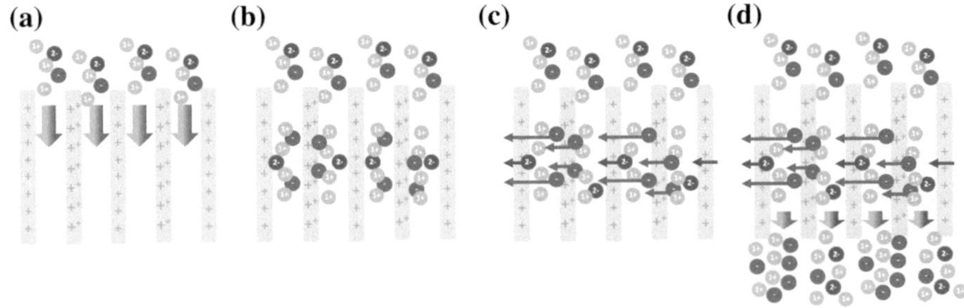

Fig. 3 Valence selectivity, particularly for monovalent ions, can be obtained by several types of membranes and used to separate higher valence species from monovalent species

Energy Use in Electrodialysis

The energy needed to drive an ED process in a unit cell with a binary salt mixture like water and NaCl, $E^{ED}_{u.c.}$, is the sum of the ohmic potential energy, *IR*, and the reversible potential of mixing the two different solutions, as summarised in Eq. 1. A unit cell is a pair of membranes and a pair of aqueous streams of high and low concentration ($C_{high/low}$). Typically the membranes have the higher resistances in an ED unit cell, but this depends much on the thickness of the flowing solutions (0.2–2 mm) and the concentration of the solution (0.01–1 M).

$$E^{ED}_{u.c.} = E^{OCP}_{u.c.} + IR_{u.c.} \tag{1}$$

where OCP is for open-circuit potential, *I* is for current, *R* is for ohmic resistance of the unit cell and

$$E^{OCP}_{u.c.} = \bar{\alpha} E^{rev}_{mix} = \bar{\alpha} \frac{2\bar{R}T}{F} ln \left[\frac{C_{high}}{C_{low}} \right] \tag{2}$$

where $\bar{\alpha}$ is the average apparent permselectivity for a pair of membranes, $\bar{R}$ is the molar gas coefficient, *T* is temperature, and *F* is the Faraday constant.

Beyond these main energy costs (concentration difference and ohmic), other energy usage matters as well. Concentration polarisation leads to higher energy spending because of the very high ohmic resistance one can get on the dilute side of the membrane and the locally increased concentration difference across the membrane. Pumping water through the membrane module also requires energy. A third, secondary energy spending, relates to fouling mitigation and clean up, where filtering and membrane module cleaning are the main factors.

ED Applications in Hydrometallurgy

Generally, we find that ED efforts in the hydrometallurgical sector fall into a category of either recovering wanted species or removing unwanted species. Unwanted species are sometimes in the category of disturbing other parts of the process industry and other times in the category of being of environmental concern.

Metallic Value Recovery

The literature has shown that it is possible to recover between 70 and 99% of valuable metals such as gold, noble metals, rare-earth metals, copper, and silver by use of ED. Intensive research has been directed to the opportunities of recovering

Cu from electroplating wastewater. In the range of initial Cu-ion concentrations from 1–4 g/L, removal rates between 70 and 85% were achieved using ED [10–12]. When coupling ED to an in situ complexation reaction step [13, 14] and an electrolysis step [15], recoveries >99% were obtained. Reig et al. [16] applied Selective ED to remove As and recover Cu and Zn from acidic metallurgical process waters. Non-selective and selective membranes were used in this approach. The recoveries were 80% for Cu(II), 87% for Zn(II), and 95% for As(V). In another report, full recovery of Ag ions from rinse water of an industrial plant with an initial concentration of 100mg/L was achieved by applying ED [17]. Scarazzato et al. used HEDP-acid to form a stable complex with 4.5 g/L copper (Cu(II)–HEDP). They reported the highest Cu^{2+} recovery of 99.7% with HEDP (94.4%) and water using ED [14]. Iizuka et al. proposed a process for separating a mixed solution of Li (0.02 M) and Co (0.02 M), based on bipolar membrane ED coupled with metal–ion chelation. The selectivity for Li and Co in the metal recovery cells was 99% [18]. Hershey et al. investigated the separation of radioactive strontium ($1.3 \cdot 10^{-7}$M) from cesium ($7 \cdot 10^{-6}$ M) based on the formation of strontium chelate complexes with EDTA and DTPA by ED. 94% of the cesium was reserved in the feed cell while 2.2% of it transferred to the anode cell with 85% of the strontium [19]. Roman et al. proposed a method based on the combination of membrane processes consisting of membrane-based solvent extraction and ED stages for the recovery of a solution of Zn and Fe with a high concentration of hydrochloric acid. 99% of the initial HCl and 75% of the initial Zn were recovered in the concentrated phase [20]. Recently, efforts have been made to produce Li from brines, ores, and spent Li-ion batteries. ED can be used to produce Li salts with purity >95% while keeping a high extraction efficiency rate (around 80%) and low energy consumption [21].

Metal Removal

ED can further be used to extract species that can be harmful to the environment, like Pb, Cu, and Cd. The selective separation of metal ions from, e.g., electroplating baths in the galvanic industry allows the reuse of both the concentrate solution in the electroplating bath and the diluate as rinse water. Santarosa et al. [22] studied the performance of different IEMs on the removal of Ni and Zn from synthetic solution and real industrial effluents and achieved removal of up to 90%. ED has also proven to be efficient for the separation of Ni and Co ions in sulfate solution [23]. The use of cadmium cyanide baths in the electroplating industry generates a strong concern due to high toxicity associated with Cd and CN. Applying ED, Cd, and CN was recovered from synthetic wastewater in concentration ranges from 1–3 g/L Cd yielding removals of 5–22% for Cd and 17–46% for CN. However, when other metal ions such as Co, Cr, or Fe were present in the solution, the concentrate was not feasible for reuse in the electroplating bath since transport of these former metal ions occurs with Cd and CN ions [1]. Lead removal from wastewater is an important challenge in the battery industry. For ED applications, removal between 80 and 90% has been reported for Pb concentrations in synthetic wastewater ranging from 100–1000 ppm [24, 25].

Electrodialytic Remediation is a process to remove contaminants disposed in solid material like harbor sediment or wastewater sludge. In different setups, 44–82% Cu, 65–81% Pb, 78–91% Zn, and 98% Pb were removed from harbor sediments [26, 27]. Extensive studies have been performed to evaluate the feasibility of ED processes in the treatment of wastewater sludge and bio-ashes from sludge incineration. These waste products from wastewater treatment are rich in phosphor. In order to use the sludge, e.g., as fertilizer, heavy metals must be removed. With ED remediation, 70% of Cd removal from sludge suspended in citric acid or in distilled water was obtained [28], while when treating the ashes 75% of Cd removal was achieved [29]. Removal of Cu, Cr, Pb, Zn, and Ni from sewage sludge [29–31] and ashes [31, 32] has also been investigated.

Outlook and Perspectives

To preserve resources and increase environmental and economical sustainability, it is of interest to design industrial processes as closed-loop systems. As a process-integrated method, ED is a promising technology for the purification of metal-containing effluents. This can be fit into several existing hydrometallurgical processes. Clearly, several efforts of ED into the hydrometallurgical sector have been made. We see several examples with room for higher recovery and for tailoring membranes with respect to selectivity and conductivity. For mixed species solutions, the selectivity of the membranes is a challenge that should be further addressed in future work. While most research focuses on pristine membrane properties like selectivity and conductivity, little is documented on the sustainability of these properties and the durability of the bulk membrane material. However, when transferring membranes to industrial applications, this research field needs exploration.

References

1. Marder L, Bernardes AM, Ferreira JZ (2004) Cadmium electroplating wastewater treatment using a laboratory-scale electrodialysis system. Sep Purif Technol 37(3):247–255
2. Juda W, McRae WA (1950) Coherent ion-exchange gels and membranes. J Am Chem Soc 72 (2):1044–1044
3. Valero F, Barceló A, Arbós R (2011) Electrodialysis technology-theory and applications. In: Desalination, trends and technologies. IntechOpen
4. Strathmann H (2010) Electrodialysis, a mature technology with a multitude of new applications. Desalination 264(3):268–288

5. Galama AH, Daubaras G, Burheim OS, Rijnaarts HHM, Post JW (2014) Seawater electrodialysis with preferential removal of divalent ions. J Membr Sci 452:219–228
6. Galama AH, Daubaras G, Burheim OS, Rijnaarts HHM, Post JW (2014) Fractioning electrodialysis: a current induced ion exchange process. Electrochim. Acta 136:257–265
7. Burheim OS (2017) Engineering energy storage. Academic Press
8. Zlotorowicz A, Strand RV, Burheim OS, Wilhelmsen Ø, Kjelstrup S (2017) The permselectivity and water transference number of ion exchange membranes in reverse electrodialysis. J Membr Sci 523:402–408
9. Jalili Z, Krakhella KW, Einarsrud KE, Burheim OS (2019) Energy generation and storage by salinity gradient power: A model-based assessment. J Energy Storage 24:100755
10. Caprarescu S, Corobea MC, Purcar V, Spataru CI, Ianchis R, Vasilievici G, Vuluga Z (2015) San copolymer membranes with ion exchangers for Cu (II) removal from synthetic wastewater by electrodialysis. J Environ Sci 35:27–37
11. Caprarescu S, Purcar V, Vaireanu D-I (2012) Separation of copper ions from synthetically prepared electroplating wastewater at different operating conditions using electrodialysis. Sep Sci Technol 47(16):2273–2280
12. Cifuentes L, Crisostomo G, Alvarez F, Casas JM, Cifuentes G (1999) The use of electrodialysis for separating and concentrating chemical species in acidic Cu–Fe–As–Sb electrolytes. In: Fourth international conference on electrorefining and electrowinning of copper, vol 99, pp 479
13. Frioui S, Oumeddour R, Lacour S (2017) Highly selective extraction of metal ions from dilute solutions by hybrid electrodialysis technology. Sep Purif Technol 174:264–274
14. Scarazzato T, Buzzi DC, Bernardes AM, Espinosa DCR (2015) Treatment of wastewaters from cyanide-free plating process by electrodialysis. J Clean Prod 91:241–250
15. Peng C, Liu Y, Bi J, Xu H, Ahmed AS (2011) Recovery of copper and water from copper-electroplating wastewater by the combination process of electrolysis and electrodialysis. J Hazard Mater 189(3):814–820
16. Reig M, Vecino X, Valderrama C, Gibert O, Cortina JL (2018) Application of selectrodialysis for the removal of as from metallurgical process waters: recovery of Cu and Zn. Sep Purif Technol 195:404–412
17. Güvenç A, Karabacakolu B (2005) Use of electrodialysis to remove silver ions from model solutions and wastewater. Desalination 172(1):7–17
18. Iizuka A, Yamashita Y, Nagasawa H, Yamasaki A, Yanagisawa Y (2013) Separation of lithium and cobalt from waste lithium-ion batteries via bipolar membrane electrodialysis coupled with chelation. Sep Purif Technol 113:33–41
19. Hershey HC, Mitchell RD, Webb WH (1966) Separation of cesium and strontium by electrodialysis. J Inorg Nucl Chem 28(2):645–649
20. San Roman MF, Gándara IO, Ibañez R, Ortiz I (2012) Hybrid membrane process for the recovery of major components (zinc, iron and HCl) from spent pickling effluents. J Membr Sci 415:616–623
21. Gmar S, Chagnes A (2019) Recent advances on electrodialysis for the recovery of lithium from primary and secondary resources. Hydrometallurgy 189:105124
22. Santarosa VE, Peretti F, Caldart V, Zoppas J, Zeni M (2002) Study of ion-selective membranes from electrodialysis removal of industrial effluent metals II: Zn and Ni. Desalination 149(1–3):389–391
23. Tzanetakis N, Taama WM, Scott K, Jachuck RJJ, Slade RS, Varcoe J (2003) Comparative performance of ion exchange membranes for electrodialysis of nickel and cobalt. Sep Purif Technol 30(2):113–127
24. Sadrzadeh M, Mohammadi T, Ivakpour J, Kasiri N (2008) Separation of lead ions from wastewater using electrodialysis: comparing mathematical and neural network modeling. Chem Eng J 144(3):431–441
25. Mohammadi T, Razmi A, Sadrzadeh M (2004) Effect of operating parameters on Pb2+ separation from wastewater using electrodialysis. Desalination 167:379–385

26. Kirkelund GM, Ottosen LM, Villumsen A (2009) Electrodialytic remediation of harbour sediment in suspension-evaluation of effects induced by changes in stirring velocity and current density on heavy metal removal and pH. J Hazard Mater 169(1–3):685–690
27. Pedersen KB, Ottosen LM, Jensen PE, Lejon T (2015) Comparison of 2-compartment, 3-compartment and stack designs for electrodialytic removal of heavy metals from harbour sediments. Electrochim Acta 181:48–57
28. Jakobsen MR, Fritt-Rasmussen J, Nielsen S, Ottosen LM (2004) Electrodialytic removal of cadmium from wastewater sludge. J Hazard Mater 106(2–3):127–132
29. Juang RS, Wang SW (2000) Metal recovery and EDTA recycling from simulated washing effluents of metal-contaminated soils. Water Res 34(15):3795–3803
30. Ebbers B, Ottosen LM, Jensen PE (2015) Comparison of two different electrodialytic cells for separation of phosphorus and heavy metals from sewage sludge ash. Chemosphere 125:122–129
31. Ottosen LM, Pedersen AJ, Hansen HK, Ribeiro AB (2007) Screening the possibility for removing cadmium and other heavy metals from wastewater sludge and bio-ashes by an electrodialytic method. Electrochim Acta 52(10):3420–3426
32. Kirkelund GM, Jensen PE, Villumsen A, Ottosen LM (2010) Test of electrodialytic upgrading of MSWI APC residue in pilot scale: focus on reduced metal and salt leaching. J Appl Electrochem 40(6):1049–1060

Leaching of Eudialyte—The Silicic Acid Challenge

Dag Øistein Eriksen, Kurt Simon Forrester and Mark Stephen Saxon

Abstract Heavy rare-earth elements (HREE) are considered as a group of critical elements of high supply risk. The most abundant mineral containing HREE is xenotime, YPO_4, but to extract Y and the substituting HREE tough handling to dissolve the phosphate is required. Eudialyte, on the other hand, is much less common but is easy to leach. The mineral is basically an alkaline zirconium silicate and it is leachable at $pH < 3$. This means that even organic acids or dilute mineral acids may be used for the dissolution. However, the challenge with eudialyte is that silicates are also dissolved and after a while these silicates form gels. Usually, the silicates are referred to as silicic acids. Such gels have a detrimental effect on chemical processes where fluid flows are imperative. This work is a review of efforts conducted by both universities and private enterprises on solving this challenge.

Keywords Eudialyte · Silicic acid · Silica gel · Xenotime · Dysprosium · Rare-earth elements · REE

Introduction

The green economy requires efficient electric generators and motors. Therefore, the need for high-strength permanent magnets is pronounced and is expected to remain high. Iron–neodymium–boron (Fe–Nd–B) magnets are the strongest permanent

D. Ø. Eriksen (✉)
Primus.inter.pares AS & Department of Chemistry, University of Oslo, Oslo, Norway
e-mail: d.o.eriksen@kjemi.uio.no

K. S. Forrester
Arn Perspective Limited, Richmond, Surrey, UK
e-mail: k.forrester@arnperspective.com

M. S. Saxon
Leading Edge Materials, Suite 1305 - 1090 West Georgia St, Vancouver, BC V6E 3V7, Canada
e-mail: msaxon@leadingedgematerials.com

G. Azimi et al. (eds.), *Rare Metal Technology 2020*, The Minerals, Metals & Materials Series, https://doi.org/10.1007/978-3-030-36758-9_16

magnets on the market today. However, the different uses require slight changes in their composition. One of the elements added is dysprosium giving a higher temperature tolerance for the magnet. The next magnet on the list is samarium–cobalt. This magnet has the advantage that it is less prone to corrosion as Fe–Nd–B magnets are.

While neodymium (Nd) together with praseodymium (Pr) belongs to the light rare-earth elements (LREE), dysprosium (Dy) is part of the heavy rare-earth elements (HREE). LREE are found in the minerals bastnäsite ($La(Ce)CO_3F$) and monazite ($La(Ce)PO_4$) where lanthanum (La) and cerium (Ce) are the major constituents. HREE are mostly produced from the yttrium phosphate, xenotime, YPO_4. These three minerals have two things in common: they are hard to leach and the content of thorium is often high. High temperatures, concentrated acids or bases are required to get the rare-earth elements (REE) into solution. Still, these minerals are the major sources for these elements.

After the moratorium on the export of REE from China in 2009/2010 the price, in particular on Dy, rose sky high and the search for new sources became intense. One of the minerals of particular interest in this respect turned out to be eudialyte. It has the following features:

- It is leachable with acids at $pH < 3$
- It consists of more HREE than of LREE
- Usually, very little thorium and uranium is related to the mineral
- It has zirconium, Zr, as its main element and is thus also of interest as a source for other metals. This also means it is a potential source of hafnium, Hf.

However, the drawback of eudialyte is that it also releases silicic acid upon exposure to acids. Silicic acid tends to form gels upon standing or when chemical potentials are changed, e.g. when mixing immiscible phases in liquid–liquid extraction.

Among the eudialyte ore deposits studied and surveyed in the years after 2010 were Kipawa in Quebec, Canada; Kvanefjeld, and Kringlerne in Greenland and Norra Kärr in Sweden. These surveys were conducted by private enterprises. The three latter deposits were also part of the EU Seventh Framework Program (FP7)-sponsored research project EURARE. The Norra Kärr deposit is owned by Tasman Metals Ltd., now a part of Leading Edge Materials Ltd.

Theory

Eudialyte has been assigned the formula [1]:

$$Na_{15}Ca_6(Fe, Mn, REE)_3Zr_3SiO(O, OH, H_2O)_3(Si_3O_9)_2(Si_9O_{27})_2(OH, Cl, F)_2 \quad (F.1)$$

As can be deduced from (F.1) there are at least three possible silicates or silicic acids which may be extracted, e.g. $SiO(OH)_2$, $Si_3O_9.nH_2O$, $Si_9O_{27}.qH_2O$, where n and q are integral numbers. Rastsvetaeva has made a comprehensive review of structural mineralogy of the eudialyte group [2]. According to Rastsvetaeva, eudialytes possess two nine-membered rings and a six-membered ring. The Na- and Ca-sites are the sites in the crystal lattice most prone to undergo a transformation. These sites are centres of the ring structures. Thus, when eudialyte is exposed to acid, it is likely that also silicates are released. However, it is also well known that silicates of calcium and aluminium can be quite stable as such minerals are abundant in the Earth's crust, e.g. feldspar, anorthite, nepheline, wollastonite, etc. Eudialyte is often associated with minerals like nepheline and feldspars, especially albite, and quartz, calcite, and aegirine. Since REE are paramagnetic eudialyte may be separated from feldspars with magnets and to make concentrates (1–10%) is relatively simple. To achieve a highly concentrated fraction (>30%) requires a huge effort and is usually not economically feasible due to the employment of different beneficiation methods, e.g. floatation in addition to magnetic separation.

Tests Performed on Norra Kärr Ore Concentrate

One strategy for avoiding dissolution of silicate is to transform the silicate to a compound which may be separated from the other elements present. An obvious possibility is to apply a carbochlorination process. In such processes carbon is used to reduce the chloride forming elements, i.e. Fe, Zr, and Si, and the chlorine gas forms the gaseous compounds, i.e. $FeCl_3$, $ZrCl_4$, and $SiCl_4$. These may be distilled off and fractionated and the rest of the elements present can be leached as dissolvable chlorides. Tasman Metals conducted one study at Elkem Technology, Kristiansand, Norway and one at RPC, Fredricton, New Brunswick, Canada. The tests were promising but not conclusive [3, 4]. The yields reported by Elkem for REE were 20–30%, which is rather low. Moreover, the yield of Si was much lower, contrary to the expectation. RPC reported improved results by reducing the carbon content from 30 to 15% by weight. They also reported 750 ºC as a more efficient temperature than 850 ºC as the latter affected the mixing efficiency due to sintering.

Presumably, one reason for low yields was due to too low mixing of carbon and ore. The tests were all performed in a batch where the solids were placed in a crucible. Thus, the chlorine gas could only attach at the surface. Fluidized bed reactors or similar are preferred for such processes. However, such a process makes an extremely corrosive environment and the operation is therefore costly. The search for less expensive processes has therefore continued.

The next simple strategy was to leach with acid and see if the silicic acid is manageable. On their Norra Kärr ore concentrate, Tasman Metals conducted several tests at relevant laboratories. At ANSTO, Australia, they tested leaching with sulphuric acid. They developed a leaching concept where the silicic acid was precipitated from the PLS (pregnant leach solution) by adding MgO for neutralization and H_2O_2. Unfortunately, this also removed the dissolved zirconium. To recover Zr a carbochlorination step on the precipitate was proposed but not tested.

At Primus.inter.pares AS, Oslo, Norway two concepts were tested: Leaching with oxalic acid and leaching with nitric acid. To employ oxalic acid also the complexing ability of the oxalate for Zr, Hf, and Nb were utilized, while the REE would remain in the residue or be re-precipitated as oxalates. The leaching were performed at a quite stable pH varying between 1.0 and 1.4. Sulphuric acid was added to maintain acidity. The results obtained were the following:

An oxalic acid concentration of 0.3 M or higher is required for achieving the complexation mentioned above and the REE in the solid phase. Also, it is beneficial to adjust the pH with oxalic acid compared to sulphuric acid. Conclusions from the test work are:

- Zirconium, hafnium and niobium are leached at reasonable yields in both sulphuric and oxalic acid. Still, it seems that the presence of sulphate enhances the yield. These metals form sulphate complexes.
- Dissolution of silicate, i.e. silicic acid, seems to be more dependent on the pH than the type of acid.
- Silicic acid is dissolved, but fresh solutions are manageable as long as the pulp density is below 12%. All leachates formed gels upon standing.
- When the oxalic acid concentration is high enough the rare-earth elements do not dissolve, or is re-precipitated as oxalates.

The reason for performing tests with nitric acid was twofold: To comply with other technology for producing high purity REE and the availability of acid at a potential production site. Also, a report on no problem with gel formation was received from Rare Earth Salts, Nebraska, USA, who had tested the concentrate from Norra Kärr. Leaching tests were performed at 30 °C and 10% pulp density for initial nitric acid concentrations of 0.1–2.7 M. Figure 1 shows chemical yields of selected elements versus the initial concentration of nitric acid.

As can be seen, the yields for the metals reach a maximum at 1.6 M HNO_3 while silicic acid yield is rather constant. Also, the nitric acid leachates formed silicic acid gels. The reason why Rare Earth Salts did not experience any problems with gel formation is presumably that they perform a pre-calcination step causing the crystal structure to be changed and second, their electrochemical process [5] produces hydrolyses of rare-earth metals forming hydroxides which also may form solid silicon hydroxides.

On behalf of Tasman Metals, Process Research Ortech (PRO) employed their proprietary process using HCl and $MgCl_2$ to leach Tasman's eudialyte concentrate. According to PRO, the process worked well with no observed gelling. PRO is reluctant to release details of their process.

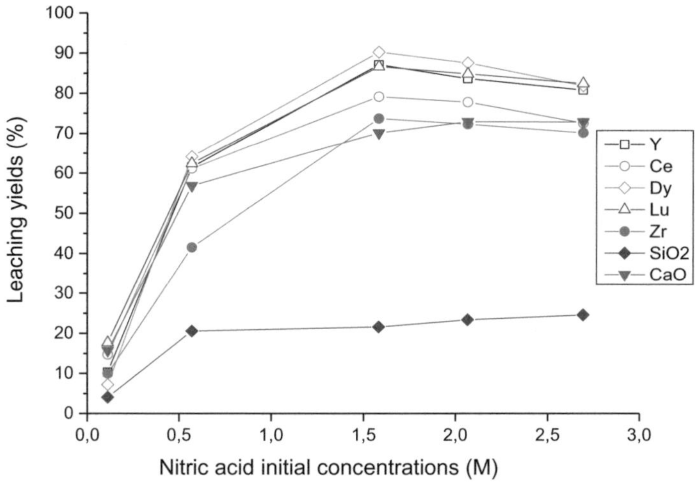

Fig. 1 Yields of selected elements from leaching of a eudialyte concentrate from Norra Kärr with nitric acid. Leaching parameters are 30 °C, pulp density of 10% and 2 h leaching time

Published Works

Pretreatment of eudialyte ore by exposing the ore to energy to avoid silica gel formation is confirmed by Chanturia et al. who used ultrasonic, electrochemical, and thermal pretreatments to improve the leaching abilities of the valuable elements in the Russian eudialyte ore they tested [6]. These authors claim the energy added to the solution promoted dispersion of colloidal silica gel and eliminated precipitation of silica gel on the mineral surfaces. The ultrasonic treatment had the best effect.

Gorrepati et al. studied the dissolution and agglomeration of silicic acid in HCl at $pH < 0$, and have shown that the process follows two stages: first mono-silicic acid is forming dimers which aggregates, then these primary aggregates flocculates [7]. They also studied the effect of added salts and showed that flocculation was enhanced by the presence of ions in the following order: $CsCl < NaCl < MgCl_2 < CaCl_2 < AlCl_3$.

Vossenkaul et al. studied the mechanism of silica gel formation with a focus on leachates from eudialyte processing [8]. These authors focus on the fact that to get polymerisation of silicic acid water molecules are needed in the process. So, by reducing the activity of water the formation of gel is prohibited. As a consequence, they performed initial leach tests with high acidities of HCl and H_2SO_4 (2–10 M) followed by washing with water. Their results show enhanced yields of REE with HCl compared to H_2SO_4, but for Si and Zr, it was the opposite. Formation of REE double sulphates is believed to be the cause of the lowered yield of REE while the formation of sulphate complexes of Zr the reason for the enhanced yield of Zr. No explanation for the difference in Si yields was given.

The same group has introduced a preprocessing step called "Fuming," which is the addition of an acidic solution to the eudialyte concentrate heated at boiling temperature. This treatment is followed by a leaching step with water at ambient temperature [9]. The group has also managed to recover Zr, Hf, and Nb from eudialyte using the same procedure with sulphuric acid and H_2O_2 as a reaction promoter [10].

Recently, Balinsky et al. have published a study on sulphuric acid leaching of eudialyte comparing direct leaching, fast leaching and water leaching of dehydrated acid/concentrate mixture. Their best results in terms of average yield of value components, i.e. REE, Zr, Hf, Mn and Nb, was obtained by direct leaching under mild conditions, i.e. 1 M, T = 60 °C. Gelling of silicic acid was avoided by keeping the pulp density below 100 kg/m^3 (10%) [11].

Discussion

Combining PRO's results with Gorrepatti's and Vossenkaul's research RPC, on behalf of Leading Edge Materials, conducted tests using HCl and metal chlorides, i.e. Mg, Al, and Fe, at a pulp density of 30%. The silicic acid dissolution was avoided and the pulp showed no formation of the gel upon standing. High yields of REE were obtained, but Zr remained in the residues, so a two-step leaching procedure is needed [12].

An industrial process requires a stable, robust means to handle the ore and recover the valuable constituents. One of the sub-requirements is often that quite concentrated solutions are processed. This reduces the size of the equipment and assures high throughput of product per time. Therefore, the 30% pulp density reported by RPC is a milestone. However, recycling of acid and added Al or Mg may turn out to be a challenge. Also, the second leach of Zr, Hf, Nb, and Ta is not yet developed.

Based on the published work and also unpublished, proprietary reports made for the private enterprises developing the eudialyte deposits, it seems clear that an initial calcination and a double leach is the safe way to establish a stable, robust method for recovery of REE, Zr, Hf, and Ta and Nb. This is, however, not the least expensive procedure as energy is required and more steps than if only one leaching is enough.

Based on the work performed on this subject by the group at RWTH Aachen University [7–9], Schreiber et al. [13] have published an article where a comparison of the environmental impacts of a hypothetical European supply chain based on Norra Kärr ore to those of today's Bayan Obo route in Inner Mongolia. To show expected differences for light and heavy rare-earth components, neodymium (Nd) and dysprosium (Dy) were chosen as representatives for each group. Based on the group's different leaching concepts, the authors find that it is technically possible to establish production in Europe with a much better environmental footprint than the Chinese productions.

Acknowledgements The authors are grateful to Leading Edge Materials for the possibility of publishing some of the corporation's proprietary data concerning eudialyte from Norra Kärr, Sweden.

References

1. Wikipedia: en.wikipedia.org/wiki/Eudialyte
2. Rastsvetaeva RK (2007) Structural mineralogy of the eudialyte group: a review. Crystallogr Rep 52(1):47–64
3. Emamifard B, Andersen E (2013) Elkem Technology, private communication
4. Botha N, Cheung L (2018) Report MIS-J2068, private communication
5. Brewer J, Method for extraction and separation of rare earth elements. US201462037714P 20140815, WO2016025928 (A1)
6. Chanturia VA, Minenko VG, Samusev AL, Chanturia EL, Koporulina EV (2017) The mechanism of influence exerted by integrated energy impacts on intensified leaching of zirconium and rare earth elements from eudialyte concentrate. J Min Sci 53(5):890–896 (2017)
7. Gorrepati EA, Wongthahan P, Raha S, Scott Fogler H (2010) Silica precipitation in acidic solutions: mechanism, pH Effect, and salt effect. Langmuir 26(13):10467–10474 (2010). https://doi.org/10.1021/la904685x
8. Voßenkaul D, Birich A, Müller N, Stoltz N, Friedrich B (2017) Hydrometallurgical processing of eudialyte bearing concentrates to recover rare earth elements via low-temperature dry digestion to prevent the silica gel formation. J Sustain Metall 3:79–89. https://doi.org/10.1007/s40831-016-0084-2
9. Davries P, Stopic S, Balomenos E, Panias D, Paspaliaris I, Friedrich B (2017) Leaching of rare earth elements from eudialyte concentrate by suppressing silica gel formation. Miner Eng 108:115–122
10. Ma Y, Stopic S, Gronen L, Friedrich B (2018) Recovery of Zr, Hf, Nb from eudialyte residue by sulfuric acid dry digestion and water leaching with H2O2 as a promoter. Hydrometallurgy 181:206–214
11. Balinski A, Atanasova P, Wiche O, Kelly N, Reuter MA, Scharf C (2019) Recovery of REEs, Zr(+Hf), Mn and Nb by H_2SO_4 leaching of eudialyte concentrate. Hydrometallurgy 186:176–186
12. Forrester, K (2019) Private communication, August 2019
13. Schreiber A, Marx J, Zapp P, Hake J-F, Voßenkaul D, Friedrich B (2016) Environmental impacts of rare earth mining and separation based on eudialyte: a new European way. Resources 5:32 (2016). https://doi.org/10.3390/resources5040032

Co-precipitation of Impurity (Ti, Fe, Al, Zr, U, Th) Phases During the Recovery of $(NH_4)_3ScF_6$ from Strip Liquors by Anti-solvent Crystallization

Edward Michael Peters, Carsten Dittrich, Bengi Yagmurlu and Kerstin Forsberg

Abstract Scandium can be extracted from waste streams of other industrial processes, particularly the bauxite residue and TiO_2 acid waste, by acidic leaching and solvent extraction of the leach solutions. Stripping of the organic phase using NH_4F solutions produces strip liquors containing Sc (>2000 mg/L). Scandium can be separated from these liquors by anti-solvent crystallization of $(NH_4)_3ScF_6$. In this study, the extent to which impurities co-precipitate as separate crystalline phases or are incorporated into the crystal lattice of $(NH_4)_3ScF_6$ was investigated. The impurity metals Fe, Zr, and U co-precipitated with the Sc phase. Moderate Ti precipitation was only observed from strip liquors containing mainly Fe and Ti impurities. Detection of these phases by powder XRD was difficult due to almost similar peak positions of the ammonium metal hexafluoride salts. However, EDS confirmed that the impurity metals were present in the precipitates in relative abundances that matched non-proportionally those of the initial strip liquors, except for Ti. SEM images showed that $(NH_4)_3ScF_6$ crystals obtained from strip liquors containing predominantly scandium were bigger (2–3 μm) compared to crystals of the mixed precipitate samples (<2 μm) obtained from strip liquors containing relatively high impurity levels. This could be attributed to surface diffusion impediment of one metal ion by other metal ions at the solid–liquid interface and surface incorporation of foreign metal ions on the growth steps or kinks of one solid phase, thereby reducing the crystal growth rate of that phase. The excess supersaturation is then consumed by crystal nucleation as observed.

Keywords Anti-solvent crystallization · Impurity co-precipitation · $(NH_4)_3ScF_6$ · $(NH_4)_3FeF_6$ · $(NH_4)_3ZrF_7$ · $(NH_4)_2UF_6$

E. M. Peters · K. Forsberg (✉)
Department of Chemical Engineering, KTH Royal Institute of Technology, SE100-44 Stockholm, Sweden
e-mail: kerstino@kth.se

C. Dittrich · B. Yagmurlu
MEAB Chemie Technik GmbH, Dennewartstraße 25, 52068 Aachen, Germany

G. Azimi et al. (eds.), *Rare Metal Technology 2020*, The Minerals, Metals & Materials Series, https://doi.org/10.1007/978-3-030-36758-9_17

Introduction

Scandium is rarely concentrated in mineral ores, and it is widely dispersed in the earth's crust at an average concentration of 22 ppm [1–3]. A few ores have been discovered and exploited mostly in the Asian world [4–6]. Due to the vast applications of Sc which include production of Al–Sc alloys and use in Solid Oxide Fuel Cells (SOFCs) amongst others [7–9], there has been growing research interest in the recovery of Sc from waste streams of other metallurgical industries. These include the bauxite residue from the alumina industry, the acid waste from the titanium dioxide industry and lateritic cobalt–nickel processing waste streams [10–13]. This would turn waste into value and minimize the environmental pollution caused by stockpiles of waste residue or storage in evaporative ponds while recovering valuable metals such as Sc whose price has been estimated at USD132 per g ingot as of 2018 [7].

The processing of such waste streams would generally involve acid leaching of the waste streams, solvent extraction using a suitable organic solvent, stripping of the metals from the organic phase into an aqueous phase, crystallization of metal salts particularly the $(NH_4)_3ScF_6$ phase from the strip liquor (SL), calcination of $(NH_4)_3ScF_6$ to ScF_3, and finally metallothermic reduction of ScF_3 to Sc metal [3, 11, 14, 15]. The strip liquors obtained often contain various metals such as Sc, Fe, Ti, Zr, Al, U, Th, Y, and V amongst others in an NH_4F solution since NH_4F solutions were observed to be effective stripping solutions [16]. Consequently, anti-solvent crystallization by addition of alcohols, particularly ethanol and methanol, produces ammonium metal fluorides of the general formula $(NH_4)_xMe^{y+}F_{(x+y)}$, where Me is the metal. Anti-solvent crystallization of $(NH_4)_3ScF_6$ was conducted successfully from synthetic strip liquors containing about 2000 mg/L Sc and some Fe and Ti impurities with a final product having purities >98.2% by mass [17, 18]. However, as the concentration of impurity metals increases in solution, the final product is likely to be a mixture of different ammonium metal fluoride phases and the relative abundance of the impurity metals in the final solid products would depend on their relative solubilities in the NH_4F–alcohol mixtures and their tendency to be incorporated in the crystal lattices of the main phases.

$(NH_4)_3ScF_6$ undergoes polymorphic phase transitions at positive temperatures, where the high temperature phase, α-$(NH_4)_3ScF_6$, with a cubic structure transforms into the β-$(NH_4)_3ScF_6$ with a tetragonal or monoclinic structure [19], although some authors only mention the tetragonal phase [20, 21]. The polymorphic transition occurs at about 47 °C, [21, 22] with some sources mentioning a temperature of 57 °C [23, 24]. However, there seems to be a disagreement in literature on whether the cubic phase transforms first into the monoclinic or the tetragonal phase at this temperature, and some sources mention a third transition at 20 °C between the monoclinic and tetragonal phases [23–25]. For iron, the phase $(NH_4)_3FeF_6$ is mostly reported in literature [20, 26]. Other impurity phases that have been reported in literature include NH_4FeF_4, $(NH_4)_3ZrF_7$, $(NH_4)_2ZrF_6$, $(NH_4)ZrF_5$, $(NH_4)_2TiF_6$, $(NH_4)_3AlF_6$, $(NH_4)_2ThF_6$, $(NH_4)_4UF_8$, $(NH_4)_2UF_6$, and $(NH_4)UF_5$ [27–31]. The crystallographic structures of these salts have been studied extensively with structural similarities being observed. For instance, α-$(NH_4)_3ScF_6$ has a cubic structure similar to that of α-$(NH_4)_3FeF_6$ [20, 22] and $(NH_4)_3ZrF_7$ is isostructural with $(NH_4)_3AlF_6$ where the [AlF_6^{3-}] complexes are replaced with [ZrF_7^{3-}] complexes [25]. The synthesis of $(NH_4)_3ZrF_7$ was accomplished by dissolving ZrF_4 in a hot concentrated NH_4F solution and allowing crystallization to occur slowly [25]. A similar procedure can be employed for the precipitation of other ammonium metal fluoride salts with alcohol addition to improve the yield [30]. The $(NH_4)_3ZrF_7$ phase was observed to undergo thermal decomposition to ZrF_4 accomplished by loss of NH_4^+ and F^- ions. It transforms to $(NH_4)_2ZrF_6$ at 297 °C and then to $(NH_4)ZrF_5$ at 357 °C and further to ZrF_4 at 410 °C [30, 32]. $(NH_4)_3ZrF_7$ was also observed to undergo six structural phase transitions in the temperature range ca. 240–290 °C [29].

The U phase $(NH_4)_4UF_8$ is the stable phase at 25 °C and high NH_4F concentrations in the range 24.2–45.1 wt%, that is, about 6.5–12.2 mol/L solution, respectively. At lower NH_4F concentrations, the phases $(NH_4)_2UF_6$ and $7NH_4F.6UF_4$ were obtained, but $(NH_4)_3UF_7$ was not observed [27]. The phase $(NH_4)_2UF_6$ was also observed to undergo polymorphic transformations near room temperature with the orthorhombic phase, γ-$(NH_4)_2UF_6$, being in equilibrium with NH_4F solutions at 25 °C. The solubility of U in NH_4F solutions could be predicted using the equation $[U^{4+}] = 655[NH_4F]^{-5} + 0.00465[NH_4F]^{0.5}$ [27], where all concentrations are molal (mol/kg solvent). Therefore, the solubility of U^{4+} is expected to decrease progressively with increase in NH_4F concentration up to about 14 mol/kg (equivalent to ca. 34.1 wt% or ca. 9.2 mol/L solution), for example, in 1 mol/kg NH_4F, U^{4+} has a solubility of about 655 mol/kg and this decreases drastically to about 2.7 mol/kg in 3 mol/kg NH_4F. The trend is due to the double common ion effect caused by increase in both NH_4^+ and F^- ions in solution. The solubility is also expected to be further reduced by the addition of ethanol to the NH_4F solutions. At NH_4F concentrations above 14 mol/kg, the solubility of U increases slightly with increase in NH_4F concentration and this could be due to the formation of one or more higher fluoride complexes, such as UF_8^{4-}.

In the presence of impurity ions in solution, the crystallization kinetics of one salt can be altered due to adsorption of the impurity ions on the crystal surfaces. This, in turn, impedes the surface diffusion rate of solute ions on the solid–liquid interface and temporary or permanent occupation of step and kink sites by impurity ions in the crystal lattice reduces the incorporation rate of solute ions into their crystal lattice. A small amount of impurity ions can be sufficient to alter the crystal growth kinetics significantly [33]. Despite the differences in ionic radii of cations, Sc atoms readily substitute Fe atoms in many compounds and Sc and Ti atoms can be interchanged in almost all proportions in several compounds [20]. Information related to the solubility of the ammonium metal fluorides in aqueous solvents is scarce. The solubilities of $(NH_4)_3ScF_6$ have been reported at different temperatures in a 3 mol/L NH_4F strip liquor and at constant temperature (25 °C) in 3 mol/L NH_4F strip liquors containing some Fe and Ti impurities and dosed with different ethanol concentrations [17]. This study focuses on the extent of co-precipitation of these impurity phases along with $(NH_4)_3ScF_6$ from NH_4F strip liquors with compositions varying from high to low Sc/impurity concentration ratios.

Methodology

Anti-solvent Crystallization

Anti-solvent crystallization of $(NH_4)_3ScF_6$ using ethanol as the anti-solvent was conducted from six strip liquors with varying compositions. Table 1 shows the composition of the strip liquors.

The strip liquors were obtained from MEAB Chemie Technik GmbH, Germany, after conducting solvent extraction of pregnant leach solutions and stripping of the metals from the organic phase using 3 mol/L NH_4F solutions. Each of the strip liquors was subjected to anti-solvent crystallization using ethanol solvent of purity 99.94 v/v% [CAS: 64-17-5]. Table 2 shows the ethanol molar concentrations employed for strip liquors A and B, while Table 3 shows the ethanol molar concentrations employed for the strip liquors C, D, E, and F as well as the corresponding ethanol to SL volumetric ratios. The amount of strip liquor used in each experiment was about 20 g. The ethanol molar concentrations are expressed as moles of ethanol per liter of total solution, whereas the volumetric ratios are expressed as volume of ethanol per volume of strip liquor.

The ethanol was added all at once (time zero) without controlling the flow rate, and the experiments were conducted for 3 h under agitation at 250 rpm. At the end

Table 1 Composition of the strip liquors determined by ICP-OES in mg/L

SL	Sc (mg/L)	Fe	Ti	Zr	Al	U	Th
A	2987.0	2.0	425.0	1.0	7.0	0.0	0.0
B	2648.0	1.0	2.0	0.0	5.0	22.0	0.0
C	361.8	230.4	312.6	480.7	0.0	30.1	13.3
D	2380.5	0.8	114.3	0.0	0.0	19.6	0.0
E	774.7	205.4	228.2	120.7	0.0	0.0	5.4
F	435.9	81.6	1791.8	2277.8	0.0	61.7	48.0

Table 2 Ethanol molar concentrations and volumetric ratios for SLs A and B

mol/L Ethanol	Ratio v/v
2.86	0.2
7.62	0.8
10.01	1.4

Table 3 Ethanol molar concentrations and volumetric ratios for SLs C, D, E, and F

mol/L Ethanol	Ratio v/v
4	0.31
6	0.54
8	0.88
10	1.41

of the experiments, a supernatant sample was obtained from each of the experiments using a syringe fitted with a 0.22 μm PVDF membrane filter. The samples were diluted accordingly and analyzed using a Thermo Fisher iCAP™ 7400 ICP-OES analyzer for elemental composition. In cases where a considerable amount of solid precipitate was obtained, solid samples were taken before and after washing with about 2 mL of ethanol, but in most cases, only solid samples before washing were taken since minute quantities of precipitate were obtained. After drying under ambient conditions, the solid samples from experiments A and B were analyzed using a PANalytical X'pert PRO powder X-ray diffractometer and the solid samples from experiments C, D, E, and F were analyzed using a SIEMENS powder X-ray diffractometer D5000 to determine the mineral phases in the precipitates. A Phillips SL 30 Environmental Scanning Electron Microscope (ESEM) equipped with EDS was used to analyze the morphology of the precipitates and their elemental surface composition. It should be noted that the metal concentrations presented in the results and discussion section are expressed as moles of metal per unit volume of the initial strip liquor, while the ethanol concentrations are presented as moles of ethanol per unit volume of the total solution.

Results and Discussion

Concentration Profiles

Figure 1 shows the concentration profiles obtained after crystallization from four strip liquors, namely, A, B, C, and F. The concentration profiles obtained for SLs C and F which had moderate to very high Ti concentrations, respectively, are representative of those for SLs D and E. In all the plots shown in Fig. 1, the concentration of Ti was generally observed to remain fairly constant, while the concentrations of major elements like Sc, Fe, and Zr decreased significantly. Other elements such as U, Th, and Al were of lower concentrations such that it is difficult to note the concentration changes in Fig. 1. For this reason, Table 4 shows the cumulative percentage decrease in concentration (ΔC%), which was computed by Eq. 1.

$$\Delta C\% = \left(\frac{C_{SL} - C_e}{C_{SL}}\right) \times 100 \tag{1}$$

where C_{SL} and C_e are the metal concentrations in the strip liquor and in the supernatant obtained at a specific ethanol concentration. Both concentrations are expressed as moles per liter of initial strip liquor.

It can be noted in Table 4 that all elements had significant concentration changes, except Ti. Based on these observations, it can be deduced that ammonium metal fluorides of Fe, Zr, Al, U, and Th co-precipitated with the $(NH_4)_3ScF_6$ phase. The most probable impurity phases include $(NH_4)_3FeF_6$, $(NH_4)_3ZrF_7$, $(NH_4)_3AlF_6$, and

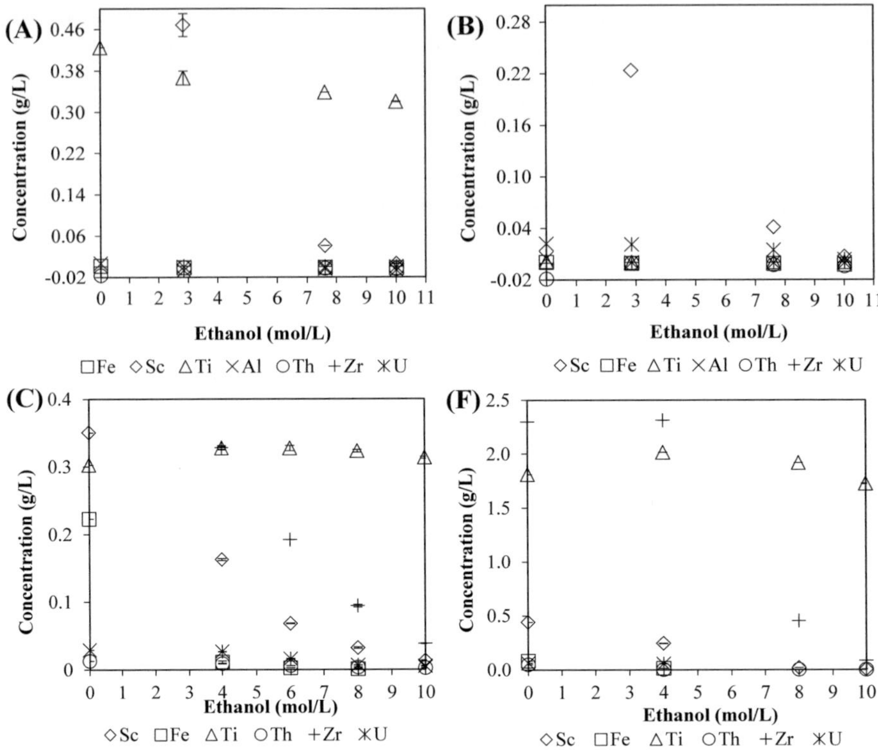

Fig. 1 Solution concentration profiles: **A**, **B**, **C**, and **F** represent strip liquors **A**, **B**, **C**, and **F**, respectively

Table 4 Cumulative percentage decrease in concentration

SL	Ethanol (mol/L)	Sc%	Ti %	Fe %	Al %	Zr %	U %	Th %
A	2.86	84.32	13.87	–	98.66	–	–	–
	7.62	98.63	20.40	–	100.00	53.30	–	–
	10.01	99.80	24.75	–	100.00	75.99	–	–
B	2.86	91.55	–	–	95.24	–	3.60	–
	7.62	98.44	–	–	100.00	–	33.88	–
	10.01	99.74	–	–	100.00	–	84.19	–
C	10.00	96.22	0.00	100.00	–	91.79	83.35	83.24
D	10.00	99.17	4.73	100.00	–	–	59.72	–
E	10.00	97.51	0.00	99.52	–	91.28	–	72.78
F	10.00	99.22	4.76	100.00	–	96.39	100.00	99.16

$(NH_4)_2UF_6$. The heptafluoride phase of Zr is stable under ambient conditions, and this transforms to the hexafluoride upon heating to 297 °C [30, 32]. In addition, the Ti phases, which could be $(NH_4)_3TiF_7$ or $(NH_4)_2TiF_6$, appear to have higher solubilities in the 3 mol/L NH_4F solution compared to phases of Fe, Zr, and Al. In strip liquor A that contained mainly Fe and Ti impurities, the Ti concentration in solution decreased by up to about 25% with increase in ethanol concentration to 10.01 mol/L, whereas a rather negligible decrease in Ti concentration was observed for strip liquors C, D, E, and F which contained considerable quantities of other metal impurities such as Zr, U, and Th. This could imply preferable precipitation of other metal impurities whose ammonium fluoride phases have lower solubilities than Ti phase(s). This is corroborated by the fact that the Ti concentration in the solids obtained increased with increase in ethanol concentration for solids obtained from strip liquor A, and minute quantities of Ti were detected by EDS in the solids obtained from strip liquors C and F (see Tables 5 and 6) where the Ti concentration was almost similar to or higher than the Sc concentration in solution, respectively. In order to predict the solubility of the respective phases, an investigation of the chemical speciation in the NH_4F–alcohol solutions is required. However, the necessary thermodynamic data to perform this evaluation is currently not available. The XRD patterns of the solid phases obtained are shown in Fig. 2, but as noted by comparing the reference patterns of these salts, it is difficult to distinguish the salts since their peak positions are almost similar as a result of almost similar crystallographic structures of these salts [20, 22]. In addition, Al, U, and Th were present in the solid precipitates in very low concentrations and could not be detected by XRD.

Phase Characterization

Since strip liquors A and B were predominantly abundant in Sc, the XRD patterns of the precipitates obtained from these strip liquors at all ethanol dosages matched that of $(NH_4)_3ScF_6$ (PDF card—00-040-0595) of monoclinic structure as shown in Fig. 2 labelled A. The plots labelled C, D, E, and F correspond to the XRD patterns

Table 5 Elemental compositions of the precipitates from strip liquors A and B as obtained by EDS (wt%)

SL	Ethanol (mol/L)	N	F	Sc	Ti	Fe	U	Total	$(NH_4)_3ScF_6$ purity%
A	2.86	16.87	53.47	29.66	0.00	0.00	0.00	100.00	100.00
	7.42	18.91	53.91	27.02	0.17	0.00	0.00	100.00	99.87
	10.01	10.49	52.01	37.05	0.36	0.09	0.00	100.00	99.74
B	2.86	20.43	61.18	18.39	0.00	0.00	0.00	99.99	100.00
	7.42	20.67	57.57	21.21	0.00	0.00	0.55	100.00	99.46
	10.01	20.06	54.40	24.92	0.00	0.00	0.62	100.00	99.48

Table 6 Elemental composition of the solids ([Ethanol] = 8 mol/L) from SLs C–F, determined by EDS (wt%)

SL	N	F	Sc	Ti	Fe	Zr	U	Total
C	15.80	57.34	8.21	0.21	6.50	11.59	0.35	100.00
D	16.64	52.42	30.94	0.00	0.00	0.00	0.00	100.00
E	15.62	54.70	20.65	0.00	6.38	2.65	0.00	100.00
F	12.04	48.68	6.21	1.23	1.58	29.15	1.11	100.00

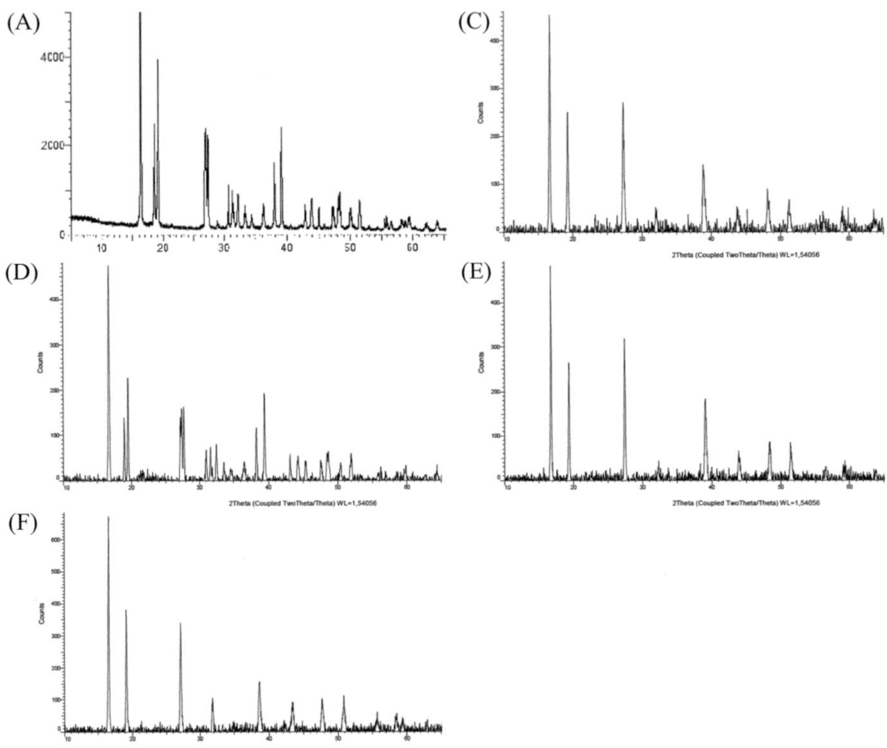

Fig. 2 XRD patterns of precipitates: **A**—precipitates from SL **A** at all ethanol concentrations; **C**, **D**, **E**, and **F**—precipitates from the respective SLs at an ethanol concentration of 10 mol/L

for the precipitates obtained from strip liquors C, D, E, and F, respectively, for the ethanol concentration of 10 mol/L. Strip liquor D was also rich in Sc with low concentrations of impurity metals and the XRD spectrum of the precipitates obtained from this strip liquor matched that of $(NH_4)_3ScF_6$ (PDF card—00-040-0595). Strip liquors C, E, and F had considerable quantities of Fe, Ti, and Zr, and it became difficult to distinguish the phases by analyzing the XRD patterns, since the high-intensity peaks of these salts have almost similar peak positions. The XRD patterns of the salts obtained from these strip liquors matched that of

$(NH_4)_3ScF_6$ (PDF card—01-074-6689), which corresponds to the cubic phase. However, it cannot be concluded with certainty that the cubic phase precipitates in the presence of considerable quantities of impurities, since it is quite possible that the double peaks at 2Θ angles of about 19 and 27° in the plot labelled A or D can be transformed into single peaks by the overlapping of peaks for the different salts as shown in plots C, E, and F. Although the Th and U phases have different peak positions from the phases of the other metals, they could not be detected by XRD.

Morphology, Crystal Size, and Purity

Figure 3 shows the morphology and crystal sizes obtained from strip liquors A and B at ethanol concentrations of 2.86, 7.62, and 10.01 mol/L, corresponding to ethanol to SL volumetric ratios of 0.2, 0.8, and 1.4, respectively. The crystal size decreased with increase in the ethanol concentration due to an increase in the initial supersaturation. The modal crystal size at the lowest ethanol concentration was about 2–3 μm, whereas it diminished to <2 μm at the highest ethanol concentration investigated. Table 5 shows the elemental composition, determined by EDS in wt %, for the precipitates obtained from strip liquors A and B at different ethanol concentrations. Strip liquor A had about 425 mg/L Ti, and the precipitates obtained from this strip liquor had Ti as the impurity in concentrations <0.4 wt%. The concentration of Ti in the solids increased with increase in ethanol concentration. This could be due to an increase in the saturation level of the Ti phase causing it to precipitate in small amounts or substitute Sc^{3+} ions in the crystal lattice or it could be due to an increase in mother liquor entrapment since the mass of the solid

Fig. 3 Micrographs of precipitates obtained from SLs (**A**) and (**B**)

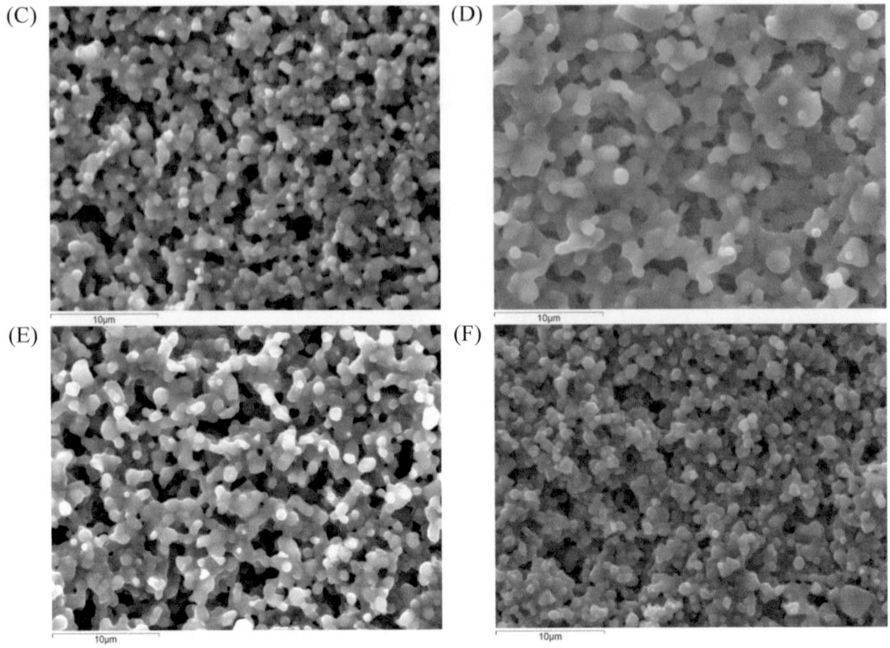

Fig. 4 Micrographs of precipitates obtained at an ethanol concentration of 10 mol/L from SLs (**C**), (**D**), (**E**), and (**F**)

precipitate increases. The purity of the precipitates with respect to the Sc phase was about 99.8% at higher ethanol concentrations. This was computed taking into account the masses of $(NH_4)_3ScF_6$ phase and elemental masses of the impurities in the solid samples. The precipitates obtained from strip liquor B had U impurity, and the U concentration in the solids increased up to 0.62 wt% with increase in ethanol concentration. Strip liquor B had a U concentration of 22 mg/L, and the purity of the precipitates with respect to the Sc phase was about 99.5% at higher ethanol concentrations. For both strip liquors, the impurities were only detected from an ethanol concentration of 7.62 mol/L and upwards.

Figure 4 shows the morphology and crystal sizes for precipitates obtained from strip liquors C, D, E, and F, respectively. It appears that the purer Sc phase obtained from strip liquor D had bigger crystals and the precipitates obtained from strip liquors C, E, and F, which contained Sc, Fe, and Zr had smaller crystal sizes. This could be due to the effect of impurities in retarding the crystal growth rates. The reduction in crystal growth rate occurs by the presence of foreign ions at the solid–liquid interface, which increases the resistance to surface diffusion of the host ions, and possibly occupy some step and kink sites on the growing crystal front, thereby reducing the surface incorporation rate of the host ions into their crystal lattice. The terms 'foreign ions' and 'host ions' are used here since the major phase in the precipitates obtained from the different strip liquors was different in each case. This implies that where the

Zr phase was the major phase precipitating, Sc^{3+} ions would be considered to be 'foreign ions' or impurities. The retardation in crystal growth rates is often accompanied by an increase in the nucleation rate due to excess available supersaturation. This can be noted in the micrographs (C, E, and F) as an increase in the number density of crystals per unit surface area as compared to micrograph D.

Table 6 shows the elemental surface composition of the precipitates determined by EDS. The relative abundance of the elements, except Ti, in the solids corresponds to their relative abundance in the strip liquors, although non-proportionally. This clearly confirms that the precipitation of Ti from these strip liquors is marginal because even in strip liquor F where Ti had a concentration of about 1792 mg/L, only 1.2 wt% Ti was detected in the solid. This is most likely due to the tendency of titanium to exist as the stable titanyl ion, TiO^{2+}, in aqueous solutions, instead of Ti^{4+} [34]. Uranium seems to be precipitating because some U is detected in the solids despite being present in the strip liquors at concentrations below 62 mg/L. Possible U phases that could precipitate from strip liquors containing three mol/L NH_4F include $(NH_4)UF_5$ and $(NH_4)_2UF_6$. However, none of these was detected by powder XRD due to their low concentrations in the solids. Thorium was not detected in the solids by EDS, but this could be due to very low concentrations that are undetectable by EDS.

Conclusions

The extent of co-precipitation of impurity phases (Zr, Fe, Ti, Al, U, Th) has been studied during anti-solvent crystallization of $(NH_4)_3ScF_6$ from strip liquors using ethanol as the anti-solvent. The concentration of Ti in solution only decreased moderately in strip liquors that had low impurity:Sc concentration ratios and remained fairly constant in strip liquors that contained high levels of Fe, Zr, U and Th relative to the Sc content. In the latter, the concentration of all metals decreased significantly with increase in ethanol concentration except for Ti. Although Th was observed to decrease in concentration, it was not detected in the solids by EDS. This implies that impurity phases of Zr, Fe, Al, and U co-precipitate, and the likely phases expected to crystallize are $(NH_4)_3ZrF_7$, $(NH_4)_3FeF_6$, $(NH_4)_3AlF_6$, and $(NH_4)_2UF_6$ or $(NH_4)UF_5$. However, the various phases were not confirmed by powder XRD due to almost similar crystallographic structures of these ammonium metal fluorides. The U and Th phases could not be detected by XRD due to their very low concentrations in the solids. It appears that the β-$(NH_4)_3ScF_6$ phase of monoclinic structure crystallizes from strip liquors containing predominantly Sc and low impurity levels, whereas the α-$(NH_4)_3ScF_6$ cubic phase crystallizes from strip liquors containing a considerable amount of impurity metals, where co-precipitation of the impurity phases occurs. In addition, large crystal sizes >2 μm were observed for the precipitates obtained from strip liquors containing low impurity levels, and smaller crystals of size <2 μm were obtained from strip liquors containing higher impurity levels. The impurity content in the precipitates increased with increase in ethanol concentration and the relative abundance of the metals in the solids corresponded non-proportionally to the relative abundance of these metals in the initial strip liquors except for Ti. It is recommended to determine the solubilities of these impurity phases in NH_4F–ethanol mixtures.

Acknowledgements This project has received funding from the European Union's Horizon 2020 research and innovation programme under grant agreement No. 730105.

References

1. Hedrick JB (2007) 2007 Minerals yearbook-Rare Earths
2. Cotton SA (1999) Recent advances in the chemistry of scandium. Polyhedron 18:1691–1715. https://doi.org/10.2533/chimia.2012.399
3. Habashi F (2013) Extractive metallurgy of rare earths. Can Metall Q 52:224–233. https://doi.org/10.1179/1879139513Y.0000000081
4. SRK Consulting, TetraTech, SMHProcessInnovation, et al (2017) Revised NI 43-101 Technical Report Feasibility Study Elk Creek Niobium Project Nebraska Revised NI 43-101 Technical Report Feasibility Study Elk Creek Niobium Project Nebraska
5. Lash LD, Ross JR (1961) Vitro chemical recovers costly scandium from uranium solutions. Min Eng 967
6. Shaoquan X, Suqing L (1996) Review of the extractive metallurgy of scandium in China (1978-1991). Hydrometallurgy 42:337–343. https://doi.org/10.1016/0304-386X(95)00086-V
7. GeologicalSurvey US (2019) Mineral commodity summaries 2019. Reston, Virginia
8. Binnemans K, Jones PT, Müller T, Yurramendi L (2018) Rare earths and the balance problem: how to deal with changing markets? J Sustain Metall 4:126–146. https://doi.org/10.1007/s40831-018-0162-8
9. Laguna-Bercero MA, Kinadjan N, Sayers R et al (2011) Performance of La2-xSrxCo0.5Ni0.5O4 $\pm$ δ as an oxygen electrode for solid oxide reversible cells. Fuel Cells 11:102–107. https://doi.org/10.1002/fuce.201000067
10. Wang W, Pranolo Y, Cheng CY (2011) Metallurgical processes for scandium recovery from various resources: a review. Hydrometallurgy 108:100–108. https://doi.org/10.1016/j.hydromet.2011.03.001
11. Li D, Wang C (1998) Solvent extraction of Scandium(III) by Cyanex 923 and Cyanex 925. Hydrometallurgy 48:301–312. https://doi.org/10.1016/S0304-386X(97)00080-7
12. Kaya Ş, Dittrich C, Stopic S, Friedrich B (2017) Concentration and separation of scandium from Ni laterite ore processing streams. Metals (Basel) 7:557. https://doi.org/10.3390/met7120557
13. Hatzilyberis K, Lymperopoulou T, Tsakanika L-A et al (2018) Process design aspects for scandium-selective leaching of bauxite residue with sulfuric acid. Minerals 8:79. https://doi.org/10.3390/min8030079
14. Smirnov DI, Molchanova TV (1997) The investigation of sulphuric acid sorption recovery of scandium and uranium from the red mud of alumina production. Hydrometallurgy 45:249–259. https://doi.org/10.1016/S0304-386X(96)00070-9
15. Krishnamurthy N, Gupta CK (2015) Extractive metallurgy of rare earths, 2nd edn. Taylor & Francis
16. Peters E, Dittrich C, Kaya S, Forsberg K (2018) Crystallization of a pure scandium phase from solvent extraction strip liquors. In: Davis B et al (ed) Extraction 2018, Proceedings of the first global conference on extractive metallurgy. Springer International Publishing, pp 2707–2713

17. Peters EM, Kaya Ş, Dittrich C, Forsberg K (2019) Recovery of scandium by crystallization techniques. J Sustain Metall 5:48–56. https://doi.org/10.1007/s40831-019-00210-4
18. Kaya Ş, Peters EM, Forsberg K et al (2018) Scandium recovery from an ammonium fluoride strip liquor by anti-solvent crystallization. Metals (Basel) 8:767. https://doi.org/10.3390/met8100767
19. Sviridova TA, Sokolova YV, Pirozhenko KY (2013) Crystal structure of (NH4)5Sc3F14. Crystallogr Rep 58:220–225. https://doi.org/10.1134/S1063774513020272
20. Gshneidner KA Jr, Melson GA, Youngblood DH, Schock HH (1975) Scandium: its occurence, chemistry, physics, metallurgy, biology and technology. Academic Press Inc., Ltd., London
21. Sobolev BP (2000) The high temperature chemistry of the rare earth trifluorides, 1st ed. Institut d'Estudis Catalans, Barcelona
22. Melson GA, Stotz RW (1971) The coordination chemistry of scandium. Coord Chem Res 7:133–160. https://doi.org/10.1002/9780470166208.ch5
23. Flerov IN, Gorev MV, Ushakova TV (1999) Calorimetric investigations of phase transitions in the cryolites $(NH_4)_3Ga(1-x)Sc_{(x)}F_6$ (x=1.0,0.1,0). Phys Solid State 41:468–473
24. Vtyurin AN, Bulou A, Krylov AS, et al (2001) The cubic-to-monoclinic phase transition in $(NH_4)_3ScF_6$ cryolite: a raman scattering study. Phys Solid State 43:2307–2310. https://doi.org/10.1134/1.1427961
25. Tressaud A, Khairoun S, Rabardel L, Kobayashi T (1986) Phase transitions in ammonium hexafluorometallates(III). Phys Status Solidi 96:407–414
26. Li Y, Yao F, Cao Y et al (2017) The mediated synthesis of FeF 3 nanocrystals through (NH 4) 3 FeF 6 precursors as the cathode material for high power lithium ion batteries. Electrochim Acta 253:545–553. https://doi.org/10.1016/j.electacta.2017.09.081
27. Penneman RA, Kruse FH, George RS, Coleman JS (1964) Studies of the ammonium fluoride-uranium tetrafluoride-water system. Properties of (NH 4)4UF8, (NH4)2UF6, 7NH4F·6UF4 and of U(IV) in aqueous NH4F. Inorg Chem 3:309–315. https://doi.org/10.1021/ic50013a001
28. Hampson GC, Pauling L (1938) The structure of ammonium heptafluozirconate and potassium heptafluozirconate and the configuration of the heptafluozirconate group. J Am Chem Soc 60:2702–2707. https://doi.org/10.1021/ja01278a046
29. Fokina VD, Gorev MV, Bogdanov EV et al (2013) Thermal properties and phase transitions in (NH4) 3ZrF7. J Fluor Chem 154:1–6. https://doi.org/10.1016/j.jfluchem.2013.07.001
30. Simons JH (1964) Fluorine chemistry, 5th edn. Academic Press Inc., Ltd., London
31. Baldea A, Axente D, Abrudean M et al (1987) Thermal decomposition of ammonium uranium fluoride. J Radioanal Nucl Chem Artic 111:423–428. https://doi.org/10.1007/BF02072874
32. Haendler HM, Wheeler CM, Robinson DW (1952) The thermal decomposition of ammonium heptafluorozirconate(IV). J Am Chem Soc 74:2352–2353. https://doi.org/10.1021/ja01129a051
33. Sangwal K (2007) Additives and crystallization processes. Wiley, England
34. Madan RL (2011) Chemistry for degree students. S. Chand Publishing Ltd., New Delhi

Impurity Uptake During Cooling Crystallization of Nickel Sulfate

Ina Beate Jenssen, Seniz Ucar, Oluf Bøckman, Ole Morten Dotterud and Jens-Petter Andreassen

Abstract Battery production for electric vehicles has increased rapidly in the last decade, and the projections show an ever-increasing demand for the next years. This leads to a need for recycling of valuable metals used in batteries, such as nickel, for a sustainable development. Nickel can be recovered from process streams or side streams in hydrometallurgical industry as nickel sulfate by crystallization, in order to obtain high-purity products for applications like battery-grade nickel sulfate. In this work, $NiSO_4 \cdot 6H_2O$ was precipitated by cooling crystallization in a temperature-controlled batch reactor, and uptake of Na, Cl, and Mg, impurities commonly found in hydrometallurgical industries, were investigated. Increasing the initial impurity concentrations in the reactor caused increased impurity content in the precipitates, and the highest uptake was provided by Mg. Characterization of final products with X-ray diffraction (XRD) showed that the precipitating polymorph was altered at high impurity concentration. The results of partial dissolution studies indicated that uptake of Cl was negligible, and Na was weakly adsorbed on the crystal surface, while Mg was also incorporated in the bulk of the crystals, possibly explained by the similar properties of Mg and Ni.

Keywords Nickel sulfate · Cooling crystallization · Impurity uptake

I. B. Jenssen · S. Ucar · J.-P. Andreassen (✉)
Norwegian University of Science and Technology, Trondheim, Norway
e-mail: jens-petter.andreassen@ntnu.no

I. B. Jenssen
e-mail: ina.b.jenssen@ntnu.no

S. Ucar
e-mail: seniz.ucar@ntnu.no

O. Bøckman · O. M. Dotterud
Glencore Nikkelverk AS, Kristiansand, Norway
e-mail: oluf.bockman@glencore.no

O. M. Dotterud
e-mail: ole.dotterud@glencore.no

G. Azimi et al. (eds.), *Rare Metal Technology 2020*, The Minerals, Metals & Materials Series, https://doi.org/10.1007/978-3-030-36758-9_18

Introduction

During the last decades, the battery production for electric vehicles has increased dramatically as a result of the increasing trend of replacement of fossil fuels to meet the environmental targets set by, amongst others, EU [1, 2]. The increasing production of batteries demands an increase of the supply chain [3, 4], hence recycling of valuable metals used in batteries. One of these valuable metals is nickel, which can be recycled from spent batteries, spent catalysts, and waste streams from hydrometallurgical industry [5, 6]. In addition, recycling of nickel contributes to reducing the environmental pollution caused by nickel-containing wastes, which would be advantageous due to the potential pathological effects of nickel [5].

In hydrometallurgical industries, nickel can be contained in, for example, bleed-off electrolytes obtained from electro-refining [5, 7, 8]. Crystallization is an effective method to separate nickel from such solutions as a high-purity product, such as battery-grade nickel sulfate [7, 8]. Nickel sulfate can be crystallized as a hexa- or heptahydrate from neutral solutions. Nickel sulfate heptahydrate ($NiSO_4{\cdot}7H_2O$) typically forms below 31.5 °C, whereas above this temperature, nickel sulfate hexahydrate ($NiSO_4{\cdot}6H_2O$) is the most commonly forming phase. $NiSO_4{\cdot}6H_2O$ can be crystallized into two polymorphs, as tetragonal α-$NiSO_4{\cdot}6H_2O$ and monoclinic β-$NiSO_4{\cdot}6H_2O$ [9–11].

Industrial solutions contain various impurities, which may have significant effects on the crystallization process and the end-product in terms of phase purity, yield, particle size, etc. [12]. In order to be able to produce high-purity nickel sulfate from industrial solutions, it is necessary to understand the mechanisms of the impurity uptake and how it affects the crystallization kinetics. Recently, cooling crystallization of nickel sulfate and the effect of impurities on the nickel recovery has been investigated, and the results indicated that cooling crystallization can be employed to obtain battery-grade $NiSO_4{\cdot}6H_2O$ from industrial solutions [6]. However, further investigations are needed to elucidate the impurity–crystal interactions, which in turn dictates the phase purity of the crystallization product. It has been observed that parameters such as chloride concentration and cooling rates may affect the forming crystal phase [6]. Therefore, in this work we investigated the effects of the initial concentration of some commonly found impurities in the metal industry, sodium, chloride, and magnesium, on the impurity uptake during cooling crystallization of nickel sulfate at comparable supersaturation.

Materials and Experimental Methods

Cooling crystallization experiments were performed using a double-walled 100 mL batch reactor equipped with baffles attached to the lid. A pH probe connected to a Mettler Toledo SevenExcellence multiparameter module that was connected to LabX direct software was used to log the pH and temperature during the experiments. The temperature and cooling rate of the reactor during the experiments was controlled by a Julabo refrigerated/heating circulator. Proper mixing was ensured by a magnetic stirrer operating at 600 rpm.

Based on the contents of impurities in a typical bleed-off electrolyte solution from the industry which is considered as potential feed for nickel sulfate production, the impurities chosen in this study were Na, Cl, and Mg. Solutions containing known concentrations of impurities were prepared using $NiSO_4{\cdot}6H_2O$ (reagent grade, $\geq$98%, Sigma-Aldrich), $MgSO_4$ (anhydrous, reagent grade, $\geq$97%, Sigma-Aldrich), Na_2SO_4 (anhydrous, EMSURE® ACS, ISO, Reag. Ph Eur, Merck), and $NiCl_2$ (98%, Sigma-Aldrich). All solutions were prepared with distilled water.

Initial tests were performed in order to find a nickel sulfate concentration that allowed for seeding at 25 °C, without spontaneous crystallization taking place. For all experiments, solutions of 0.53 g $NiSO_4{\cdot}6H_2O$/g solution were prepared. Amounts of Na_2SO_4, $NiCl_2$, and $MgSO_4$ were added in order to obtain four different initial impurity concentrations (0, 3, 5, and 9 mg impurity/g solution), referred to as experiment no. 1–4, where exp no. 1 had the lowest initial impurity concentrations, and exp no. 4 contained the highest. Three replicates were performed for each initial impurity concentration. Calculations of solution speciation with the thermodynamic software Visual Minteq showed that the presence of additives did not have any significant effects on solution supersaturation. The solutions were heated to 60 °C to completely dissolve the salts and then added to the preheated reactor. All experiments were performed by cooling from 60 °C to 25 °C with a cooling time of 2 h, hence keeping the supersaturation profile the same for the experiments. Once the temperature of the solution reached 25 °C, 0.2 g of seeds (<40 μm, prepared from commercial $NiSO_4{\cdot}6H_2O$ by sieving) was added to the reactor, and the suspension was left for crystal growth and subsequent stabilization for 30 min. $NiSO_4{\cdot}6H_2O$ was used in order to prevent crystallization of $NiSO_4{\cdot}7H_2O$ which may be unstable at room temperature [13] and have an unfavorable crystal shape [14].

At the end of each experiment, samples of the suspension were taken out from the reactor, in order to characterize the precipitated nickel sulfate and measure the concentrations of impurities. The samples were filtered, washed with ethanol, and dried at both 25 °C and 40 °C for 48 h. Characterization of solid phases were conducted via powder X-ray Diffraction (XRD) (D8 Da Vinci, Bruker AXS GmBH) in the 2θ range of 5–75° with a step size of 0.013° and a step time of 0.67 s. Fourier transform infrared (FTIR) (Tensor, Bruker AXS GmBH) spectra of powder samples were collected between 4000 and 550 cm-1. Raman microspectroscopy (InVia Reflex, Renishaw) was performed (2 s integration time, 30 accumulations) using a 535 nm laser through a 10× lens. Energy-dispersive X-ray (EDX) spectroscopy (Xmax, Oxford) was performed at an accelerating voltage of 10 kV. The morphology of the nickel sulfate crystals was examined by scanning electron microscopy (SEM) (Hitachi S-3400 N). Prior to SEM analyses, the samples were placed on carbon tape and sputter coated with gold. High-resolution inductively coupled plasma mass spectrometry (HR-ICP-MS) was used to determine the concentrations of the impurities Na, Cl, and Mg in the nickel sulfate crystals. Samples for HR-ICP-MS were prepared by dissolving solid samples in 0.1 M HNO_3.

In order to investigate if the impurities were adsorbed on the surface of the nickel sulfate crystals or incorporated in the crystal lattice, samples of the dried nickel sulfate obtained from the cooling crystallization experiments were partly dissolved in ultrapure water, then filtered and dried, and the impurity concentrations in the solid were again measured by HR-ICP-MS.

Results and Discussion

XRD analyses showed that for all experiments the final product was $NiSO_4{\cdot}6H_2O$, and that the products did not undergo transformation during drying at 40 °C. However, the precipitating polymorph varied with the increasing initial concentration of impurities in the reaction solution. The tetragonal α-polymorph of $NiSO_4{\cdot}6H_2O$ was precipitated in experiment no. 1–3, while in experiment no. 4, with highest impurity concentration, monoclinic β-$NiSO_4{\cdot}6H_2O$ was formed (Fig. 1a).

The XRD patterns obtained from experiments no. 1–3 show a peak broadening and decreased peak intensity with increasing initial impurity concentrations (Fig. 1b). A small shift of the peaks is also observed for the spectra obtained for precipitate from experiment no. 3. These phenomena might indicate lattice strain in the crystal structure caused by the uptake of impurities in the precipitated nickel sulfate [15]. The formation of β-$NiSO_4{\cdot}6H_2O$ could be a result of the high concentration of impurities that may influence which of the polymorphs is formed. Precipitation of nickel sulfate in the presence of high concentrations of Na and Cl (9 mg/g), without Mg, resulted in crystallization of α-$NiSO_4{\cdot}6H_2O$, which indicates that Cl might not be the reason for formation of β-$NiSO_4{\cdot}6H_2O$, as previously reported [6]. One possible explanation for the formation of β-$NiSO_4{\cdot}6H_2O$ is that

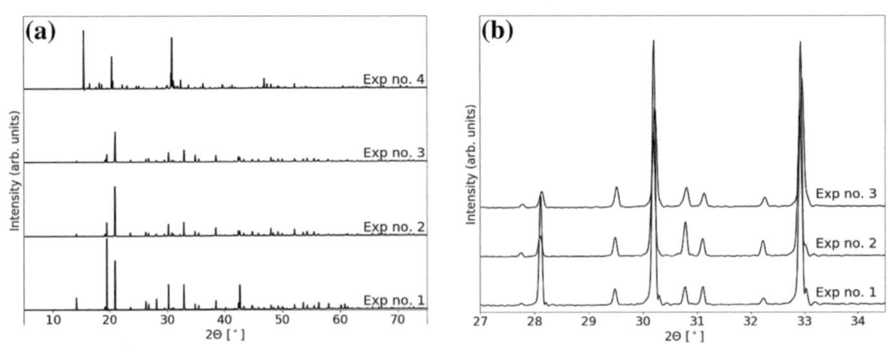

Fig. 1 XRD patterns of the precipitated nickel sulfate for each experiment with varying amounts of impurities, total spectra (**a**) and a magnified in (**b**)

incorporation of Mg alters the solubility of one of the polymorphs, which then results in a change in the final precipitating phase. The same phenomenon has been observed in $CaCO_3$ crystallization in the presence of Mg [16]. Another possible explanation for the formation of this polymorph could be that nucleation may have taken place before reaching 25 °C, at a higher temperature than for experiments no. 1–3, due to higher ionic strength of this solution caused by high impurity concentration, since β-$NiSO_4 \cdot 6H_2O$ has been reported to form at higher temperatures [9]. However, this was difficult to observe, due to the dark green color of the solution in the reactor.

Observations from scanning electron microscopy (SEM) images (Fig. 2) show a clear difference in the crystal shape of the nickel sulfate precipitates from experiment no. 1–3 (α-$NiSO_4 \cdot 6H_2O$), compared to the precipitate from experiment no. 4 (β-$NiSO_4 \cdot 6H_2O$), containing the highest initial impurity concentration, which is in accordance with the observations from XRD. This difference observed for the two polymorphs is also in agreement with morphology observations found recently [6].

Concentrations of the impurities Na, Cl, and Mg in the precipitates were analyzed by HR-ICP-MS, and the amounts of impurity per mass of solid were

Fig. 2 SEM images of the precipitated $NiSO_4 \cdot 6H_2O$ from each experiment containing different amounts of impurities, **a** Exp no. 1, **b** Exp no. 2, **c** Exp no. 3, **d** Exp no. 4

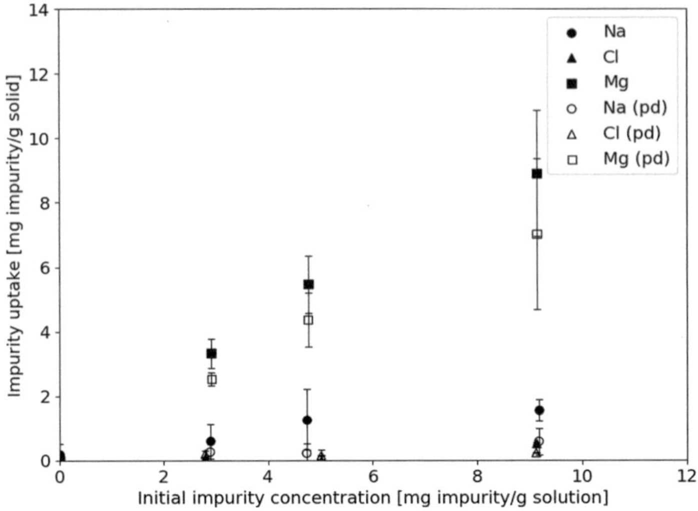

Fig. 3 Influence of initial impurity concentrations in solution on the uptake of impurities Na, Cl, and Mg in the nickel sulfate precipitated during cooling crystallization. The filled markers represent the concentrations of impurities in the precipitated nickel sulfate crystals, and the ones marked (pd) represent the impurity concentrations in the nickel sulfate crystals after partly dissolving the crystals in ultrapure water. The error bars represent standard deviation for three replicates

calculated from Eq. (1) and plotted against the initial impurity concentration in the solutions (Fig. 3).

$$\text{Impurity uptake} = \frac{m_{impurity}}{m_{solid}} \tag{1}$$

The concentrations of Cl in the solutions analyzed by HR-ICP-MS were lower than or close to the detection limit of the instrument (300 μg/L), and thus uptake of Cl can be regarded as negligible. Na and Mg uptakes were within the detectable range, where Mg showed the highest uptake and a linear dependency to the initial impurity concentration in solution. These results imply that Cl is completely removed during the filtration and washing of the crystals, while Na and Mg are more difficult to wash off from the precipitated nickel sulfate.

In order to investigate the mechanism for the uptake of impurities, small amounts of the precipitated samples were partly dissolved in ultrapure water, dissolving approx. 50 wt% of the crystals. The samples were then filtered and dried and analyzed for their impurity contents again. The changes in the impurity contents before and after this partial dissolution/washing of the samples may give an indication if the impurities are incorporated in the bulk of the crystals or adsorbed on the surface. From Fig. 3, it is seen that in all samples, Na is almost completely washed off, indicating that it was loosely adsorbed on the surface of the nickel sulfate and easily removed during this second washing procedure. The concentrations of Mg detected in the solid also decreased; however, significant amounts of Mg were still present in the samples, indicating that Mg could have been both adsorbed on the crystal surface and incorporated in the bulk, in agreement with the previous reports of Mg uptake and removal from nickel sulfates [6, 17]. It must be noted that the seeds added in the experiments may have an impact on the results for the impurity concentrations of the nickel sulfate. Because the seeds are pure nickel sulfate, the impurities will not be equally distributed in the total nickel sulfate crystal, and by partly dissolving the crystals, a certain amount of the outer part (containing impurities) will be removed, and thus the impurity/solid ratio will decrease.

EDX spectroscopy of samples from exp no. 4 verified the Mg content along with the precursor ions of $NiSO_4{\cdot}6H_2O$ (Fig. 4a), which further demonstrated the presence of Mg in the bulk of the crystal since the depth of beam penetration is estimated to be between 0.5 and 1.0 μm at the corresponding analysis conditions

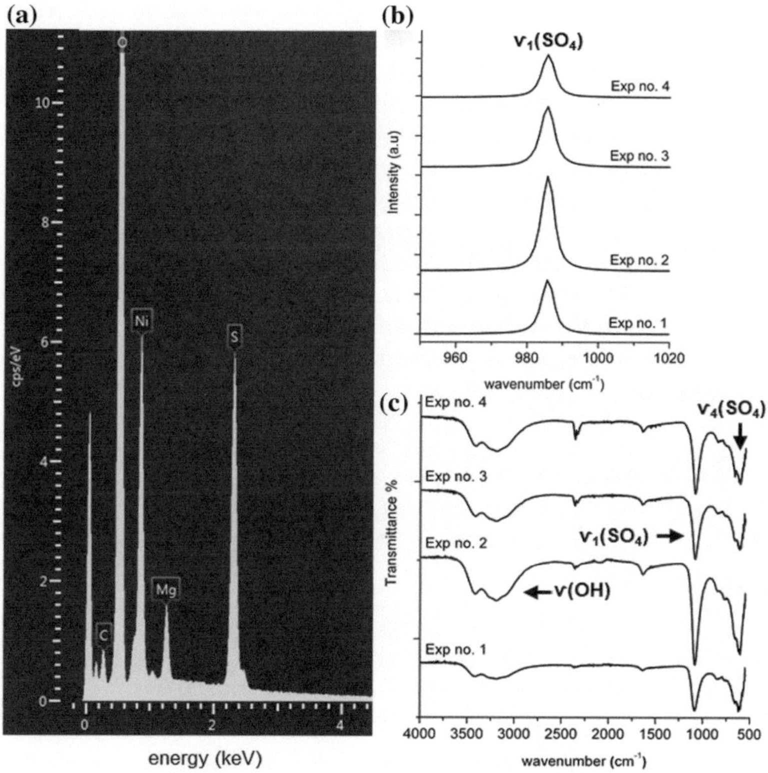

Fig. 4 **a** EDX spectrum of precipitate from exp no. 4, and **b** Raman and **c** FTIR spectra of the precipitated $NiSO_4{\cdot}6H_2O$ from each experiment. Characteristic Raman and FTIR peaks of $NiSO_4{\cdot}6H_2O$ are labeled on the spectra

[18]. The similar ionic radii (86 pm and 83 pm, respectively) and divalent charge of Mg and Ni can explain the higher uptake and stronger inclusion of Mg in the precipitates compared to Na^+ and Cl^- ions.

Precipitates were further characterized with FTIR and Raman spectroscopy to investigate any disorder or structural changes due to the incorporation of an impurity ion into the host lattice. However, neither the characteristic peaks nor the relative peak intensities showed any change in correlation with the impurity concentration (Fig. 4). These results suggest that Mg presence did not induce a significant strain in the crystal, in contrast to the indications from the XRD data. The broadening of the XRD peaks can also be explained by changes in crystallinity; however, further explanations cannot be offered within the experimental data of this study.

Conclusions

$NiSO_4{\cdot}6H_2O$ was precipitated by cooling crystallization from 60 °C to 25 °C, and the effect of impurity concentration of Na, Cl, and Mg on the impurity uptake was investigated. Two polymorphs of nickel sulfate were formed, depending on the initial experimental conditions. In experiments with no impurities or low initial concentrations of impurities, α-$NiSO_4{\cdot}6H_2O$ was precipitated, while for high initial impurity concentration β-$NiSO_4{\cdot}6H_2O$ was the resulting phase. The preliminary results indicate that Cl uptake was negligible, whereas significant amounts of Na and Mg were detected in the final crystallization products. Partial dissolution studies showed that Na was adsorbed on the surface of the crystals and could easily be removed with further washing, while Mg was more difficult to remove, and found both as adsorbed on the crystal surface and incorporated in the crystal bulk. However, further studies are necessary to fully understand the mechanisms for uptake of impurities in $NiSO_4{\cdot}6H_2O$.

Acknowledgements The authors thank the Norwegian Research Council, Yara, Glencore Nikkelverk AS, and Boliden Odda for financing the project (project number 236674).

References

1. Berckmans G et al (2017) Cost projection of state of the art lithium-ion batteries for electric vehicles up to 2030. Energies 10(9):1314
2. Communities, C.o.t.E., White paper: European transport policy for 2010: Time to decide. 2001: The Commission
3. Bloomberg New Energy Finance (2017) Electric vehicle outlook 2017. Bloomberg Finance L. P, Technical Report, 16 July 2019. http://data.bloomberglp.com/bnef/sites/14/2017/07/BNEF_EVO_2017_ExecutiveSummary.pdf
4. Nitta N et al (2015) Li-ion battery materials: present and future. Mater Today 18(5):252–264

5. Coman V, Robotin B, Ilea P (2013) Nickel recovery/removal from industrial wastes: a review. Resour Conserv Recycl 73:229–238
6. Han B et al (2019) Purification of nickel sulfate by batch cooling crystallization. Chem Eng Technol 42(7):1475–1480
7. Agrawal A et al (2008) Extractive separation of copper and nickel from copper bleed stream by solvent extraction route. Miner Eng 21(15):1126–1130
8. Nyirenda RL, Phiri WS (1998) The removal of nickel from copper electrorefining bleed-off electrolyte. Miner Eng 11(1):23–37
9. Steele BD, Johnson F (1904) XIII—The solubility curves of the hydrates of nickel sulphate. J Chem Soc Trans 85, 113–120
10. Angel R, Finger L (1988) Polymorphism of nickel sulfate hexahydrate. Acta Crystallogr C 44 (11):1869–1873
11. Seidell A (1940) Solubilities of inorganic and metal organic compounds: a compilation of quantitative solubility data. D. van Nostrand
12. Mullin JW (2001) Crystallization. Elsevier Science
13. Friesen M, Burt HM, Mitchell AG (1980) The dehydration of nickel sulfate. Thermochim Acta 41(2):167–174
14. Madsen RSK et al (2019) Novel membrane percrystallisation process for nickel sulphate production. Hydrometallurgy 185:210–217
15. Che M, Vedrine JC (2012) Characterization of solid materials and heterogeneous catalysts: from structure to surface reactivity. Wiley
16. Berner RA (1975) The role of magnesium in the crystal growth of calcite and aragonite from sea water. Geochim Cosmochim Acta 39(4):489–504
17. Lu X (2014) Eutectic freeze crystallization of nickel sulfate from an industrial stream: a comparison with evaporative crystallization. In: Novel applications of eutectic freeze crystallization, p 13
18. Nagy IZ et al (1977) Energy dispersive X-ray microanalysis of the electrolytes in biological bulk specimen: I. specimen preparation, beam penetration, and quantitative analysis. J Ultrastruct Res 58(1):22–33

Potential of a Nigerian Cassiterite Ore for Industrial Steel Coatings

Alafara A. Baba, Abdulrasheed O. Yusuf, Mustapha A. Raji, Kuranga I. Ayinla, Abdullah S. Ibrahim, Folahan A. Adekola, Abdul G. F. Alabi, Christianah O. Adeyemi, Sadisu Girigisu, Rasaki A. Gbadamosi and Aishat Y. Abdulkareem

Abstract As the world demand for pure tin and its compounds are quite steady with an estimated projection of 5% per annum and coupled with its wide array of applications in iron and steel industries in Sub-Sahara Africa, the need for continuous ore purification has become paramount. Thus, the treatment of Nigerian cassiterite ore for industrial value addition by selective acid leaching and solvent extraction was investigated. The effects of leachant concentration, temperature, and particle size on ore dissolution rates were examined. At optimal leaching conditions, 76.7% ore reaction was achieved within 120 min. Tin recovery from the leach liquor by 0.5 mol/L tributyl phosphate (TBP) in kerosene at 27 ± 2 °C gave 89.31% efficiency. Stripping of tin from organic-loaded phase recorded 98.9% purity and beneficiated to obtain high-grade tin chloride ($SnCl_2$, *melting point*: 239.7 °C; *density*: 3.89 g/cm^3), suitable as steel coating indices in some defined indigenous steel industry.

Keywords Cassiterite ore · Nigeria · Pure tin · Tin chloride · Steel coatings Hydrometallurgical process

---

A. A. Baba (✉) · A. O. Yusuf · M. A. Raji · K. I. Ayinla · A. S. Ibrahim · F. A. Adekola · C. O. Adeyemi · R. A. Gbadamosi
Department of Industrial Chemistry, University of Ilorin, P.M.B. 1515, Ilorin 240003, Nigeria
e-mail: baalafara@yahoo.com

A. G. F. Alabi
Department of Material Science and Engineering, Kwara State University, P.M.B. 1530, Malete, Nigeria

C. O. Adeyemi · S. Girigisu
Science Laboratory Technology Department, Federal Polytechnic Offa, P.M.B. 420, Offa, Nigeria

A. Y. Abdulkareem
National Mathematical Centre, Sheda-Kwali, P.M.B. 118, Abuja, Nigeria

G. Azimi et al. (eds.), *Rare Metal Technology 2020*, The Minerals, Metals & Materials Series, https://doi.org/10.1007/978-3-030-36758-9_19

Introduction

The fast development of science and technology coupled with some distinct properties of tin (Sn) such as malleability, ductility, and resistance to corrosion makes pure tin metal useful for a wide array of utilizations in tin plating, phone pads, alloying, and metals coating among others. In addition to these, the consumption of pure tin metal and its compounds are increasing tremendously as a result of its versatile end uses [1] in which the global price is equivalent to about US $22,000 per metric ton, affirming it about seven times as valuable as refined zinc [2]. In recent times, the major uses of tin are in industrial coatings on steel (27% Sn) most importantly in canning foods and drinks, electrical and electronic industries (23% Sn), in alloys such as bronze (8.5–16% Sn) as well as other pertinent engineering alloys. In placing tin metal values to address the aforementioned applications, efforts should be geared on sourcing for high-grade tin varieties such as cassiterite ore as its vast deposits are widely spread globally especially in Africa and most importantly in Nigeria where its proven reserves are currently put at US$370/ton [3].

Thus, cassiterite, the primary mineral, occurs in the form of stannic oxide (SnO_2) with varying colors: black, brown, reddish-brown, or grayish-white is an important ore from which pure tin can be economically extracted [4]. This variation in color could be due to the presence of iron as the major gangue in combination with other impurities such as tantalum, tungsten, and niobium [5]. These gangues associated with cassiterite often have an adverse effect on the quality of tin concentrates which if not adequately removed may alter its industrial utilities [6].

However, in order to meet its defined industrial demands, most especially in the steelmaking industry, the cassiterite ore has to be purified from the aforementioned associated impurities. A number of techniques have been carefully investigated for the extraction and purification of cassiterite ore and varieties for industrial value additions through roasting–acid leaching [7–9], beneficiation–metallurgy route [10–13], reduction–smelting electrolysis [14], vacuum carbothermal reduction [15], and froth flotation [16]. These techniques present various setbacks, such as fatal arsenic pollution leading to high energy consumption as well as difficulty in separating metal values of interest from the associated impurities among others.

Compared to the aforementioned techniques, acid leaching appears to be cleaner, low-cost, eco-friendly, and effective and has been a 'win-win' technique utilized in the extraction and processing of tin [17] and addressing some of the aforementioned drawbacks. Considering its effectiveness, authors have employed acid leaching in the purification of tin ore varieties at optimized conditions in different media. For example, Baba et al. [3] reported that an increase in the HCl concentration gave rise to increase in the fraction of cassiterite ore reacted at various time intervals and that the reaction order obeys first order at defined leachant concentration (≤ 4 M HCl). Also, Yang et al. [18] have also successfully removed nearly 99.4% iron and other associated impurities from a low-grade tin concentrate at a leaching temperature of 35 °C within 2 h for extraction optimization investigations.

In this study, utilization of acid leaching by hydrochloric acid solution to selectively extract and purify tin from indigenous cassiterite ore for possible removal of associated impurities from the cassiterite leach pregnant solution was examined. Combinations of precipitation, solvent extraction, and crystallization processes were adopted to obtain high-grade tin chloride of industrial utilities.

Materials and Method

The cassiterite ore used for this investigation was sourced from *Lafiagi* (8°52'0''N 5°25'0''E), Edu Local Government Area of Kwara State, Nigeria. The sample was pulverized and sieved into three different particle sizes (−112 + 90, −90 + 75, and −75 + 43 μm) with ASTM standard sieves. Considering the sample surface area, all experiments were carried out with the smallest particle (−75 + 43 μm) assumed to have a larger surface area, unless otherwise stated [19]. Characterization of the raw sample and some selected products at optimized conditions were accordingly examined by X-ray diffraction (XRD), X-ray fluorescence (XRF), scanning electron microscopy (SEM), and atomic absorption spectroscopy (AAS).

Leaching Tests

A 600 ml Pyrex leaching reactor equipped with a thermometer to monitor the reaction temperature and a stirrer was used in this study. For each run, 10 g/L of the pulverized ore was added to the solution at defined lixiviant concentration and desired temperature at varied time intervals (5–120 min). The lixiviant concentration which gave the maximum dissolution efficiency was used for optimization of other leaching parameters such as reaction temperature and particle size. The fraction of ore reacted in the lixiviant was calculated from the initial difference between the mass of the reacted and un-reacted at specified conditions [20, 21]. Some selected products at 70 °C by 2.0 mol/L HCl solutions were accordingly examined through XRD, EDS, and SEM techniques for better assessment of the un-reacted species for ease of predicting the tin optimal extraction conditions.

Solvent Extraction and Beneficiation Tests

The amount of tin in the pregnant solution was spectrophotometrically determined [22]. The organic and aqueous phases were kept at a ratio of 1:1. A 30 ml of the pregnant solution (aqueous phase) and organic phase (TBP in kerosene) were appropriately mixed in a stoppered bottle and shaken for 20 min. After disengagement, the stripping of pure tin ion from organic-loaded phase recorded 98.9% purity and subsequently beneficiated to obtain a well-characterized high-grade tin chloride of industrial value [23].

Results and Discussion

Characterization Studies

The cassiterite ore examined by XRF contain 65.61 wt% Sn, 2.13 wt% W, 5.11 wt% Ca, 0.93 wt% Fe, 2.15 wt% Pb, and 2.50 wt% Nb. The mineral purity test by X-ray diffraction (XRD) affirmed admixture of *Cassiterite* ($Sn_{2.00}O_{4.00}$: 96-900-7534), *Galena* ($Pb_{4.00}S_{4.00}$: 96-901-3404), and *Rutile* ($Ti_{1.98}O_{4.00}$: 96-900-1682) as major compounds present in cassiterite ore body before acid treatment.

Leaching Results

The leaching investigations showed that the cassiterite ore dissolution in hydrochloric acid solution increases appreciably with increasing lixiviant concentration, reaction temperature, and decreasing particle diameter at moderate stirring as follows:

(i) **Effect of HCl concentration**: HCl concentration was varied between 0.1 and 2.5 mol/L at 50 °C within 5–120 min using the smallest particle size (−75 + 43 μm). Increasing the lixiviant (HCl) concentration appreciably increases the ore dissolution up till 2.0 mol/L HCl solutions. During this time, 48% of the ore reacted within 120 min. Further increase in lixiviant concentration to 2.5 mol/L resulted to a decrease in ore dissolution rate with reaction reaching 45%, which could be due to precipitation phenomenon at larger lixiviant concentration as summarized in Table 1a.

(ii) **Effect of reaction temperature**: The effect of reaction temperature on the amount of ore dissolution was carried out from 27 to 70 °C. Other conditions considered were 2.0 mol/L HCl, 5–120 min, and −75 + 43 μm particle size. Thus, increasing the reaction temperature from 27 to 70 °C appreciably increases the degree of dissolution from 19.0 to 77% as depicted in Table 1b, though the reaction temperature was not set beyond 70 °C because of undue evaporation at higher temperatures.

(iii) **Effect of Particle size**: The dissolution results for particle size variations: −75 + 43 μm −90 + 75 and −112 + 90, and yielded 76.7, 67.1, and 59.1%, respectively, at optimized conditions (2.0 mol/L HCl, 70 °C) for 120 min (Table 1c). The un-reacted product (∼23.3%) analyzed by XRD was found

Table 1 Fraction of cassiterite ore reacted at various conditions

Leaching time (min.)	0.1	0.5	1.0	1.5	2.0	2.5
	(a) HCl concentration (mol/L) at 50 °C					
0	0	0	0	0	0	0
5	0.04	0.07	0.10	0.12	0.15	0.14
10	0.08	0.11	0.14	0.16	0.19	0.18
30	0.10	0.14	0.18	0.21	0.26	0.23
60	0.12	0.16	0.20	0.27	0.34	0.33
120	0.15	0.18	0.25	0.35	0.48	0.45
	(b) Reaction temperature (°C)					
	27	40	50	60	70	
0	0	0	0	0	0	
5	0.05	0.07	0.15	0.19	0.25	
10	0.08	0.13	0.19	0.28	0.34	
30	0.13	0.19	0.26	0.36	0.46	
60	0.15	0.24	0.34	0.47	0.58	
120	0.19	0.29	0.48	0.59	0.76	
	(c) Particle size (μm)					
	−75 + 43	−90 + 75	−112 + 75			
0	0	0	0			
5	0.25	0.21	0.17			
10	0.34	0.29	0.23			
30	0.46	0.40	0.35			
60	0.58	0.52	0.46			
120	0.76	0.67	0.59			

to contain siliceous compound and could serve as a viable raw material for some defined industries.

The experimental data from the leaching studies were accordingly fitted into the well-established shrinking core models (chemical, diffusion, and mixed controls) to obtain the activation energy of 33.5 kJ/mol, supporting the proposed diffusion control mechanism for the dissolution process.

Solvent Extraction Investigations

The metal ion composition of the pregnant solution by 0.5 mol/L TBP in kerosene diluent at 27 ± 2 °C at optimal leaching conditions gave 2109.4 mg/L Sn, 118.1 mg/L Fe, 24.6 mg/L W, and 7.9 mg/L Nb. Prior to obtaining purified tin solution, iron and other associated impurities were initially precipitated at pH 3.5 to improve the extraction and beneficiation operations [21]. About 89.31% pure tin

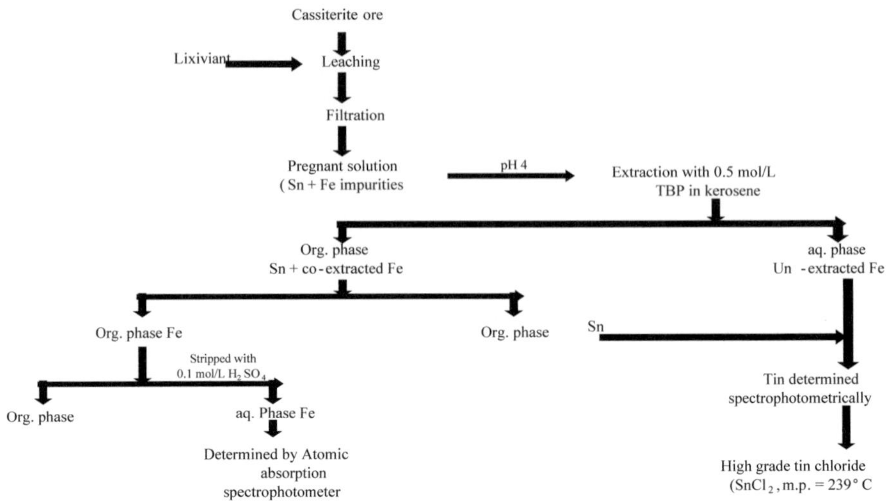

Fig. 1 Hydrometallurgical flow sheet for processing and treating a Nigerian cassiterite ore to obtain high-grade tin chloride of industrial utilities

solution was successfully extracted at pH range 4.0. The pure tin solution obtained was subsequently purified by precipitation and crystallization to obtain tin chloride of high purity (98.9%). The purified solution was accordingly beneficiated and well characterized to produce high-grade tin chloride ($SnCl_2$, melting point: 239.7 °C, density: 3.89 g/cm^3) suitable as steel coating indices in some defined indigenous steel industries. The hydrometallurgical technique for processing and purifying a typical Nigerian cassiterite ore to obtain high-grade tin chloride of industrial utilities is summarized in Fig. 1.

Conclusion

From this investigation, it can be affirmed that a cassiterite ore of Nigerian origin could be transformed to tin chloride of industrial utilities through combinations of leaching, solvent extraction, precipitation, and crystallization techniques from HCl acid pregnant solution by tributyl phosphate (TBP) extractant with an extraction efficiency of 89.31%. Hence, this investigation shows the possibility of producing high-grade, domestic, and industrially acceptable tin chloride ($SnCl_2$: 96-901-9553; *melting point* = *239.7* °C, *density* = *3.89* g/cm^3) suitable as steel coating material in some defined industries by hydrometallurgical technique. Thus, this study would contribute to the country's economy, industrialization as well as poverty reduction to the barest minimum.

Acknowledgements The authors wish to thank CSIR—Institute of Minerals and Materials Technology, Bhubaneswar-751013, India for their benevolence in supplying the TBP extractant used for this study; and the Central Research Laboratories (CRL), University of Ilorin, Nigeria for assisting in aqueous metal analyses by AAS. One of the authors (Prof. Alafara Abdullahi Baba, FCSN, FMSN) sincerely grateful to the Director Training, Mining Cadastral Office, Abuja—Nigeria; KAM Industries (Nig.) Ltd., Ilorin, Kwara State: One of the leading indigenous Steel Industries in Nigeria; Saolad Nig. Ltd. (Building Materials Merchants), Palm Avenue, Mushin, Lagos; and S. S. Amao and Sons Limited (Importer and Exporters of Motor Spare parts and facilities), Ijora, Lagos, Nigeria for their immeasurable supports to attend the TMS 2020 149th Annual Meeting and Exhibition in San Diego, California, U.S.A. between 23 and 27 February 2020.

References

1. Angadi SI, Sreenivas T, Jeon H, Baek S, Mishra BK (2015) A review of cassiterite beneficiation fundamentals and plant practices. Miner Eng 70:178–200
2. Ltd DC (2019, 08 04). Indium tin oxide touch screen. http://www.diamondcoatings.co.uk/indium-tin-oxide-touch-screen/
3. Baba AA, Adekola FA, Bale RB (2009) Study of dissolution kinetics of a Nigerian cassiterite ore by hydrochloric acid. Sci Focus 14(2):198–207
4. Houchin MR (1985) Surface titrations and electrokinetic measurements on Australian cassiterites. Colloids Surf 16:117–126
5. Bulatovic S, De Silvio E (2000) Process development for impurity removal from a tin gravity concentrate. Miner Eng 13(8–9):871–879
6. Feng Q, Wen S, Zhao W, Chen Y (2018) Effect of calcium ions on adsorption of sodium oleate onto cassiterite and quartz surfaces and implications for their flotation separation. Sep Purif Technol 200:300–306
7. Lu H, Yin J, Zhang Q, Zhang M (2013) Thermodynamics of separation stibium from tin anode slime by sulfidation roasting. Chin. J. Rare Metal. 37:845–859
8. Havuz T, Donmez B, Celik C (2010) Optimization of removal of lead from bearing-lead anode slime. J Ind Eng Chem 16:355–358
9. Li D, Guo X, Xu Z, Tian Q, Feng Q (2015) Leaching behaviour of metals from copper anode slime using an alkali fusion-leaching process. Hydrometallurgy 157:9–12
10. Cao H, Chen J, Yuan H, Zheng G (2010) Preparation of pure $SbCl_3$ from lead anode slime bearing high antimony and low silver. Trans Nonferr Mater Soc 20:2397–2403
11. Han J, Liu W, Wang D, Jiao F, Qin W (2016) Selective sulfidation of lead smelter slag with sulfur. Metall Mater Trans B 47:344–354
12. Su Z, Zhang Y, Liu B, Zhou Y, Jiang T, Li G (2016) Reduction behaviour of SnO_2 in the tin-bearing iron concentrates under CO-CO_2 atmosphere. Part 1: effect of magnetite. Powder Technol 292:251–259
13. Han J, Jiao F, Liu W, Qin W, Xu T, Xue K, Zhang T (2016) Innovative methodology for comprehensive utilization of spent MgO-Cr_2O_3 bricks: copper flotation. ACS Sustain Chem Eng 4:5503–5510
14. Belousova TI, Sotnikov KY, Psikov VS, Vasenin KY (2002) Reducing the amount of tin lost with the slime from a sulfamine tinning electrolyte. Metallurgist 46:350–353
15. Li W, Guo W, Qin K (2013) Vacuum carbothermal reduction for treating tin anode slime. JOM 65:1608–1614
16. Lalasari, L.H., Suharyanto, A., Firdiyomo, F. (2016). The effect of pretreatments on the dissolutions of impurities from Indonesian cassiterite mineral. IOP Conf Ser J Phys Conf. Ser 817. https://doi.org/10.1088/1742-6596/817/1/012064

17. Han J, Liang C, Liu W, Qin W, Jiao F, Li W (2017) Pretreatment of tin anode slime using alkaline pressure oxidative leaching. Sep Purif Technol 174:389–395
18. Yang J, Wu Y, Zhang X (2014) Study on separation of tin from a low-grade tin concentrate through leaching and low-temperature smelting processes. Miner Process Extr Metall (Trans Ins Met C) 123(4):228–233
19. Gerald O, Priscillia U, Martin O, Emmanuel E, Chukwuka N, Hwang JY (2012) Model for prediction of the concentration of extracted tin during leaching of cassiterite in potassium hydroxide solution. J Miner Mater Charact Eng 11:730–734
20. Baba AA, Raji MA, Muhammed OM, Abdulkareem AY, Olasinde FT, Ayinla IK, Adekola FA, Bale RB (2019) Potential of a Nigerian biotite-rich kaolinite ore to industrial alumina by hydrometallurgical process. J Metall Res Technol (Formerly Revue de Metallurgie) 116:222. https://doi.org/10.1051/metal/2018076
21. Baba AA, Muhammed OM, Raji MA, Ayinla IK, Abdulkareem AY, Lawal M, Adekola FA, Alabi AGF, Bale RB (2018) Purification of a Nigerian wolframite ore for improved industrial applications. In: Rare metal technology 2018, pp 265–272
22. Falk H (2007) Analytical capabilities of atomic spectrometric methods using tunable lasers: a theoretical approach. Prog Anal Spectrosc 3(181):208
23. Baba AA, Adekola FA, Bale RB (2009) Development of a combined pyro- and hydro-metallurgical route to treat spent zinc-carbon batteries. J Hazard Mater 171:838–844

The Iron Precipitate from Primary Zinc Production: A Potential Future Source for Indium and Silver

Stefan Steinlechner and Lukas Höber

Abstract Indium as a so-called hitchhiker metal is extracted almost entirely in course of the primary zinc production. Thus, a strong dependence on the output of the corresponding base metal is given. Taking the annual growth rate of approximately 3% in zinc production and the one of indium, respectively, around 15% into account, it is obvious that a shortage in indium is faced. Because of the relative new application area of indium in high-tech products, a great potential of dumped metal is present in the residues of the past production. Therefore, the present paper describes the characterization and the possible treatment of such an iron precipitation residue, including the thermodynamic background for a pyrometallurgical extraction of indium and silver.

Keywords Jarosite · Silver · Indium · Vaporization · Combined recycling

Introduction

History illustrates the importance of metals very well by the fact that in the past entire periods of time were named after the mainly utilized metals and alloys, like bronze or iron [1, 2]. Although it is not perceived by the broad society, base metals are still of significant importance today and more than ever; they form the basis of our standard of living.

A wide variety of scarce metals that are of considerable relevance to the European high-tech industry, rapidly grown in their consumption in the last decades, are supplied as a by-product by the nonferrous base metal industry. For instance, the global energy system, IT infrastructures, automobiles, aircraft, fiber optics, or mobile phones are just a few technologies in modern life utilizing various

S. Steinlechner (✉) · L. Höber
Chair of Nonferrous Metallurgy, Montanuniversität Leoben, Franz-Josef-Str. 18, 8700 Leoben, Austria
e-mail: stefan.steinlechner@unileoben.ac.at

G. Azimi et al. (eds.), *Rare Metal Technology 2020*, The Minerals, Metals & Materials Series, https://doi.org/10.1007/978-3-030-36758-9_20

nonferrous metals such as copper, zinc, rare earth elements, precious metals, indium, or cobalt to ensure their technical functions [3, 4].

Most of those scarce elements are produced in course of a base metal production, which acts as a carrier metal, thus having no own specific mineral mined. In general, they are of less economic value for the metallurgical plant than the carrier metal. Typically, they do not constitute the main product of the production site with regard to revenue and thus only indirectly influence the production quantity. As a consequence, the amount of by-product produced is directly linked to the quantity of carrier metal, so it can be difficult to react to demand peaks for such scarce metals, which in turn can lead to a critical supply situation, like in case of indium [4].

Since the implementation of the production technology for the extraction of such minor metals from complex concentrates is a rather new field, industrial residues were landfilled for decades and by-products as well as intermediate process streams are still generated today without the extraction of these elements. Further reasons for this lagging behind of extraction technology are the comparable complex structure, partly missing know-how about their element distribution in the phases of the material, and with this required challenging technology development for the successful treatment of residue or industrial intermediates.

One prevailing example, described in this paper, is the iron precipitate residue from primary zinc production. The predominant applied technology is the jarosite precipitation, aside of goethite and relatively rare applied haematite process. Considering the technology and chemical reactions of an iron removal, potentially present indium in the treated zinc concentrate is accumulated in the jarosite precipitate. Based on the formation process of jarosite also silver can be present, due to the introduced calcined concentrate utilized for pH adjustment. Although both metals are typically present in an amount of only some hundred ppm, their potential contribution to the economic viability of a possible recycling process is significant. In this context, the paper describes the formation of the iron precipitate—jarosite—the fundamentals of a possible process for the extraction of indium and silver via selective vaporization and hand in hand with that outlining the future potential of making this residue a secondary raw material resource for indium and accompanied silver.

Materials and Methods

Before 1950, zinc production was dominated by the pyrometallurgical horizontal retort process. A turning point came at the end of the 60 s, when the iron precipitation processes were developed, which offered the possibility to leach zinc and iron simultaneously followed by a selective extraction of the disturbing element iron [5, 6]. As early as the end of the twentieth century, more than 90% of the zinc was produced by hydrometallurgical routes utilizing this development, thus generating huge amount of iron precipitation residue as well [7]. In the following, the formation process is outlined, showing in which of the phases of the residue indium and silver theoretically can be found. Additionally, the thermodynamic background of an investigated selective vaporization process for indium and silver is described.

Formation Process of the Investigated Material—Jarosite

One of the most utilized precipitation processes is the so-called jarosite process, which is outlined in Fig. 1 more in detail.

Jarosite is a complex basic iron sulphate with the chemical compound $R_2Fe_6(OH)_{12}(SO_4)_4$, where R is a cation [7]. Typical cations are K^+, Na^+, NH_4^+, Ag^+, H_3O^+, or R_2 that can be Pb^{2+}. Equations 1–3 summarize the reactions during the formation [8]:

$$3Fe_2(SO_4)_3 + 6H_2O \rightarrow 6Fe(OH)SO_4 + 3H_2SO_4 \quad (1)$$

$$4Fe(OH)SO_4 + 4H_2O \rightarrow 2Fe_2(OH)_4SO_4 + 2H_2SO_4 \quad (2)$$

$$2Fe(OH)SO_4 + 2Fe_2(OH)_4SO_4 + 2NH_4OH \rightarrow (NH_4)_2Fe_6(SO_4)_4(OH)_{12} \quad (3)$$

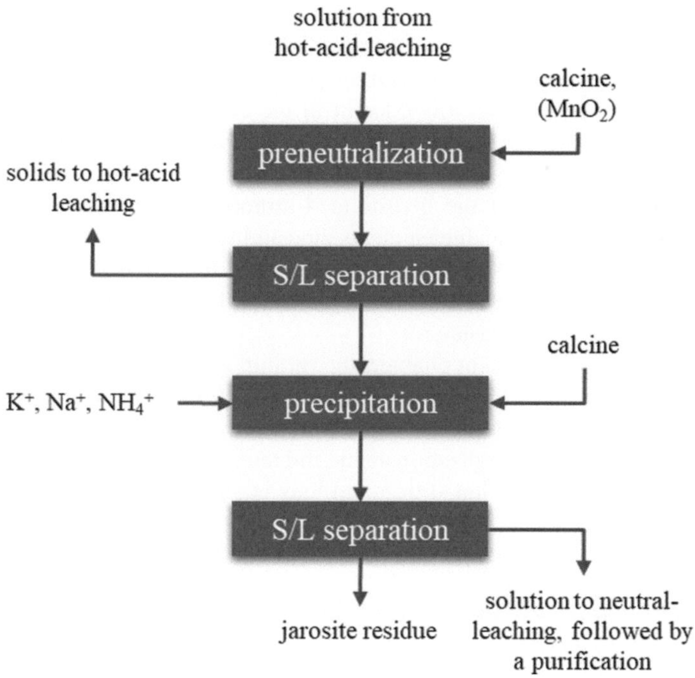

Fig. 1 Block diagram of the jarosite precipitation in course of the primary zinc production

Table 1 Elemental analysis of investigated jarosite

Jarosite		
Element	Concentration [wt%]	Method
Ag	0.018	DIN EN ISO 11885
In	0.023	DIN EN ISO 11885
Fe	27.1	DIN EN ISO 11885
Zn	6.5	DIN EN ISO 11885
Pb	6.2	DIN EN ISO 11885
S	8.4	DIN EN ISO 15350
Cl	<0.1	DIN 38405 D1-2
F	0.01	DIN 38405 Teil1

Combining the above-mentioned reactions, the overall reaction can be stated as shown in Eq. 4 [8].

$$3Fe_2(SO_4)_3 + 10H_2O + 2NH_4OH \rightarrow (NH_4)_2Fe_6(SO_4)_4(OH)_{12} + 5H_2SO_4 \quad (4)$$

It can be seen that sulfuric acid is formed during this reaction, leading to the necessity of the addition of calcine, an additional input stream in the leaching circuit, for neutralization. This, together with the necessity to increase the pH to perform the precipitation process, is the reason why the jarosite—and the same holds true for goethite—always carries a certain amount of impurities like zinc ferrite, Pb, Ag, In, traces of gold, and gangue compounds introduced by the calcine or concentrate, which dissolves only partly. Additionally, the circumstance that in some production plants, no separation of the Pb–Ag residue from the hot-acid leaching is carried out, due to low content in the concentrate or also in favour of lower investment costs, results in an increased lead and silver content in the iron precipitate [7].

An elemental analysis of the investigated ammonium jarosite, summarized in Table 1, shows the main valuable elements zinc and lead, with a content of 6.5 wt% and 6.2 wt%, respectively. Additionally, 180 ppm of silver and 200 ppm of indium were detected. The main elements are represented by iron and sulphur as main constituents of the jarosite structure.

So as outlined in the present chapter, the indium as well as the silver partly can substitute the cation place in the jarosite structure. Additionally, they are present partially in the leached calcine as well, which was utilized for pH adjustment. Because of the usage as neutralization agent, the remaining part of the calcine cannot be separated from the precipitated iron and thus is dumped together as a sludge.

Concept for the Separation of Indium and Silver

The following considerations are based on the fact that indium is extracted mainly (>96%) [4] in course of the primary zinc production. As a result, this industry sector, at least selected plants, already operate an extraction process for indium from the acidic zinc sulphate solution. Another common by-product is the well-known lead–silver residue from the hot-acid-leaching step, thus representing again an already available infrastructure for the separation of lead and silver. Therefore, the main idea is to form a product, which is utilizable in the existing zinc industry, closing the loop and recycling zinc, lead aside of indium, and silver.

A very common method to extract zinc from various kinds of material is the vaporization as metallic zinc. Therefore, the boiling point of 907 °C is exploited and the again oxidized material is collected as a zinc oxide product in the off-gas system. A side effect is the vaporization of volatile lead compounds, similar than in the Waelz process. Considering this as a base, the main target of the treatment concept is to form a volatile indium and silver compound, so that both can be collected together with the zinc and lead, resulting in an added value.

Knowing that the state-of-the-art pre-treatment of Waelz oxide before entering the primary zinc industry is a soda washing to remove possible present chlorine compounds, also theoretically introduced chlorine compounds in the obtained jarosite recycling process do not prevent the utilization of the product in primary zinc industry.

Based on the framework described, the investigated concept for the treatment of jarosite deals with the pyrometallurgical volatilization of zinc and lead and the formation of volatile indium and silver compounds in the form of chlorides, thus forming one product accumulating four valuable elements recovered from the jarosite.

Thermal Decomposition of Jarosite

To release the indium and silver, which are bound in the jarosite structure, the first step of the selective vaporization is the decomposition of the iron precipitate. Therefore, the behaviour during a pyrometallurgical treatment for decomposition of the ammonium jarosite was investigated by a thermogravimetric analysis. The obtained plot of the experiment utilizing synthetic air as purging gas is shown in Fig. 2.

Considering the chemical formula of ammonium jarosite $(NH_4)Fe_3(SO_4)_2(OH)_6$, a mass loss of 3.55 wt% for ammonia NH_3, 11.27 wt% for the hydroxyl group as H_2O, and 26.7 wt% for the sulphate group can be calculated. In course of the heating-up process, the following decomposition reactions for ammonium jarosite, shown in Eqs. 5–8, can be summarized:

$$(NH_4)Fe_3(SO_4)_2(OH)_6 \cdot xH_2O \rightarrow (NH_4)Fe_3(SO_4)_2(OH)_6 + x\{H_2O\} \quad \text{(Temp. : 120 °C)} \tag{5}$$

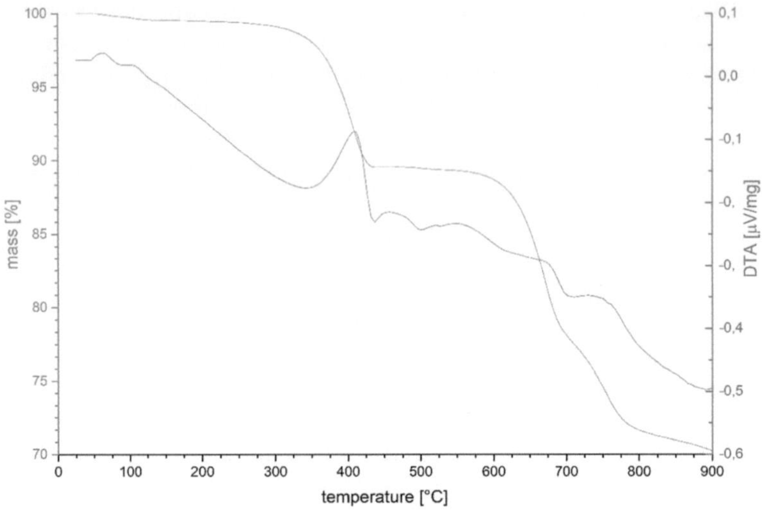

Fig. 2 Thermogravimetric analysis of jarosite under oxidizing conditions (synthetic air) [9]

$$(NH_4)Fe_3(SO_4)_2(OH)_6 \rightarrow (NH_4)(FeO)_3(SO_4)_2 + 3\{H_2O\} \quad (\text{Temp.}: 250 - 300\ ^\circ C) \tag{6}$$

$$6(NH_4)(FeO)_3(SO_4)_2 \rightarrow 6\{NH_3\} + 3\{H_2O\} + 5Fe_2O_3 + 4Fe_2(SO_4)_3 \quad (\text{Temp.}: 350 - 400\ ^\circ C) \tag{7}$$

$$2Fe_2(SO_4)_3 \rightarrow Fe_2O_3 + Fe_2(SO_4)_3 + 3\{SO_2\} \quad (\text{Temp.}: > 500\ ^\circ C) \tag{8}$$

The removal of humidity (Eq. 5) or adherent water can be neglected, as the sample was dried before. The measured mass loss of 10.4 wt%, starting at a temperature of around 310 °C up to around 400 °C, caused by the dihydroxylation (Eq. 6) as well as the separation of ammonia and water (Eq. 7) does not exactly correlate with the theoretical calculated values of 11.27 wt% and 3.55 wt%, respectively. The difference may be caused either by trapped water in the crystal structure or alternatively also by the circumstance that an industrial jarosite sample still can carry parts of the iron as goethite, for instance. The additional measured mass loss of 21 wt% can be attributed to the loss of sulphate as SO_2, which correlates with the theoretical 26.7 wt%. It can be seen that the sample mass is still decreasing slightly, leading to the assumption that either the remaining sulphate is entrapped or the kinetic of this reaction is slowed down at the end of the desulphurization.

Thermodynamic Fundamentals for the Selective Vaporization of In and Ag

As described earlier in this paper, for an effective recovery of In and Ag from jarosite waste especially the evaporation combined with metallic Zn can be seen. Therefore, the formation of halides can be exploited to enhance this process due to significantly higher vapor pressures. Both pure indium and silver can only be evaporated at temperatures higher than 900 °C and show relevant vapor pressure in a temperature range below.

As it is clearly depicted in Fig. 3, the compounds of In and Ag with chlorine or fluorine show a significant vapor pressure at lower temperatures and thus imply that a volatilization leading to a high extraction rate is possible. This is of essential interest as zinc is typically vaporized above 907 °C but has notable vapor pressure at even lower temperatures. Chlorine is less disturbing in the primary zinc industry and can be removed more easily from crude zinc oxide via a soda-leaching process, which is state of the art. Therefore, it is, compared to fluorine, the preferred species. It is worth mentioning that up to 450 °C, InCl theoretically has a higher vapor pressure than $InCl_3$—this behaviour changes at higher temperatures.

The dependence of stable phases in the systems of In and Ag with chlorine and oxygen can be helpful for conclusions and an enhancement of the evaporation procedure. Figure 4 shows the predominance diagrams of In and Ag as a function of the partial pressures of Cl_2 and O_2. The left plot illustrates that for a certain

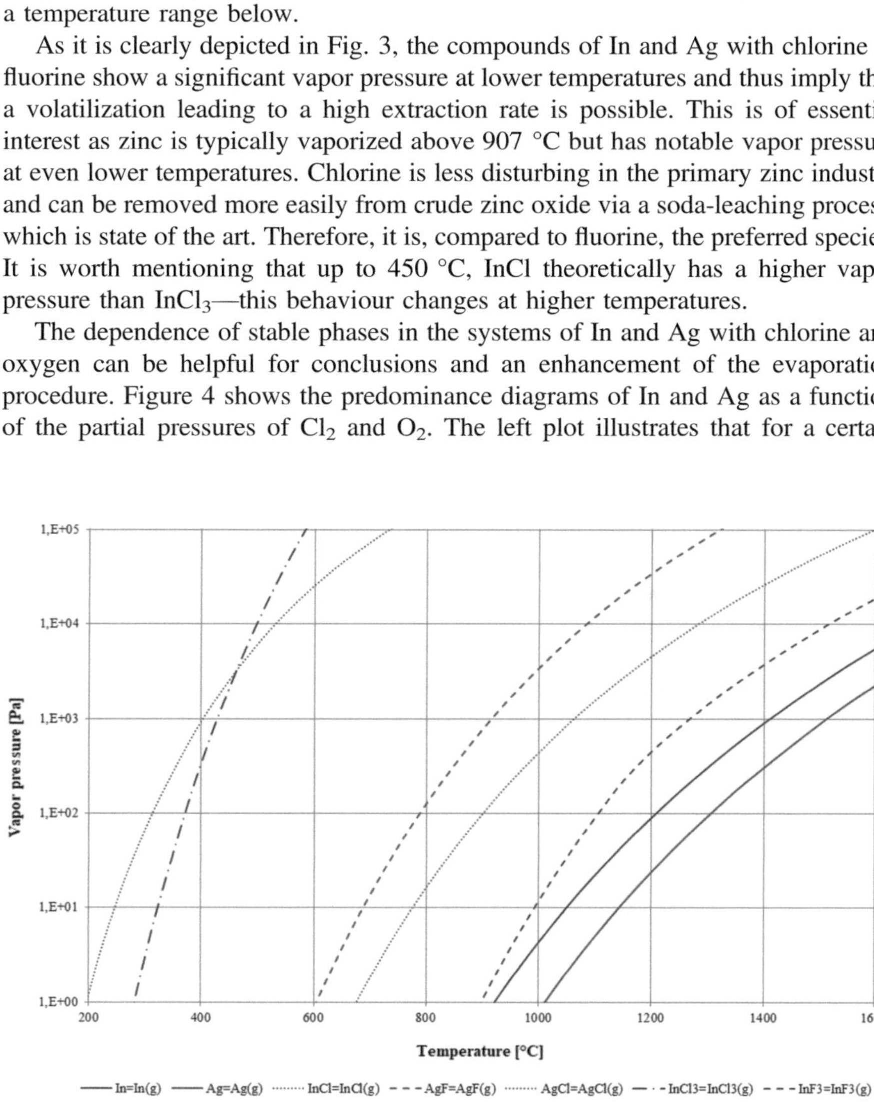

Fig. 3 Selected vapor pressure curves for indium and silver halides [10]

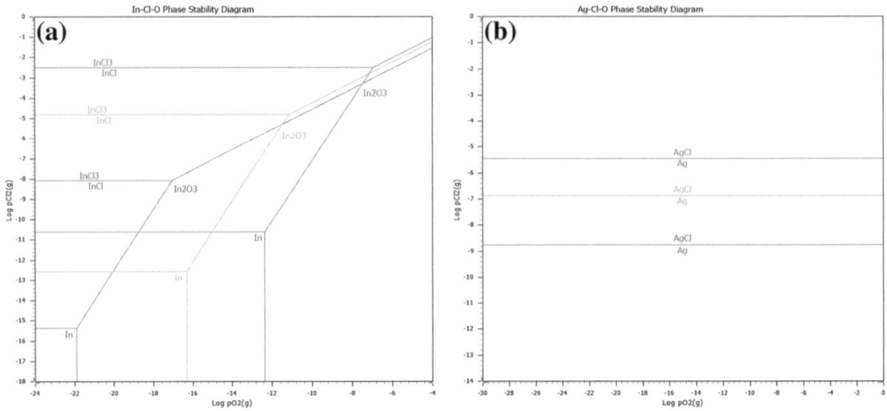

Fig. 4 **a** Predominance diagram (LPP) for the system O–Cl–In at three temperatures (green: 700 °C; yellow: 900 °C; red: 1100 °C;) **b** Predominance diagram (LPP) for the system O–Cl–Ag at three temperatures (green: 700 °C; yellow: 900 °C; red: 1100 °C) [10]

oxygen concentration, InCl can be formed from In_2O_3 with increasing chlorine partial pressure. This has the positive effect that for the evaporation of In, only one-third of chlorine is needed when InCl is formed instead of $InCl_3$. To ensure a rapid and adequate recovery process, account should be taken to promote this mechanism which is also influenced by the process temperature and partial pressure of oxygen. With rising temperatures, a higher chlorine partial pressure is needed to ensure the formation of InCl. At low temperatures, a significantly lower partial pressure of oxygen has to be provided, which is ensured by a reducing atmosphere. An interplay between temperature, chlorine supply, and reducing capability is therefore crucial for an effective indium chlorination.

As shown in the right chart, Ag phases are independent of the oxygen partial pressure as at the given temperatures ranging from 700 to 1100 °C Ag_2O is unstable. As with indium, it is evident that with decreasing temperature a lower chlorine partial pressure is needed for the chlorination. Nevertheless, temperature-dependent kinetic aspects are of importance for a thriving recovery process.

To ensure the availability of chlorine during the recovery process, three strategies are feasible. Chlorine gas could be brought into contact with the waste material which would ensure a high chlorination efficiency due to the gaseous phase. This on the other hand entails higher process costs and risky drawbacks in process safety. Another, easier and safer way, is to use halide additives which enable the essential chlorine supply. A further enhancement is the third possibility which aims to mix Jarosite waste with other waste materials containing sufficient quantities of chlorine compounds. This enables considerable ecologically and economically advantages.

As a first step, appropriate halide compounds were identified with thermodynamic calculations using the software package FactSage 7.2 with the database FactPS. The chemical reactions as stated in Eqs. 9 and 10 were implemented and its behaviour investigated at 800 °C. The amount of the chloride species was increased stepwise as an equivalent to Cl_2 from 0 to 1 mol. This allows a reasonable comparison of the different species as the same amount of chlorine is brought into reaction.

$$In_2O_3 + 3C \rightarrow 2In(l) \tag{9}$$

$$2In(l) + \frac{2}{x}(MCl_x) \rightarrow 2InCl + \frac{2}{x}M \tag{10}$$

$$In_2O_3 + 3C + \frac{2}{x}(MCl_x) \rightarrow 2InCl + 3CO + \frac{2}{x}M \tag{11}$$

Figure 5 illustrates the findings of the calculation done for six different chloride compounds. After the reduction of In_2O_3 with carbon (Eq. 9), the occurring indium is theoretical in liquid state at the present temperature and can further react to its chlorine compound, which is volatile. The overall reaction is shown in Eq. 11. The remaining amount of liquid indium after the reduction and chlorination is plotted in dependence of the ratio between Cl_2 and In_2O_3 in the mixture, representing a kind of likelihood of reaction. The lower the percentage of indium the better, as it points out a successful evaporation. It is evident that both NaCl and KCl show no effect and are therefore unsuitable for a recycling process. With $CaCl_2$, an improvement to a theoretical recovery of indium of about 30% was achieved. Although the

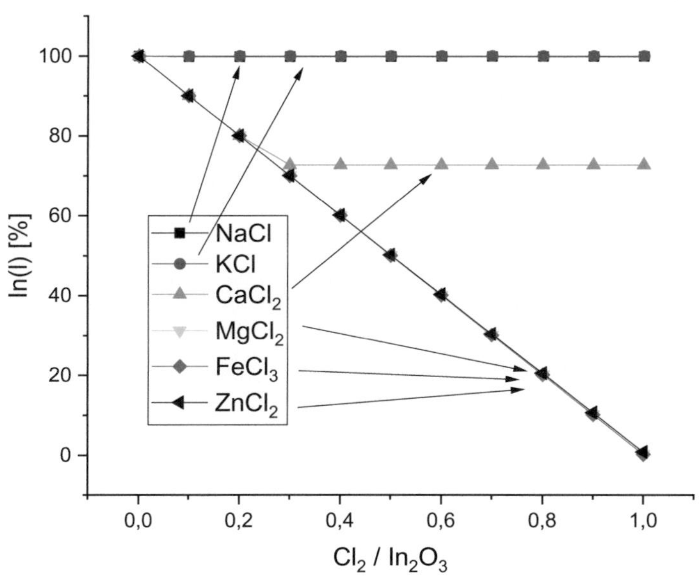

Fig. 5 Comparison of selected chloride compounds in regard to their theoretical applicability as an additive for the indium-chloride formation

evaluation was done by calculating an extraction rate, this value can only give a comparison between the investigated additive as the partial pressure of the reaction product is different in a potential industrial application, for example. Anyhow, an excellent result was obtained with the three remaining species $MgCl_2$, $FeCl_3$, and $ZnCl_2$ which enabled a potential recovery of 100%. As FactSage does not consider the kinetic aspects, further investigations are planned to verify the theoretical findings including research on the kinetic behaviour. Furthermore, different mixtures of chlorine compounds can be probed to draw conclusions on potential combined waste material recycling.

Supply/Demand Situation for Indium and Silver

Indium and silver can clearly be named as key elements for future technologies, especially in the field of IT and digitalization as well as advanced renewable energy and mobility systems. Due to an increasing demand and a stagnation or even a decrease in production of indium, a defile supply situation can be a realistic scenario. It is estimated that the global reserve of indium is about 15.000 tons. In relation to the indium production of 750 tons in 2018, the global reserves could be depleted within 20 years [11, 12].

The global indium consumption in 2016 is numbered with 1.430 tons where in the same year the primary production is listed with 655 tons. This shows that in 2016 more than 50% of the global indium production was carried out by processing secondary resources. Exact statistics of the secondary indium production were unavailable, but it is estimated that about 1.200 tons were potentially recyclable from ITOs in 2016 [12].

Around one half of the production of the by-product indium is carried in China. This entails the risk for volatile prices for the whole indium market. As seen in November 2017 when Chinese metal exchange Fanya reportedly held back 3.600 tons of indium with lacking information regarding a release into the market. Further big producers are Canada, Japan, and especially South Korea with an increasing yearly production of 230 tons in 2018 [12].

As mentioned, the indium price was quite unstable with its peak at $745 per kg in 2014 followed by a significant drop to $275 in 2015 remaining low for 2 years with an average $210 throughout 2017. Since 2018, prices are recovering [12]. Indium is named one of the most critical elements on earth due to missing alternatives of production routes and the concentration of its producers in Asia [13–15].

In comparison to indium, the situation for silver is not as critical due to several possibilities in production and a low geographic concentration. The global primary silver production was 26.616 tons compared to 4.707 tons from secondary resources. The global demand increased by 4% to 32.146 tons, driven by a recovery in retail investment [13, 16]. The silver price dropped 8.3% in 2018 in which the trade dispute between USA and China was the main influence. In the last year, the silver prices increased significantly by far over 30% and is by September 2019 listed with $620 per kg. With a possible continuation of this trend, silver can be a profitable by-product in the recycling of Jarosite waste as well, even when present in small amount [16, 17].

Results and Discussion

Thermodynamic calculations show clearly that the evaporation of indium and silver chlorides takes place at significantly lower temperatures than of the metallic phases of indium and silver and is therefore more suitable in terms of energy efficiency and also process yield. Different chloride species were investigated, and thereby $MgCl_2$, $FeCl_3$, and $ZnCl_2$ additives were identified as most suitable in terms of the thermodynamic calculated chlorination potential. The carried characterization points out that for a selective vaporization process the jarosite structure has to be decomposed to set the indium and silver free for a possible chlorination process. The thermogravimetric evaluation showed that a temperature of at least 550 °C is suitable for a decomposition but from the point of view of industrial throughput, meaning increased kinetics, a somewhat higher treatment temperature is meaningful.

Based on the market situation and the strong dependence on the primary zinc production, a huge potential to cover a part of the future indium demand can be found in the recycling of jarosite waste. The high concentration of primary indium production in Asia and America enhances the necessity for the European Union to take secondary resource, such as the jarosite, into account. Millions of tons of jarosite were dumped for decades without a noteworthy extraction of indium, underlining the potential of this material as well, assuming a viable extraction process.

References

1. Paschen P (2012) Kupfer-Aphrodite-Venus. World Metall ERZMETALL, 380–387
2. Paschen P (2011) Eisen-Ares-Mars. World Metall ERZMETALL 64:348–353
3. Schweizerische Akademie der Technischen Wissenschaften: Seltene Metalle. https://www.satw.ch/fileadmin/user_upload/documents/02_Themen/06_Rohstoffe/SATW-Seltene-Metalle_DE.pdf. Accessed 24 Nov 2017
4. Chapman A, Arendorf J, Castella T, Thompson P, Willis P, Espinoza LT, Klug S, Wichmann E (2013) Study on critical raw materials at EU level. http://ec.europa.eu/DocsRoom/documents/5605/attachments/1/translations/en/renditions/native. Accessed 24 Nov 2017
5. Krüger J, Püllenberg R (1980) Das Eisenproblem bei der hydrometallurgischen Zinkgewinnung. Erzmetall, 70–76
6. Röpenack A (1978) Die Bedeutung der Eisenfällung für die hydrometallurgische Zinkgewinnung. Erzmetall, 272–276

7. Sinclair RJ (2005) The extractive metallurgy of zinc. In: Spectrum series. Australasian Institute of Mining and Metallurgy, 13. AusIMM, Carlton, Vic (Band)
8. Pawlek F (1982) Metallhüttenkunde. Walter de Gruyter, Berlin
9. Leuchtenmüller M.: Untersuchungen zum Verhalten von Silber und Indium in Rückständen der Zinkindustrie in einem pyrometallurgischen Aufarbeitungsschritt. MSc thesis, Montanuniversitaet Leoben, Austria
10. Steinlechner S, Antrekowitsch J (2018) Thermodynamic considerations for a pyrometallurgical extraction of indium and silver from a Jarosite residue. Metals 8:335. https://doi.org/10.3390/met8050335
11. Lokanc M, Eggert R, Redlinger M (2015) Colorado School of Mines, Golden, Colorado: the availability of indium: the present, medium term, and long term
12. USGS National Minerals Information Center: Indium 2019 (2015–2019)
13. Innovationsforschung F.-I. f. S.-u. et al.: Rohstoffe für Zukunftstechnologien 2016. In: DERA Rohstoffinformationen, Band: 28. DERA, Hannover (2016)
14. Shanks WP III et al (2017) Germanium and indium. Professional Paper, Reston, VA
15. Hayes SM, McCullough EA (2018) Critical minerals: a review of elemental trends in comprehensive criticality studies. Resour Policy 59:192–199
16. World Silver Survey (2019) Silver institute. Washington, D.C
17. silverprice.org. https://silverprice.org/spot-silver.html

Recovery of Platinum Group Metals From Secondary Sources by Selective Chlorination from Molten Salt Media

Ana Maria Martinez, Karen Sende Osen and Anne Støre

Abstract The use of molten salts as "solvents" offers interesting possibilities for the treatment of ores, industrial by-products as well as scraps, allowing the development of processes that are mid-way between pyro- and hydrometallurgy, with the advantage of salts offering a wide choice of chemical and electrochemical properties. The molten mixtures can be fed to industrial electrolytic cells (electrometallurgy) where the reactions can take place. One example of this type of reactions is the selective chlorination using gaseous mixtures in a molten chloride salt mixture. Within the frame of the EU-financed project PLATIRUS (GA 730224), the possibility of selective chlorination of platinum group metals (PGM) contained in spent automobile catalyst samples, using a molten salt as reaction media was investigated. To predict the selective separating conditions of the PGMs from the sample matrix, the Pourbaix-type E-pO^{2-} diagrams of the most relevant components in the spent catalyst were compared with those of relevant chlorinating gaseous mixtures, i.e., Cl_2/O_2. This allowed to predict the optimal chlorinating conditions, which were tested and confirmed experimentally using Cl_2 gas.

Keywords Platinum group metals PGM · Molten chlorides · Selective chlorination · Secondary resources · Spent catalyst · Recycling

Introduction

Platinum group metals (PGM) are among the least abundant elements in the Earth's crust, and they are classified as critical raw materials (CRMs). They have large range of applications from jewelry to automotive and electronics. Considering autocatalysts only, Europe is the world's largest consumer of PGMs, with an annual demand of 40 tons of platinum, worth 1137 MEUR [1].

A. M. Martinez (✉) · K. S. Osen · A. Støre
SINTEF Industry, 7465, Trondheim, Norway
e-mail: anamaria.martinez@sintef.no

G. Azimi et al. (eds.), *Rare Metal Technology 2020*, The Minerals, Metals & Materials Series, https://doi.org/10.1007/978-3-030-36758-9_21

The supply of PGMs is currently ensured mostly by primary sources (72%), and the overall supplied PGMs that comes from both mines and recycled products cannot meet the global demand. Then, the recycling of secondary resources must be promoted to tackle sustainability challenges associated with PGMs. If 100% of the PGM available in end-of-life autocatalysts is recovered, an additional 29 tons of PGM will become available in the global market, which is about 1.5 times of the global supply–demand gap [2].

Within the frame of the EU-financed project PLATIRUS (GA 730224), SINTEF has investigated the possibility of recovering PGM from spent automobile catalysts by selective chlorination from a molten salt.

The use of molten salts as "solvents" offers interesting possibilities for the treatment of ores, industrial by-products as well as scrap. By using salts in the temperature range of 200–500 °C, it is possible to develop processes that are mid-way between pyro- and hydrometallurgy, with the advantage of salts offering a wide choice of chemical and electrochemical properties. One example of this type of reactions is the selective chlorination using gaseous mixtures in a molten chloride salt mixture.

Chlorination of metallic oxide mixtures (or even ores or industrial by-products) in molten chlorides media can take place at moderate temperatures, leading to the selective separation of different compounds by choosing the appropriate conditions (salt, temperature, chlorinating agent, etc.). The dissolution reaction is favored by significant solvation of the oxide ion (O^{2-}) and the produced metal cation (M^{n+}) giving stable complexes with the anions (Cl^-) of the solvent. This is a considerable advantage compared with classical gas–solid reaction systems using solid reductants (i.e., carbo–chlorination), where the rate and conversion levels are limited by the intimate contact of the reductant and the metal oxide particles, and the reaction is occurring non-selectively at much higher working temperatures (ca. 1000 °C).

Selective separation conditions of the PGMs from the spent catalyst matrix were predicted by comparing the Pourbaix-type E-pO^{2-} predominance diagrams of the most relevant components in the material with those of the utilized chlorinating gaseous mixture. The most promising conditions were tested and confirmed experimentally using Cl_2 gas in two molten salt media with different oxoacidic properties, namely, the eutectic LiCl–KCl mixture and the equimolar $CaCl_2$–NaCl melt. The effect of the working temperature in the chlorination process was also studied, and the PGM recovery rates were assessed in each case. The results allow to establish the optimal operating conditions for the recovery of PGMs from spent automobile catalyst waste.

Molten Salts as Reaction Media

Molten salts are ionized solvents that, as any other liquid solvent (aqueous or non-aqueous), can dissolve substances and make them react with one another, following reactions in solution at high temperature.

Chemical reactions in liquid media at high temperatures offer two main advantages: (i) Reactions kinetics are more enhanced than at ambient temperature and nearly always obey the laws of chemical equilibrium; (ii) The reactions can be controlled, as the temperature can be easily controlled, and then reaction selectivity is as large as those offered in other liquid solvents.

The structure of the molten salt is looser and more disordered than in the crystalline state, thus allowing high solubility (complete miscibility) of the solute added, which is subjected to the phenomenon of solvation by the species of the molten salt used as solvent. The solvation involves interactions of the electrostatic-type (Coulombic) and, frequently, cation–anion coordination. The differences in the solvation state have the most important consequence of modification of reactivity.

The O^{2-} Anion: Acidity in Molten Salts

The oxide anion, O^{2-}, frequently exists in the free state, i.e., without the interaction of solvation other than the Coulombic interaction with the surrounding ions of the molten salt.

The O^{2-} ion plays a particularly important role in molten salts, which arises from the natural abundance of compounds that contain it, constituting the major part of mineral resources. On the other hand, the O^{2-} ion is the most remarkable base (electron–pair donor), capable of combining with all acceptors of electron pairs (Lewis acid), and particularly, metallic cations. It is then possible to develop a system of reactions in molten salt solvents involving the oxide ion O^{2-} following the acid–base concept, as in analogy with the case of the H^+ ion in water and other molecular solvents [3, 4], as in Reaction 1. The level of oxoacidity in a molten salt can be measured by the pO^{2-} value, analogous to pH and playing the same role, as defined in Reaction 2.

$$\text{oxobase} \rightarrow \text{oxoacid} + O^{2-} \qquad \text{(Reaction 1)}$$

$$pO^{2-} = -log\, a_{(O_{2-})} = log\, m_{(O_{2-})} \qquad \text{(Reaction 2)}$$

The E-$_{P}O^{2-}$ Diagrams

The E-pO^{2-} diagrams in molten salts are like the E-pH diagrams (Pourbaix diagrams) in aqueous systems. They represent the variation of the equilibrium potentials of the Red-Ox systems as a function of pO^{2-}, at fixed concentrations of the oxidizing and reducing substances.

In the case of molten alkaline chloride solvents, the higher value of E corresponds to the oxidation of Cl^- ions to chlorine, at a partial pressure that, in ordinary conditions, does not exceed 1 atm. The higher value corresponds to the normal chlorine electrode (NCE) taken as reference electrode, with E = 0 by convention. On the other hand, the lower E value corresponds to the alkaline metal, which can be determined by thermochemically tabulated data with respect to the NCE at the given temperature.

Concerning the pO^{2-} values, the only theoretical limit is imposed by the solubility of the alkali oxide.

Solubilization Conditions of Insoluble Metallic Oxides

The solubilization of an insoluble oxide (MO) in a molten salt solvent needs the establishment of a sufficiently high pO^{2-} in the molten salt. This can be obtained either by the addition of a strong enough oxoacid reactant, which will combine with the O^{2-} anion giving the conjugated oxobase, or by the oxidation of the O^{2-} into oxygen by the addition of a sufficiently strong oxidizing reactant, like chlorine gas. Maximum efficiency can be obtained when these two actions are combined.

In the case of molten chloride solvents, the insoluble oxides are converted to soluble chlorides, and this solubilization process (dissolution or vaporization) is often called chlorination, which can be selective by choosing the appropriate conditions, i.e., pO^{2-} value, and then separations of metallic oxide mixtures can take place.

The chlorination conditions can be predicted by comparing the E-pO^{2-} predominance diagrams of the M-O-Cl substances with that of the chlorinating agent, in the molten chloride at the operating temperature. When the zone of the gaseous mixture covers an insoluble oxide, the reactant does not have a solubilizing action. In the contrary, when a zone covers a chloride (soluble or volatile) salt, it indicates that the reactant has a chlorinating action. Comparing several elements indicates if these actions can be selective.

Experimental

Experimental Setup

90 g of a molten salt mixture (eutectic LiCl–KCl or equimolar $CaCl_2$–NaCl) was mixed with 10 g of spent catalyst powder in a quartz crucible, which was placed in a quartz cell under inert atmosphere (Ar 99.999%) with a water-cooled quartz lid that allowed the insertion of a thermocouple, and gas and sampling tubes. The glass reactor was placed in a vertical furnace with a mullite liner.

All salts and the catalyst waste material were dried at 200 and 100 °C, respectively, for at least 48 h prior to use. The working temperature was measured using a thermocouple type S (Pt–Pt 10%Rh) shielded by a closed-end alumina or quartz tube. The temperature was monitored continuously by means of a multi-channel Keithley 2000 Multimeter.

Samples of the molten bath were taken prior to and after the chlorination experiments, using a quartz tube with a quartz frit to avoid solid particles. The samples were cooled down in a desiccator and then stored until analyzed by ICP-SFMS (Inductively Coupled Plasma Sector Field Mass Spectrometry).

After the chlorination trial, the furnace was cooled down under argon to room temperature, and the next day the cell was opened and investigated.

The furnace, gas system, valves, gas control panel, electronic mass flow controllers–control system, pressure gauge, and transmitter and pressure relief valve are schematically shown in Fig. 1. The complete equipment was installed in a walk-in hood, and only materials and apparatus able to stand high temperature and chlorine atmosphere were used, i.e., glass, special valves, mass flow controllers (MFC), pressure gauges, etc.

To absorb the unreacted chlorine gas coming from the cell, 6 M sodium hydroxide solutions in gas washing bottles were placed in the gas outlet. This proved to be a very efficient scrubber, and as long as the sodium hydroxide was not saturated, no chlorine gas passed these bottles. Also, a liquid N_2 (LN2) cooling trap was positioned at the end of the gas line to account for safety if volatile chlorides were released from the reactor. A U-tube was connected in parallel with the gas outlet to ensure an alternative escape route for the off-gas in case of clogging of the NaOH scrubbers or LN2 trap to prevent pressure build-up in the reactor.

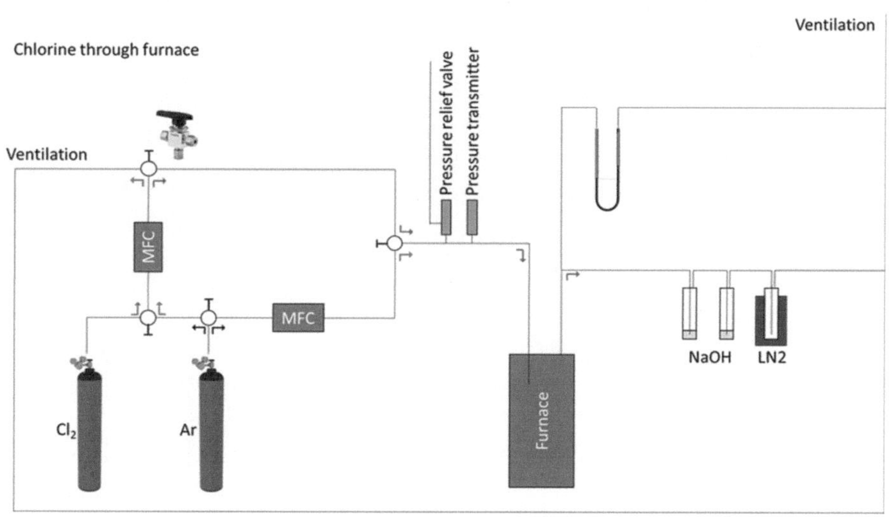

Fig. 1 Schematics of the chlorination apparatus

Methodology

The reactor was kept completely gastight when it was under chlorine gas atmosphere. When the set temperature was obtained, still under argon addition, the gas inlet tube was lowered into the melt. The gas was then switched to chlorine via the gas control panel so that chlorine gas was bubbled directly into the molten salt. The flow of Cl_2 gas was 100 ml min^{-1}, which corresponds to ca. 17.5 g h^{-1}. This value was chosen following the data from the literature [5]. When the chlorination was finished, argon gas was introduced to strip the melt of chlorine and remove the chlorine from the gas line before opening the cell for further manipulation, i.e., taking melt samples, etc.

Samples of the molten bath were taken prior to and after each of the chlorination tests, using a quartz tube with a quartz frit to avoid solid particles. The samples were cooled down in a desiccator and then stored until analyzed by ICP-MS/OES (Inductively Coupled Plasma Mass Spectrometry/Optical Emission Spectrometry).

Once the experiment was finished, the furnace was cooled down under argon to room temperature, and the next day the cell was opened and investigated. The catalyst material left at the bottom of the reactor was recovered, after washing the entrapped salt with water, and then analyzed by ICP-MS/OES.

Results and Discussion

Choice of the Molten Salt Media

The most attractive molten salt media are the molten halides, as there is no limitation of high pO^{2-}, and particularly molten chlorides, where the insoluble oxides are converted to soluble chlorides.

In general, molten chlorides are not recommended to be used at working temperatures higher than 850–900 °C, due to the high vapor pressures at those temperatures, leading to large vaporization loss of the reaction media.

The most common molten chloride media are the eutectic LiCl–KCl molten mixture and the equimolar NaCl–KCl mixture. Both have been widely studied, and many data on the chemical and electrochemical properties of metallic elements in those media can be found in the literature. Moreover, their oxoacidic properties are also well known (e.g. [6, 7]).

The eutectic mixture LiCl–KCl is very attractive, due to its low melting point (ca. 348 °C), thus allowing a wide range of working temperatures, going from relatively low, i.e., 400 °C, to relatively high, i.e., 800 °C. The equimolar mixture NaCl–KCl has a higher melting point (ca. 645 °C); thus, 727 °C (1000 K) is the standard temperature used in this type of molten salt.

Another attractive and well-studied molten chloride is the eutectic $CaCl_2$–NaCl, of composition close to that of the equimolar mixture. The melting temperature is ca. 500 °C; thus, 550 °C is the usual working temperature used in the literature. The equimolar $CaCl_2$–NaCl mixture possess a higher oxoacidity than the above-mentioned molten chlorides [8], thus affecting the O^{2-} exchange reactions and the solvation effect of the metallic ions. In addition, it is known that PGMs are easily forming complexes with Ca^{2+} ions, thus perhaps influencing the solubilization reactions of PGMs in $CaCl_2$-based media.

The eutectic LiCl–KCl and equimolar $CaCl_2$–NaCl mixtures were used in the chlorination tests. The effect of temperature in the chlorination kinetics was also studied in both cases.

Characterization of the Raw Material

The spent catalyst material was provided in the form of fine particles by Monolithos Catalysts Ltd. [9]. X-ray diffraction (XRD) analysis of the particles showed that it consists mainly of SiO_2, Al_2O_3, and MgO (indialite/cordierite ceramic monolith). Some hydroxide phases as well as silicon were also found (cf. Fig. 2). Quantification of the different phases was done using Bruker TOPAS Software, being Indialite: 73.09%; Cordierite: 17.44%; magnesium silicon oxide hydroxide hydrate: 4.56%; and silicon: 4.92%.

ICP-MS/OES analysis of the catalyst sample is shown in Table 1. The material has a total PGM content of ca. 2500 ppm.

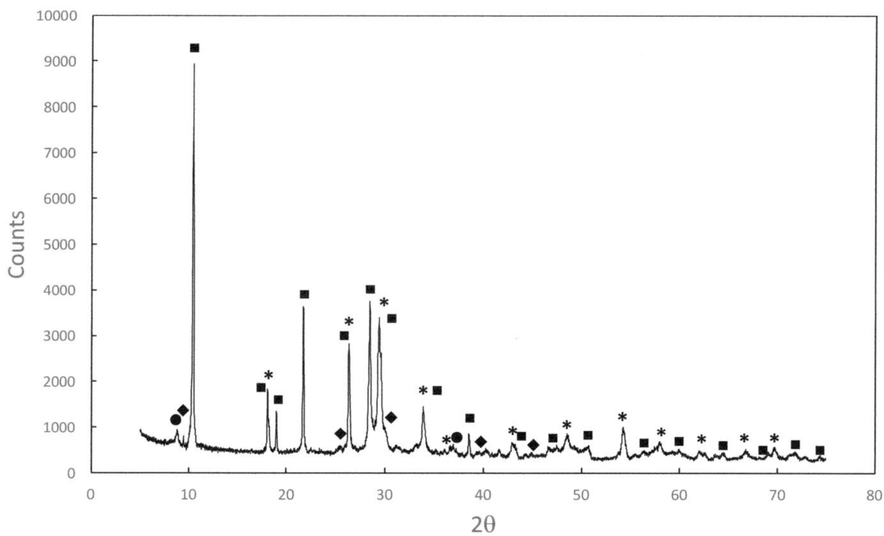

Fig. 2 XRD diffractogram of the spent catalyst sample. (*) $Mg_2(Al_4Si_5O_{18})$ Indialite; (■) $Mg_2Al_4Si_5O_{18}$ Cordierite; (●) $Mg_3Si_4O_{10}(OH)_2(H_2O)$; (♦) $Na_{3.7}Al_{3.7}Si_{8.3}O_{24}(H_2O)_{11}$

Table 1 Elemental composition of the spent catalyst material analyzed by ICP-OES. The PGM content is analyzed by ICP-MS

Pd	Pt	Rh	Al	Si	Mg	Zr	Ce	Ba	Ca	Fe	La	Pb	Ti	Zn
ppm	ppm	ppm	wt%	wt%	wt%	wt%	wt%	ppm	ppm	ppm	ppm	ppm	ppm	ppm
1453	819	279	20.0	17.9	5.8	2.7	4.4	8011	4337	7699	6848	1969	7055	2667

Chlorination in the Eutectic LiCl–KCl Mixture

The E-pO^{2-} diagrams of the most relevant elements present in the waste catalyst material (Si, Al, and Mg) in the eutectic LiCl–KCl mixture are available in the literature [10, 11]. Their comparison with that of the Cl_2/O_2 mixture is also available in the literature [12]. This information enables predicting the solubilization conditions of the different elements, either by dissolution as chloride complexes or by vaporization.

Even though chlorine possesses the strongest oxidizing capacity in molten chlorides, this reactant is not capable of solubilizing the most stable metallic oxides. Also, it is expected that SiO_2 cannot be chlorinated at the experimental conditions, most likely remaining as insoluble silicates. Al and Mg could be easily chlorinated as AlO^+ and Mg(II) complexes, if kinetics of the chlorination reaction are high enough.

Moreover, the theoretical regions of the PGM-O-Cl compounds as a function of the partial pressures of Cl_2 and O_2 gases could be determined from thermodynamic data (cf. Fig. 3). The thermodynamic data show that all PGMs can be chlorinated with pure chlorine and that Rh-O compounds are very stable even at low oxygen partial pressures. However, the solvation effect of the chloride complexes in the molten salt (activity) must also be considered.

In order to confirm the theoretical predictions, chlorination experiments were carried out using chlorine gas at two different working temperatures, i.e., 450 and 750 °C, and at different chlorination times, i.e., 1, 2, 3, and 5 h.

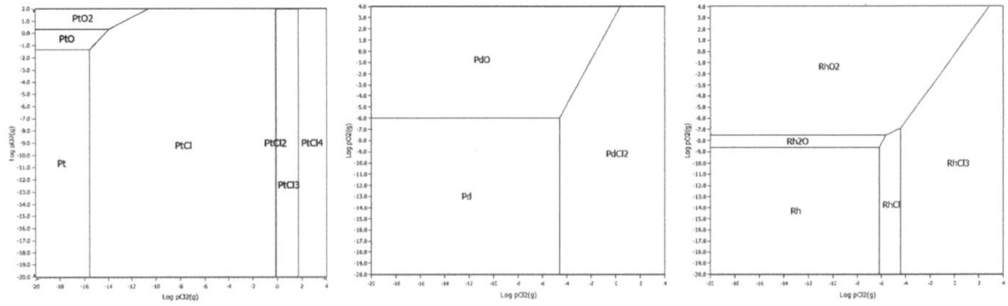

Fig. 3 Theoretical stability regions of the PGM-O-Cl compounds as a function of the partial pressures of Cl_2 and O_2 gases. Temperature 450 °C Data obtained from HSC Database

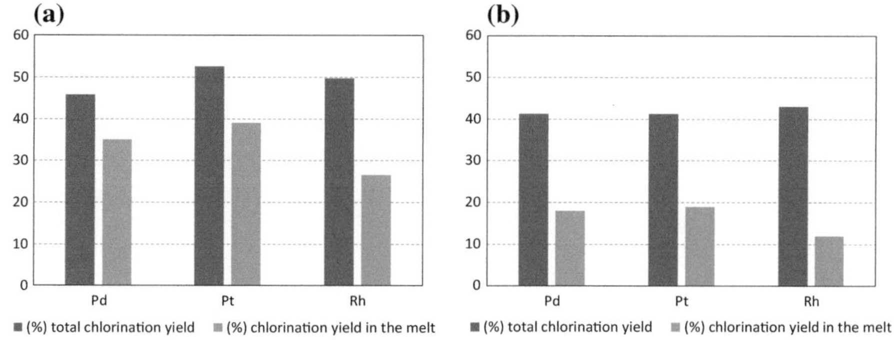

Fig. 4 PGM content in the remaining catalyst material and in the molten salt samples after the chlorination runs in the eutectic LiCl–KCl mixture at two different working temperatures: **a** 450 °C; and **b** 750 °C

Figure 4 shows the total chlorination yield of PGMs, which was calculated by analyzing the remaining catalyst sample after the chlorination runs compared with the initial PGM content in the waste catalyst material, as a function of temperature. The chlorination yield in the melt was determined by analysis of the melt samples. Indirectly, the amount of PGM chlorocomplexes lost in the vapor phase can also be extracted from the results obtained. The results showed that increasing the working temperature did not affect the total chlorination yield, but only the content of PGM in the vapor phase.

Moreover, Fig. 5 shows the amount of PGMs in the molten salt as a function of the chlorination time. At lower temperatures, it was found that after 2–3 h chlorination time, the amount of PGM oxo-complexes in the molten bath is constant. In the case of high working temperatures, the effect of the elapsed time in the losses of the dissolved PGM-chloride complexes by evaporation is clear.

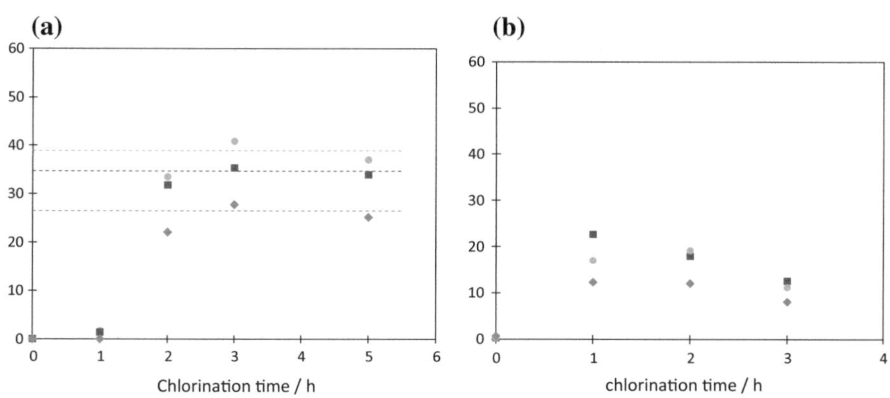

Fig. 5 PGM yield in melt (%) in the eutectic LiCl–KCl melt as a function of the chlorination time at **a** 450 °C; and **b** 750 °C. (●) Pt; (■) Pd; and (♦) Rh

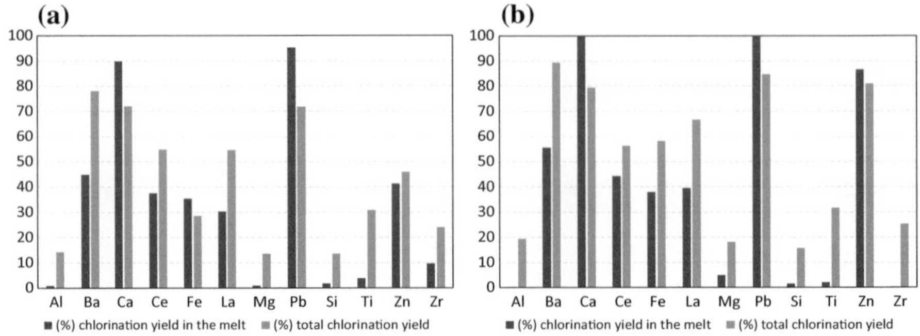

Fig. 6 Element analysis of the remaining catalyst material and molten salt samples after the chlorination runs in the eutectic LiCl–KCl mixture at two different working temperatures: **a** 450 °C; and **b** 750 °C

The results obtained for other elements are shown in Fig. 6. In the case of the metallic elements from the main phase indialite/cordierite, i.e., Al, Mg, and Si, the results showed that they remained in the catalyst material after chlorination, most likely as silicates, which are not possible to solubilize using chlorine gas at the working conditions.

Lanthanides (La and Ce), alkaline earth (Ca and Ba), and base metals (Zn, Pb, Fe) are chlorinated and remain in the molten bath, in some cases with an efficiency up to almost 80–100%, as in the case of Ca, Ba, Pb, and Zn at 750 °C. At the operating conditions, the gaseous $FeCl_3$ formed during the chlorination reaction and reacted with the chloride ions of the molten bath, resulting in stable Fe(II) species dissolved in the melt, according to Reaction 3:

$$FeCl_3(g) + Cl^- \rightarrow FeCl_2(diss) + Cl_2(g) \qquad \text{(Reaction 3)}$$

Surprisingly, total chlorination yields of Zr and Ti reached 20–30%. This is probably due to the fact that the catalyst waste material contained some carbon acting as auxiliary reductant, and thus it underwent a carbo–chlorination reaction in which higher pO^{2-} values can be achieved [6, 13], and then gaseous $TiCl_4$ and $ZrCl_4$ formed according to Reaction 4. This was confirmed by the fact that, especially at the higher temperatures, i.e., 750 °C, both elements were found in the vapor phase. However, a reaction similar to Reaction 3 cannot be ruled out, explaining the small amounts of dissolved Zr- and Ti-chlorocomplexes obtained in the bath samples at lower temperatures.

$$MO_2(s) + C + 2Cl_2(g) \rightarrow MCl_4(g) + CO_2(g) \qquad \text{(Reaction 4)}$$

Under these conditions, it is possible to predict the separation of the PGMs from other metal chlorocomplexes dissolved in the molten LiCl–KCl by electrolysis, due to the differences in standard potentials of the elements involved (cf. Table 2). It seems to be difficult to separate the different PGMs by electrochemical methods, due to their similar standard potential. However, a PGM alloy could be selectively separated from the rest of the dissolved species present in the electrolyte.

Chlorination in the Equimolar $CaCl_2$–NaCl Melt

In this case, chlorination experiments were carried out at two different working temperatures, i.e., 550 and 790 °C, and at different chlorination times, consecutively during the same run. Figure 7 presents the ICP-MS analysis results of molten salt samples taken during the test.

As in the case of the LiCl–KCl melt, the effect of the elapsed time in the loss of the dissolved PGM-chloride complexes by evaporation is clear at higher temperatures. The rest of the elements showed a similar behavior, except for the lanthanide elements, i.e., Ce and La, that gave chlorination yields in melt close to 70% at the higher working temperature of 790 °C.

Table 2 Theoretical standard potentials of the reduction systems of relevant elements dissolved in the molten chloride, including the matrix (eutectic LiCl–KCl melt). Temperature 450 °C

Reduction reactions	E^o/V versus Cl_2/Cl^- reference
$PtCl_4$ (s) + $4e^-$ = Pt (s) + $4Cl^-$	−0.212
$PdCl_2$ (s) + $2e^-$ = Pd (s) + $2Cl^-$	−0.328
$RhCl_3$(s) + $3e^-$ = Rh(s) + $3Cl^-$	−0.357
$RhCl_2$ (s) + $2e^-$ = Rh (s) + $2Cl^-$	−0.369
$PtCl_2$ (s) + $2e^-$ = Pt (s) + $2Cl^-$	−0.404
$FeCl_2$ (s) + $2e^-$ = Fe (s) + $2Cl^-$	−1.292
$PbCl_2$ (s) + $2e^-$ = Pb (l) + $2Cl^-$	−1.308
$ZnCl_2$ (s) + $2e^-$ = Zn (l) + $2Cl^-$	−1.595
$CeCl_3$ (s) + $3e^-$ = Ce (s) + $3Cl^-$	−3.016
$LaCl_3$ (s) + $3e^-$ = La (s) + $3Cl^-$	−3.083
$CaCl_2$ (s) + $2e^-$ = Ca (l) + $2Cl^-$	−3.567
LiCl (s) + e^- = Li (l) + Cl^-	−3.680
$BaCl_2$ (s) + $2e^-$ = Ba (l) + $2Cl^-$	−3.848
KCl (s) + e^- = K (l) + Cl^-	−3.878

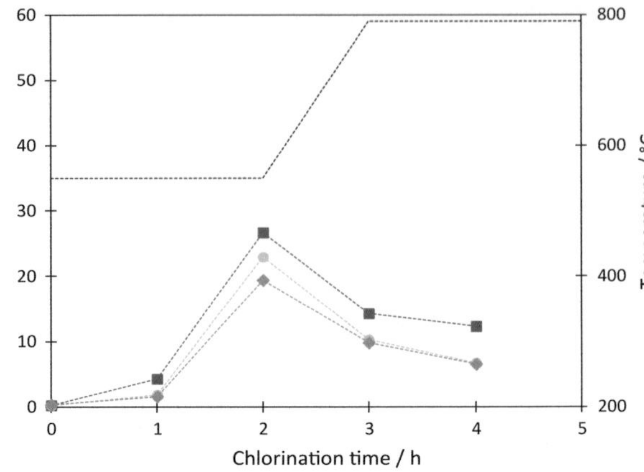

Fig. 7 PGM yield in the melt (%) in the equimolar $CaCl_2$–NaCl mixture as a function of the chlorination time and at the working temperature showed by the broken line, i.e., 550 °C and 790 °C. (●) Pt; (■) Pd; and (♦) Rh. The x-axes do not reflect the real elapsed time during the test

Summary and Concluding Remarks

The possibility of selective chlorination of PGMs contained in spent automobile catalyst samples, using a molten salt as reaction media, was investigated. The selective separating conditions of the PGMs from the spent catalyst matrix were estimated by comparing the Pourbaix-type E-pO^{2-} predominance diagrams of the most relevant components in the material with those of the utilized chlorinating gaseous mixture. The most promising conditions were tested and confirmed experimentally that using Cl_2 gas in two molten salt media with different oxoacidic properties, namely, the eutectic LiCl–KCl mixture and the equimolar $CaCl_2$–NaCl melt is possible. The effect of the working temperature in the chlorination process was also studied, and the PGM recovery rates were assessed in each case.

Similar results were obtained in both melts, and chlorination yields of the PGMs were found to be 40–50% after 2–3 h of passing chlorine gas through the molten salt at 100 ml min^{-1}, i.e., ca. 17.5 g h^{-1}. Increasing the working temperature did not lead to better chlorination yields, but the evaporation losses were significantly higher.

Elemental analysis of the remaining catalyst material as well as that of the salt samples taken during the chlorination runs allowed to determine the elements dissolved in the molten chloride as chlorocomplexes. Lanthanides (La and Ce), alkaline earth (Ca and Ba), and base metals (Zn, Pb, Fe) are chlorinated in some cases with an efficiency of up to almost 80–100%, as for Ca, Ba, Pb, and Zn at the higher working temperatures.

It was also found some extent of carbo–chlorination reaction, due to small amounts of carbon existing in the waste catalyst material. This explains the certain chlorination of Ti and Zr, which were found to escape in the vapor phase as gaseous $TiCl_4$ and $ZrCl_4$.

The results of this study show that the most suitable molten chloride system is the eutectic LiCl–KCl melt, allowing working at relatively low temperature of 450 °C.

Under these conditions, the PGMs can be selectively recovered from the waste catalyst material by chlorination and subsequent electrolysis from the same molten chloride acting as an electrolyte.

Acknowledgements Zhaohui Wang is acknowledged for performing XRD/TOPAS analysis, Tecnalia Research and Innovation for the ICP-MS/OES analysis of the catalyst sample and Kai Tang for providing thermodynamic data of PGM-Cl-O compounds. This work has received funding from the European Union's H2020 Programme under GA No. 730224 (www.platirus.eu).

References

1. World Platinum Investment Council, Platinum Quarterly Q3 2015
2. Ryan A, Clarke B, Atilan O, Watts J, Trickey S (2010) Recycling: volume, momentum and consequences for PGMs. In: The 4th international platinum conference on platinum in transition 'boom or bust'. The Southern African Institute of Mining and Metallurgy
3. Lux H (1939) Acids and bases in fused salts: the determination of the oxygen ion concentration. Z Für Elektrochem 45(4):303–309 (in German language)
4. Flood H, Förland T (1947) The acid and basic properties of oxides. Acta Chem Scand 1:592–604
5. Process for Recovering Platinum Group Metals from Ores and Concentrates, US 2010/0257978 A1
6. Combes R, Vedel J, Tremillon B (1975) Measure of pO^{2-} by means of a zirconia-stabilized membrane electrode and potentiometric determination of the equilibrium constants of O^{2-} exchange reactions in the equimolar NaCl-KCl molten mixture. Electrochim Acta 20(3):191–200 (in French language)
7. Picard G (1993/94) Three-dimensional reactivity diagrams for reaction chemistry in molten chlorides. Molten Salt Forum 1(2):25–40
8. Castrillejo Y, Martinez AM, Haarberg GM, Børresen B, Osen KS, Tunold R (1997) Oxoacidity reactions in the equimolar molten $CaCl_2$-NaCl Mixture at 575 °C. Electrochim Acta 42(10):1489–1494
9. https://monolithos-catalysts.gr/en/
10. Seon F (1981) O^{2-} exchange reactions in the eutectic LiCl-KCl mixture at 470 °C. Application to the selective chlorination of metallic oxides in molten salts media. PhD thesis, University Pierre et Marie Curie, Paris VI (in French language)
11. Shenin-King J (1994) Processes involved in the chlorination of metallic oxides in molten salts media. PhD thesis, University Pierre et Marie Curie, Paris VI, (in French language)
12. Tremillon B (1997) Reactions in solution. An applied and analytical approach. Wiley
13. Littlewood R (1962) Diagrammatic representation of the thermodynamics of metal-fused chloride systems. Journal of Electrochemical Society 109(6):525–234

Study on the Mechanisms for Vanadium Phases' Transformation of Vanadium Slag Non-salt Roasting Process

Junfan Yuan, Yijun Cao, Guixia Fan, Hao Du, David Dreisinger, Guihong Han and Meng Li

Abstract Sodium and calcium salts roasting are main processes for the production of vanadium compounds using vanadium slag as raw material. These processes suffer from heavy pollution and operation difficulty, respectively. Based on the principle regarding ferrovanadium spinel oxidation, as well as investigation of vanadium phases' in situ transformation, a clean and efficient vanadium extraction process featuring non-salt vanadium slag roasting, ammonium meta-vanadate conversion, was established. The new process is expected to be able to realize high vanadium recovery and reduce exhaust gas and wastewater at source. Further, the tailing can be comprehensively utilized in conjunction with ironmaking process. This process may provide scientific and technical support for the green upgrading of vanadium industry.

Keywords Non-salt roasting · Vanadium slag · Ammonium leaching · Roasting and leaching mechanisms

---

J. Yuan · Y. Cao · G. Fan · G. Han · M. Li
School of Chemical Engineering, Zhengzhou University, Zhengzhou 450001, China

H. Du · M. Li
National Engineering Laboratory for Hydrometallurgical Cleaner Production Technology, Key Laboratory of Green Process and Engineering, Institute of Process Engineering, Chinese Academy of Sciences, Beijing 100190, China

Y. Cao
Henan Province Industrial Technology Research Institute of Resources and Materials, Zhengzhou University, Zhengzhou 450001, China

D. Dreisinger · M. Li (✉)
Department of Materials Engineering, University of British Columbia, Vancouver V6T1Z4, Canada
e-mail: limengubc@outlook.com; li.meng@zzu.edu.cn

G. Azimi et al. (eds.), *Rare Metal Technology 2020*, The Minerals, Metals & Materials Series, https://doi.org/10.1007/978-3-030-36758-9_22

Introduction

Vanadium is a strategic metal existing naturally in about 65 different minerals. Vanadium compounds have been widely used in many industries [1, 2], due to their unique physical and chemical properties [3]. Vanadium slag, a steel by-product slag produced during metallurgical treatment of vanadium titanomagnetite ore, is the direct raw material for vanadium extraction, accounting for more than 60% of the world's total production of vanadium [4, 5].

The most common industrial practice to recover vanadium from vanadium slag is by alkaline salt roasting followed by water leaching [3]. In this process, vanadium slag is mixed with sodium salts (NaCl, $NaCO_3$, and Na_2SO_4) and then roasted at 800–850 °C. However, in the alkaline roasting process, large amount of corrosive gases (HCl, Cl_2, SO_2, and SO_3) are generated due to the decomposition of sodium salts, which can seriously damage the equipment and pollute the environment [6]. Further, chromium spinel in the vanadium slag will be partially oxidized to hexavalent chromate salts under alkaline roasting condition, resulting in the accumulation of toxic CrO_4^{2-} in the process water and the leaching residues [7]. Furthermore, in order to precipitate the vanadium, around 30–50 t of high salinity ammonia wastewater containing Na_2SO_4, $(NH_4)_2SO_4$ and/or NH_4Cl will be generated per ton of V_2O_5 product.

In order to solve the problems associated with sodium salt roasting technology, calcium oxide or calcium carbonate is used as additive to replace sodium salts. During the roasting process, calcium vanadates ($Ca(VO_3)_2$, $Ca_3(VO_4)_4$, and $Ca_2V_2O_7$) are generated [8], which dissolve in dilute sulfuric acid solution, followed by precipitation and calcination to obtain V_2O_5 [9]. Compared with the sodium salt roasting technology, calcium roasting does not release harmful kiln gases. However, impurities like P and Fe dissolve into the leaching liquor, creating significant burden for subsequent product quality control. More importantly, during acid leaching, calcium sulfate is generated and accumulated in the leaching residue, causing significant difficulty for further residue utilization as an ironmaking feed [10]. Efforts are made to develop a new approach using ammonium carbonate as an alternative leaching reagent to replace sulfuric acid [9, 11]; however, the leaching reagent $(NH_4)_2CO_3$ is volatile and may decompose easily at temperatures above 35 °C [12–14]. According to Li and Vlek's research, ammonia losses were 20% for $(NH_4)_2CO_3$ solution containing 100 ppm N at around 46 °C, complicating exhaust gas recycling [13].

In this study, the principle of the non-salt roasting process was proposed. First, under the roasting temperature, chromium spinel in the raw vanadium slag cannot be converted to carcinogenic chromate salts. Second, after cooling crystallization, NH_4VO_3 can be easily separated as a product, and the mother liquor can be recycled for further leaching operations, avoiding the discharge of process water. More importantly, the leaching residue does not contain alkali metals; it thus can be utilized as ironmaking feed or additives.

Experiment

Materials

The vanadium slag, supplied by Chengde Iron and Steel Group Co., Ltd., Hebei, China, was dried in an oven overnight at 80 °C. The chemical compositions of the slag were determined by inductively coupled plasma optical emission spectrometry (ICP-OES, PerkinElmer Optima 7,300 V, USA), and the results are presented in supplementary Table 1. Phase compositions were determined by X-ray diffraction analysis (XRD, PANalytical Empyrean, Netherlands), with results shown in Fig. 1. The scanning electron microscopy (SEM, FEI Quanta 250, USA) and energy-disperse X-ray spectrometry (EDS, EDAX Genesis, USA) analysis results of the slag were shown in Fig. 2. In addition, vanadium slag was dry sieved, giving particle size fraction 48–75 μm for this work. Supplementary Table 1 shows that Fe, Si, V, Ti, and Mn are the major elements in the slag.

Experimental Procedure

During the non-salt roasting process, 50 g of vanadium slag was loaded in a corundum boat and roasted in a tube furnace. A number of different gas atmospheres were studied in the tube furnace. The gas flow rate was controlled by a flow meter. After roasting, the slag was air-cooled to room temperature followed by

Table 1 Chemical composition of vanadium slag (<200 mesh)

Compound	V_2O_5	FeO	SiO_2	CaO	Al_2O_3	MgO	Cr_2O_3	TiO_2	MnO
wt%	10.65	39.58	20.21	1.28	2.56	1.89	4.20	13.67	5.96

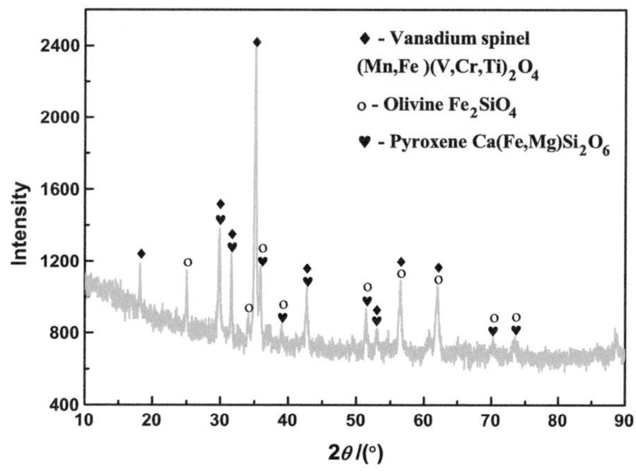

Fig. 1 XRD pattern of the vanadium slag

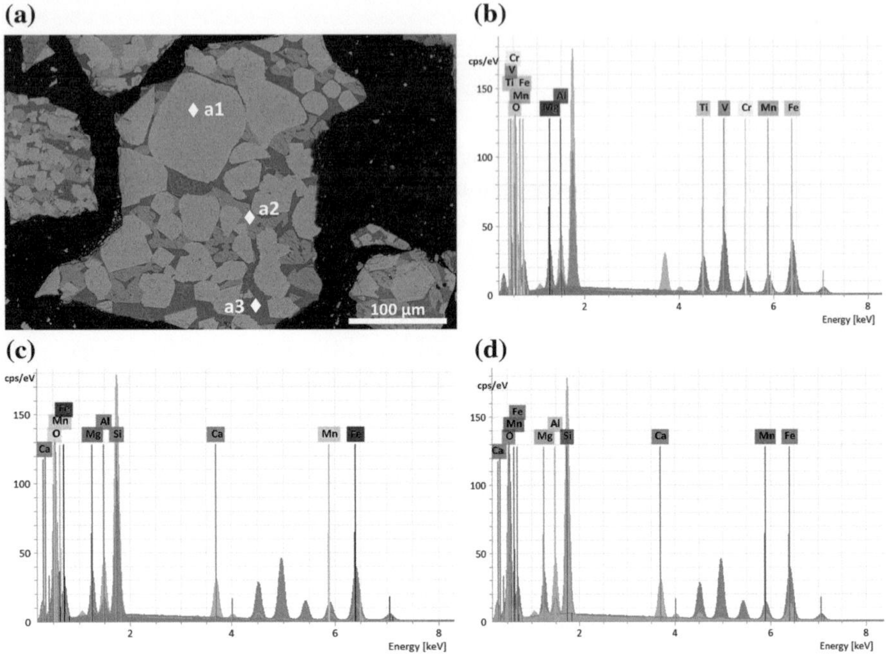

Fig. 2 The backscattered micrograph and the EDS elements mapping of the vanadium slag

milling by an agate mortar and dry sieved to obtain fine powders of 48–75 μm for XRD analysis and leaching experiments.

The roasted slag was leached in a glass vessel positioned in a thermostatic water bath (DF-101S, Bejing Xingde Instrument & Equipment Co., Ltd), equipped with a reflux condenser, mechanical stirring protected by Teflon. The leaching conditions were as follows: roasted vanadium slag particle size of 48–75 μm, ammonium oxalate concentrations of 5–15% by weight, solid-to-liquid (S/L) ratio of 1:4 (50 g slag and 200 g solution), temperature of 30–90 °C, time of 1–120 min, and stirring speed of 300 rpm. After leaching, the slurry was filtered, and the filter cake was washed with deionized water.

The recovery of vanadium was calculated using formula (1).

$$X = (1 - [V]r/[V]o) \cdot 100\% \tag{1}$$

where X represents the vanadium recovery, $[V]_o$ and $[V]_r$ are the contents of vanadium metal in the original vanadium slag and the leaching residue, respectively.

Results and Discussion

Roasting Mechanisms

After roasting at 400 °C for 2 h (Fig. 3a), the appearance of the diffraction peaks of Fe_2SiO_4, $Fe_{2.932}O_4$, and Fe_3O_4 suggested that fayalite was only partially decomposed and oxidized. However, after roasting at 700 °C for 2 h, as can be seen in Fig. 3a, the diffraction peaks of fayalite disappear and the diffraction peaks of Fe_2O_3 appear, indicating that fayalite was completely oxidized to Fe_2O_3. The oxidation of fayalite at the temperature from 400 to 700 °C can be expressed by the following reactions:

$$3Fe_2SiO_4 + O_2 = 2Fe_3O_4 + 3SiO_2 \quad (2)$$

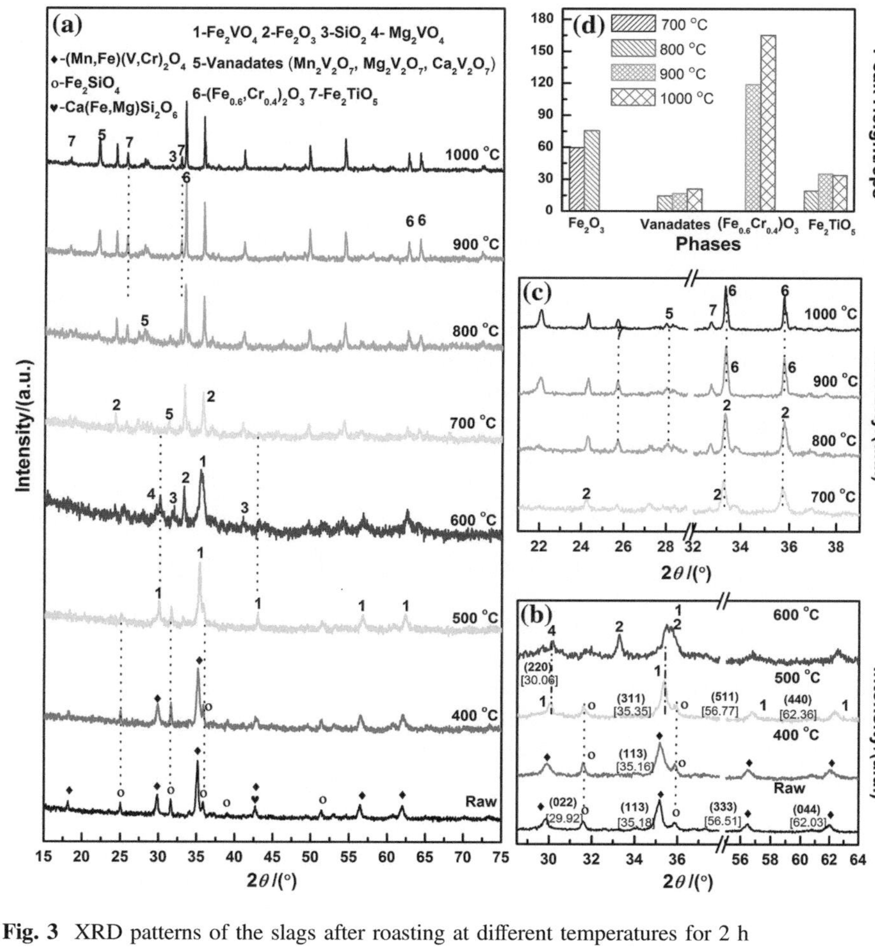

Fig. 3 XRD patterns of the slags after roasting at different temperatures for 2 h

$$4Fe_3O_4 + O_2 = 6Fe_2O_3 \tag{3}$$

$$2Fe_2SiO_4 + O_2 = 2Fe_2O_3 + 2SiO_2 \tag{4}$$

Further, after roasting at 700 °C (Fig. 2a) for 2 h, the characteristic diffraction peaks of vanadium spinel disappeared and the characteristic peaks of $Fe_2O_3{\cdot}V_2O_3$ and Fe_2O_3 were observed, suggesting that vanadium spinel was decomposed, but vanadium was not oxidized to the highest valence of V^{5+}. According to the study of Fotiev et al. [15], V_2O_3 can be easily oxidized to V_2O_5 in the temperature range of 100–370 °C. However, the diffraction peaks of V^{4+} still did not appear at 700 °C in this study, indicating that some types of solid solution $Fe_2O_3{\cdot}V_2O_3$ were formed. These solid solutions may inhibit the further oxidation of V^{3+} at the initial stage of the reaction. These results suggested that the oxidation of vanadium spinel can be presented by the following reaction:

$$FeV_2O_4 + O_2 = Fe_2O_3 \cdot V_2O_3 \tag{5}$$

After roasting at 800 °C for 2 h, the characteristic diffraction peaks of $Fe_2O_3{\cdot}V_2O_4$ and CaV_2O_6 were observed (Fig. 3a), indicating that V^{3+} in the solid solution began to be oxidized to V^{4+} and V^{5+}. The characteristic diffraction peaks of MnV_2O_6 and MgV_2O_6 were also observed, indicating that V_2O_5 could react with MgO and MnO. The presence of $Fe_2O_3{\cdot}V_2O_3$ and $Fe_2O_3{\cdot}V_2O_4$ diffraction peaks indicated that some vanadium cannot be fully oxidized to V^{5+}. The incomplete oxidation of vanadium slag can be explained that the melting point of most vanadate compounds (MnV_2O_6 and MgV_2O_6) is around 820–1,093 °C [16, 17], and may also form vanadate crust, affecting the further reaction of O_2 and vanadium spinel. The oxidation of vanadium compounds at the temperature from 800 to 1,000 °C mentioned above can be described by the following reactions:

$$2Fe_2O_3 \cdot V_2O_3 + O_2 = 2Fe_2O_3 \cdot V_2O_4 \tag{6}$$

$$2Fe_2O_3 \cdot V_2O_4 + 2CaO + O_2 = 2CaV_2O_6 + 2Fe_2O_3 \tag{7}$$

$$2Fe_2O_3 \cdot V_2O_4 + 2MnO + O_2 = 2MnV_2O_6 + 2Fe_2O_3 \tag{8}$$

$$2Fe_2O_3 \cdot V_2O_4 + 2MgO + O_2 = 2MgV_2O_6 + 2Fe_2O_3 \tag{9}$$

The formation of $Fe_2O_3{\cdot}V_2O_4$ was also detected by other researchers [18–21].

In the whole oxidation process, the chromium spinel was very stable and cannot be oxidized to Cr^{6+} by O_2 at these conditions. As shown in Fig. 3a, after roasting at 1,000 °C for 2 h, the chromium exists in either chromium spinel or solid solutions R_2O_3 (R = Cr, Fe).

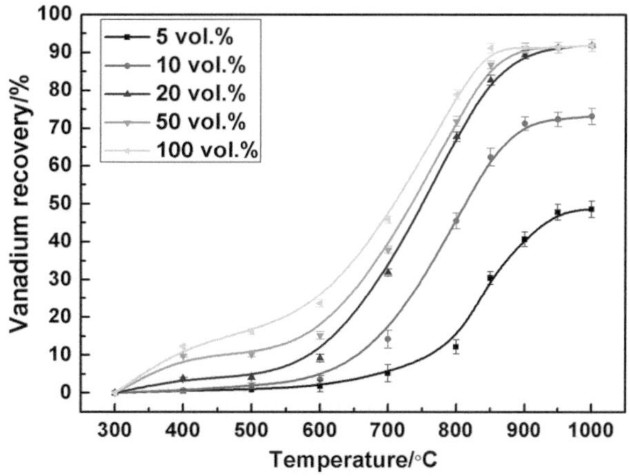

Fig. 4 Effect of oxygen concentration on vanadium recovery

Effect of Oxygen Concentration on the Vanadium Recovery

Figure 4 illustrates the effect of oxygen concentration in the roasting atmosphere on vanadium recovery. The holding time was 2 h, and the particle size was 48–75 μm. From Fig. 4, it is clear that oxygen concentration has a significant effect on the vanadium recovery, and at all the temperatures considered, an increase of oxygen concentration was beneficial for the oxidation of vanadium compounds. Such observation is expected since oxygen is the main oxidant for oxidation of vanadium slag, and the increase of oxygen concentration favors the oxidation process both thermodynamically and kinetically. However, further observation of Fig. 4 suggested that, at oxygen concentrations higher than 20 vol.%, the final recovery of vanadium did not show significant change at temperatures higher than 900 °C, indicating that roasting using air would provide enough oxygen for the reaction to proceed if other conditions were well optimized.

Conclusions

In this work, the non-salt roasting process of vanadium slag was investigated. Further, it was found that the vanadium recovery increased with an increase in the roasting temperature. The oxygen concentration and the holding temperature have been proven to play a critical role in the roasting process.

Acknowledgements This work was financially supported by the National Key R&D Program of China (Grant No. 2018YFC1901601), the Scientific Research Starting Allowance for the Outstanding PhD Talent of ZZU (Grant No. 32211254), the Key Research Project of Henan Province Colleges and Universities (Grant No. 20A440011), and Program for Innovative Research Team (in Science and Technology) in University of Henan Province (Grant No. 19IRTSTHN028).

References

1. Liu Y, Liu Z, Mnichowicz B, Harinath AV, Li H, Bahrami B (2016) Chem Eng J 287:680
2. Mase A, Sugita T, Mori M, Iwamoto S, Tokutome T, Katayama K, Itabashi H (2013) Chem Eng J 225:440
3. Moskalyk R, Alfantazi A (2003) Miner Eng 16:793
4. Gustafsson S, Zhong WW (1985) Int J Miner Process 15:103
5. Zhang Y-M, Bao S-X, Liu T, Chen T-J, Huang J (2011) Hydrometallurgy 109:116
6. Barolin H (1982) Oxidation of vanadium slag, 1st edn. Metallurgical Industry Press, Beijing pp, 121–131
7. Fang L (2012) Ferro-Alloys 227:41
8. Zhonghao Z, Yanheng W (2000) Chemical World 6:290
9. Li H-Y, Wang K, Hua W-H, Yang Z, Zhou W, Xie B (2016) Hydrometallurgy 160:18
10. Tavakoli M, Dreisinger D (2014) Hydrometallurgy 147:83
11. Li M, Liu B, Zheng S, Wang S, Du H, Dreisinger DB, Zhang Y (2017) J Clean Prod 149:206
12. Meng X, Han KN (1996) Min Proc Ext Met Rev 16:23
13. Vlek P, Stumpe J (1978) Soil Sci Soc Am J 42:416
14. Li M, Zheng S, Liu B, Wang S, Dreisinger D, Zhang Y, Du H, Zhang Y (2017) Min Proc Ext Met Rev 38:228
15. A. Fotiev and V. Strelkov, *Russ. J. Inorg. Chem*, 26, (1981)
16. Chung S, Shin S, Andriiko A, Rudenok P (1999) J Mater Res 14:2929
17. Clark G, Morley R (1976) J Solid State Chem 16:429
18. Basu S, Taniguchi M (1984) J Therm Anal 29:1209
19. Maciejewski M, Reller A, Baiker A (1985) Thermochim Acta 96:81
20. Stander P, Van Vuuren C (1990) Thermochim Acta 157:347
21. Van Vuuren C, Stander P (1995) Thermochim Acta 254:227

Reclamation of Precious Metals from Small Electronic Components of Computer Hard Disks

Rekha Panda, Manis Kumar Jha, Om Shankar Dinkar and Devendra Deo Pathak

Abstract Present research work focuses on the recovery of precious metals (Ag, Au, Pd, and Pt) from the leach liquor of small populated chips present in the hard disk of computers. Initially, the hard disks were dismantled to separate the printed circuit boards (PCBs) followed by its depopulation to liberate the mounted small electronic components. The liberated black chips were pulverized to ~100 mesh and chemically analysed. The powdered black chips containing ~0.6% Ag, ~0.3% Au, ~0.01% Pd, ~0.0003% Pt, and 20% Cu on mass basis were first leached in nitrate medium for maximum dissolution of non-ferrous metals along with Ag leaving Au, Pd, and Pt in the residue. About 99.99% of precious metals were leached out from the residue using suitable lixiviant. The obtained leach liquor was purified using advanced separation techniques (solvent extraction/ ion-exchange/ precipitation) from which marketable products (metals/ salts) could be produced.

Keywords Hydrometallurgy · Precious metals · Hard disk · Small black chips

Introduction

In recent years, the electronic industries have emerged as a boon to mankind providing phenomenal benefits, which has made life very luxurious and comfortable [1]. Electronic devices have become essential in various spheres of life, among which telecommunication, medical, trade and commerce, and defence are key areas [2–4]. Exponential increase in the demand along with cheaper cost has resulted in

R. Panda · M. K. Jha (✉) · O. S. Dinkar
Metal Extraction and Recycling Division, CSIR-National Metallurgical Laboratory, Jamshedpur 831007, India
e-mail: mkjha@nmlindia.org; maniskrjha@gmail.com

R. Panda · D. D. Pathak
Department of Applied Chemistry, Indian Institute of Technology (ISM), Dhanbad 826004, India

G. Azimi et al. (eds.), *Rare Metal Technology 2020*, The Minerals, Metals & Materials Series, https://doi.org/10.1007/978-3-030-36758-9_23

the penetration of different electronic items in the markets of both developed and developing nations [5]. Consumer electronics such as computers, laptops, mobile phones, televisions, and navigation devices have made communication and information sharing easy, comfortable, and proficient in economical ways. Massive demand for electronic goods in the market, reduced lifespan and constant replacement of older with new models due to technological transformation, has resulted in the generation of huge amounts of e-waste at their end of life [5, 6]. These electronic items are made of different valuable, precious as well as toxic materials; thus the constant generation of large volume of e-waste to the environment at a disquieting rate is a serious topic of concern globally.

All electronic devices consist of printed circuit boards (PCBs), which is made of variety of metals and materials. The manufacturing of PCBs consumes large amount of rare and precious elements as raw materials, which are very scarcely present in the Earth's crust. PCBs are made of various metals such as Cu, Ni, Zn, Pb, Sn, Au, Ag, and Pd. Copper is present in plenty along with Au and Ag in lesser quantity with other metals. Even though the presence of Au is $\sim$0.0016 wt% of the total weight of PCBs, it alone makes up for more than 65% value of metals present in PCBs [7]. About 9.3 million tonnes of e-waste was discarded in 2014 by the information and communication technology (ICT) sector worldwide [8]. About 44.7 million tonnes of e-waste was generated in 2016, which is expected to reach about 52.2 million tonnes by 2021 [9]. Keeping in view the above mentioned percentage of Au (0.0016 wt%), it can be assumed that about 148.8 tonnes of gold was disposed into the landfills, converting billions of dollars into waste. The increase in demand for electronic devices and their short life span is exerting immense load on mining that will lead to exhaustion of available primary resources, which are limited in nature. Thus, proper management of e-waste and their effective recycling is imperative to address this growing problem. Recovery of valuable metals from e-waste will not only prevent negative environmental footprints, but it will also help in sustainable utilization of metallic values, thereby minimizing the load on mining.

Different metallurgical processes including hydrometallurgy and pyrometallurgy have been used for the recycling of e-waste [10]. Pyrometallurgy has been employed for the recovery of metals such as iron, copper, aluminium, lead, and precious metals from e-waste [11–13]. As e-waste is heterogeneous in nature with complex chemical composition [14], pyrometallurgical route is not efficient because it requires high temperature with high energy consumption as well as emission of toxic gases (furans, dioxins, and greenhouse gases). Thus, hydrometallurgy still remains the preferred method to process e-waste because of its exceptional characteristics like selectivity, predictability, ease of control as well as low energy consumption [15].

Hydrometallurgy has been widely used to recover metallic values from the PCBs of e-waste, but negligible research and development efforts have been made till date to recover valuable metals from small electronic components (EC) mounted on the surface of PCBs such as microprocessors, transistors, black ceramic chips, capacitors, and inductors. Large quantities of EC are generated during the manufacturing of PCBs as well as during their end of life, which forms the part of waste stream.

The EC is rich in precious metals, and recovery of such valuable metals from the EC is of immense economic importance. The present paper reports a systematic hydrometallurgical study for the separation of non-ferrous (Cu, Ni, and Ag) and the recovery of precious metals (Au, Pd, and Pt) from the liberated black chips of the PCBs. The developed process is economical, low energy consuming, and complies with strict environmental rules and regulations. The obtained results from the present study will help to simulate the process in a continuous mode.

Experimental

Raw Material

Black chips mounted on the PCBs of scrap computer hard disk were used as feed material for the leaching experiments. All chemical reagents (nitric acid, sulfuric acid, hydrochloric acid, sodium hypochlorite, and sodium cyanide) were of laboratory grade (LR) supplied by Merck, India.

Methods

Leaching experiments were carried out in a temperature-controlled three-necked Pyrex glass reactor well fitted with condenser where the raw material along with suitable leachant at particular S/L ratio was heated to a desired temperature on a hot plate. Continuous mixing of the sample was maintained during leaching using magnetic stirring to maximize the mass transfer effect during leaching. Satisfactory mass balance was obtained for each set of experiments. The concentration of metals were analyzed using atomic absorption spectrometer (AAS) (Perkin Elmer model, Analyst 200; USA).

Results and Discussion

Pre-treatment and Physical Beneficiation

Initially, the hard disks were dismantled, and the mounted small ECs were separated from the PCBs. Further, all the populated ECs were liberated from the PCBs by thermal treatment. From the obtained liberated components, black ceramic chips were separated and pulverised to $\sim$100 mesh size (Fig. 1). The powdered black

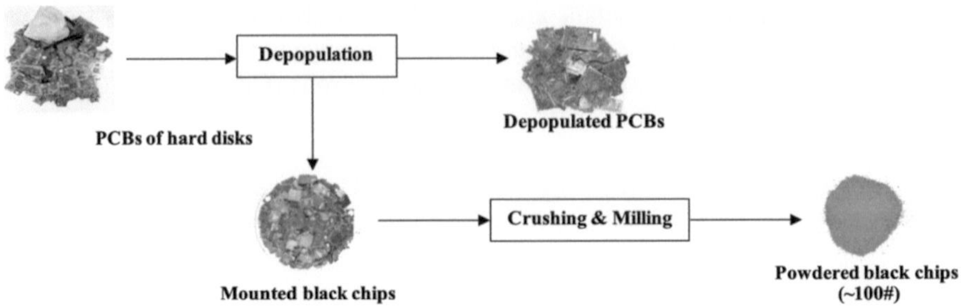

Fig. 1 Flow-sheet for sample preparation

chips containing ~0.6 wt% Ag, ~0.3 wt% Au, ~0.01 wt% Pd, ~0.0003 wt% Pt, and 20 wt% Cu were used for experimental purpose.

Leaching Studies

Systematic leaching studies were carried out for maximum dissolution of metals (non-ferrous and precious) from the beneficiated metallic fractions of black chips present in the PCBs of hard disk. Leaching of metals was carried out in two steps. In the first step, leaching studies were carried out to remove maximum non-ferrous metals (Cu, Ni, etc.) along with Ag, leaving other precious metals like Au, Pd, and Pt in the residue. In the second step, the residue enriched with Au, Pd, and Pt was leached using appropriate leachant to achieve maximum precious metals in solution.

Leaching of Non-ferrous Metals Along with Ag

The feed material (powdered black chips) contains precious as well as non-ferrous metals (mainly Cu); therefore, at first, leaching was carried out to remove the impurities (non-ferrous metals) and enrich the concentration of precious metals (Au, Pd, and Pt) in the residue. Different lixiviants, such as HCl, H_2SO_4, and HNO_3 were used to select the suitable lixiviant to dissolve all non-ferrous metals along with Ag from the black chips powder leaving precious metals in the residue. HNO_3 proved to be the most suitable leachant among all acids (Fig. 2) and was selected for further experiments. Experiments were carried out at various concentrations of HNO_3 from 1.5 to 4.5 M at 80 °C, keeping pulp density at 50 g/L for 30 min, to observe the behaviour of metal dissolution. Leaching of non-ferrous metals (Cu, Ni, and Ag) was found to increase with the increase in acid concentration (Fig. 3).

About 99.99% of Cu, Ni, and Ag were leached under the above experimental condition. Further increase in acid concentration showed negligible increase in

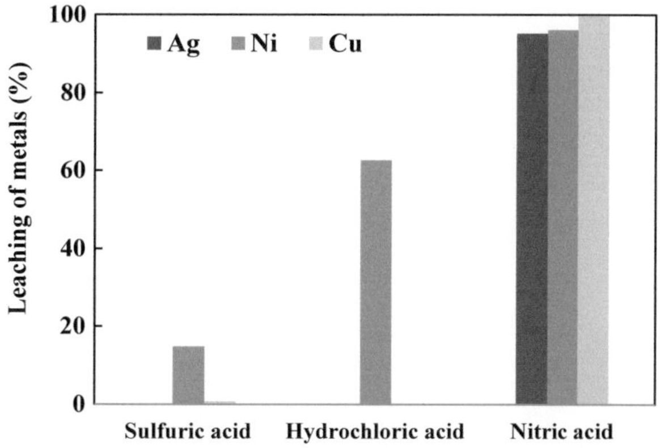

Fig. 2 Selection of leachant (Pulp density: 50 g/L, Temp: 80 °C, Time: 30 min)

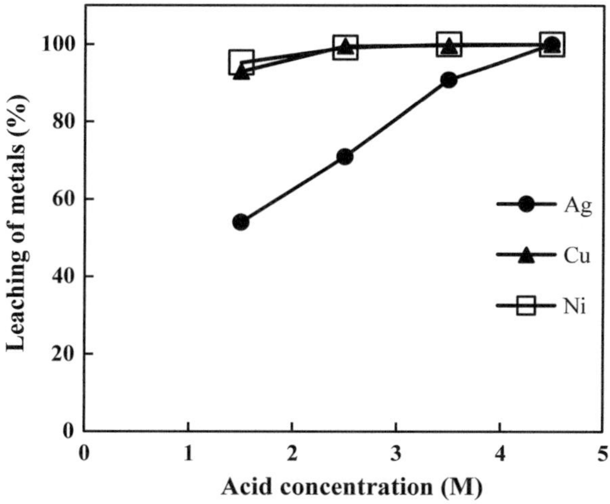

Fig. 3 Effect of acid concentration (Pulp density: 50 g/L, Temp: 80 °C, Time: 30 min)

leaching of metals. Therefore, 4.5 M HNO_3 was considered optimum for leaching of non-ferrous metals from the black chips powder. Experiments were also performed at 30 °C (room temperature), 70 °C, and 80 °C using 4.5 M HNO_3 maintaining 50 g/L pulp density and 30 min mixing time for maximum leaching of impurities (Cu, Ni, and Ag) as shown in Fig. 4. Complete removal of non-ferrous metals along with Ag (99.99%) was achieved using 4.5 M HNO_3 at 80 °C, keeping pulp density at 50 g/L for 30 min mixing time. The leached solution was filtered to separate the residue containing Au, Pd, and Pt from the leach liquor. The residue was washed and dried under vacuum for 2 h at 80 °C and further used for leaching of precious metals in the second step.

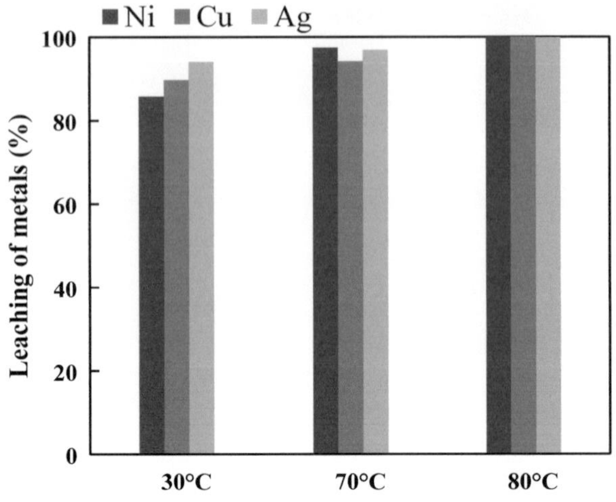

Fig. 4 Effect of temperature (Acid: 4.5 M HNO_3; Pulp density: 50 g/L, Time: 30 min)

Leaching of Precious Metals from the Leached Residue

The leached residue obtained after nitric acid leaching was found to contain ~0.41 wt% Au, ~0.014 wt% Pd, and ~0.0004 wt% Pt. Experiment was done for the leaching of precious metals using two different lixiviants, i.e., aqua regia and a mixture of hydrochloric acid and sodium hypochlorite. Maximum leaching of Au, Pd, and Pt were achieved using both the leachants at 80 °C for 1 h residence time and 100 g/L pulp density. More than 99% of precious metals were leached out which will be purified using advance separation techniques to obtain pure solutions from which their value-added products (metal/salts) could be produced. All leaching experiments were carried out in fume hood with proper ventilation following all safety measures.

Recovery of Metals as Value-Added Products

The leach liquor obtained in both leaching steps was further purified using advance separation techniques to obtain pure solutions from which metals were recovered as marketable products. The leach liquor containing Cu, Ni, and Ag was initially processed for the precipitation of Ag followed by the selective separation of Cu and Ni using solvent extraction technique. About 99.99% of Ag precipitated out as silver chloride (AgCl) using NaCl at 15 min residence time. The filtrate after Ag precipitation was processed with solvent extraction using organic extractant LIX 84IC to selectively separate Cu from Ni at pH of 2.2 maintaining phase ratio (O/A) of 1:1 for 10 min residence time. About 99.99% of Cu was extracted from the filtrate leaving 99.99% of Ni in the raffinate. The Cu-loaded organic was stripped using 10 wt% H_2SO_4 to obtain pure solution of Cu. From the obtained pure solutions of Cu and Ni,

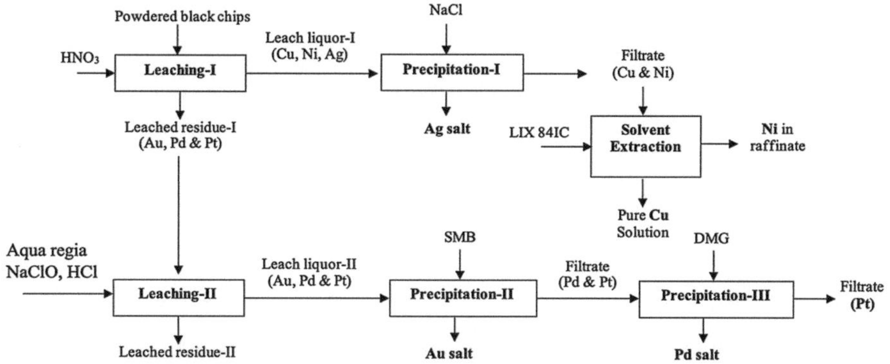

Fig. 5 Process block flow diagram for the recovery of precious metals from small electronic components of hard disk

both metals could be recovered as marketable salts (with high purity) using evaporation/ precipitation techniques.

Precipitation was also employed to recover Au, Pd, and Pt from the leach liquor obtained in the second step. Initially, sodium metabisulphite (SMB) was used to precipitate Au. The precipitate was filtered and the filtrate obtained was processed for Pd precipitation using dimethyl glyoxime (DMG). The filtrate obtained after precipitation of Pd was evaporated to obtain Pt as a salt. Thus, all metals (Cu, Ni, Ag, Au, Pd, and Pt) present in the leach liquor were recovered as value-added marketable product.

Conclusions

The developed process to recover precious metals from black chips of computer hard disks will not only minimize the loss of precious metals present in this waste stream, but it will also help to reduce the costs associated with mining and processing of their primary ores. Therefore, present hydrometallurgical flow-sheet (Fig. 5) developed for the separation of non-ferrous metals and recovery of precious metals from the liberated black chips is economical and feasible. It is a energy-saving process and complies with strict environmental rules and regulations.

Acknowledgements The authors are thankful to the Director, CSIR-National Metallurgical Laboratory, Jamshedpur for giving the permission to publish this paper. One of the authors Ms. Rekha Panda would also like to extend her sincere gratitude to CSIR, New Delhi (Grant: 31/10 (64)/ 2017-EMR-I) for providing Senior Research Fellowship for carrying out this research work.

References

1. Veit HM, Bernardes AM (2015) Electronic waste: generation and management, in: electron, waste recycle Tech. Springer International Publishing, Cham pp. 3–13
2. Mikre F (2011) The roles of information communication technologies in education–review article with emphasis to the computer and internet. Ethopian J Educ Sci 6:109–126
3. UNCTAD (2008) Measuring the impact of ICT use in business: the case of manufacturing in Thailand, pp. 20–21
4. Ogu EC, Oyerinde OD (2014) ICT and national security in developing and underdeveloped countries–the Good, the bad and the ugly: a case study of Nigeria' s cyberspace. Int J Comput Sci Inf Technol 5:5625–5633
5. Lundgren K (2012) The global impact of e-waste-addressing the challenge. Geneva
6. Namias J (2013) The future of electronic waste recycling in the United States: obstacles and domestic solutions, Columbia University
7. Dimitrijević S, Vlastimir T, Aleksandra I, Stevan D, Mladen M (2013) Recycling of precious metals from e-scrap, Iran. J Chem Chem Eng 32:17–23
8. Baldé CP, Wang F, Kuehr R, Huisman J (2014) The global e-waste monitor: quantities, flows and resources. p. 24
9. Baldé CP, Forti V, Gray V, Kuehr R, Stegmann P (2017) The global e-waste monitor: quantities, flows and resources. p. 5
10. Tuncuk A, Stazi V, Akcil A, Yazici EY, Deveci H (2012) Aqueous metal recovery techniques from E-scrap: hydrometallurgy in recycling. Miner Eng 25:28–37
11. Prabhakaran D, Kannadasan T (2013) Recovery of noble metals from electronic waste by pyrometallurgy process arensen aier. pp. 209–212
12. Syed S (2012) Recovery of gold from secondary sources-a review. Hydrometallurgy 115–116:30–51
13. Canda L, Heput T, Ardelean E (2016). Methods for recovering precious metals from industrial waste, IOP Conf Ser Mater Sci Eng 106
14. Ahmed S, Panwar RM (2014) Analysis of barriers of e-waste management using interpretive structural modeling (ISM) methodology. Int Conf Innov Trends Appl Phys Chem Math Sci Emerg Energy Technol Sustain Dev pp. 94–96
15. Cui J, Zhang L (2008) Metallurgical recovery of metals from electronic waste: a review. J Hazard Mater 158:228–256

Part IV
V, Mn, Co, Zn, Mo, Cu, REEs

Study on Vanadium Phase Evolution Law in Vanadium Slag During the Interface Reaction Process of Sodium Roasting

Dan-Qing Li, Yang Yang, Hong-Yi Li and Bing Xie

Abstract Vanadium slag is composed of many complex phases with various kinds of elements, occurrence forms, and chemical forms. At present, sodium roasting–water leaching can effectively extract vanadium; this technique has been widely used in industrial production, but the current researches to evolution law of the V-concentrating phase in V-slag during the roasting process focus on the macro-level, not directly on micro-level and especially at interface reaction process. Consequently, the microstructure, phase compositions, mineralogical morphology, and elements distribution on the surface of vanadium slag during the roasting process were investigated by various surface analysis techniques. It is essential to truly understand and control the sodium roasting–water extraction process, so the atomic evolution law can be of a great significance to optimize operating conditions in order to raise vanadium recovery rate.

Keywords Vanadium slag Sodium roasting Interface reaction · Surface analysis

Introduction

As an important nonferrous metal and strategic resource, vanadium is widely used in metallurgy, chemical, and material industries because of its chemical activity and particular assist in improving performance, such as refining the structure and grain of steel and reducing the strength and toughness of steel [1–3]. The major raw source of vanadium worldwide is from vanadium titano-magnetite ore that has proven tremendous reserves at Panzhihua mineral region in China [4]. At present,

D.-Q. Li · Y. Yang · H.-Y. Li (✉) · B. Xie
College of Materials Science and Engineering, Chongqing University,
Chongqing 400044, China
e-mail: hongyi.li@cqu.edu.cn

D.-Q. Li · Y. Yang · H.-Y. Li · B. Xie
Chongqing Key Laboratory of Vanadium-Titanium Metallurgy and New Materials,
Chongqing 400044, China

G. Azimi et al. (eds.), *Rare Metal Technology 2020*, The Minerals,
Metals & Materials Series, https://doi.org/10.1007/978-3-030-36758-9_24

vanadium is extracted mainly from vanadium titano-magnetite ore by vanadium hydrometallurgy technology, by enriching into solid slag and then leaching with water after oxidative roast with sodium slats, so-called sodium roasting–water extraction technique [5]. This technique has been widely used in industrial production [6], but the current researches to evolution law of the V-containing phase in V-slag during the roasting process focus on the macro-level phase changes, not directly on Vanadium itself. Consequently, micro-level research to state of V including valence, distribution, and phase is becoming increasingly essential to truly understand and control the sodium roasting–water extraction process [7–9]. The atomic evolution law can be of a great significance to optimize operating conditions in order to raise vanadium recovery rate.

For the vanadium slag selected in this experiment, the optimal sodium roasting scheme has been given in the literature: Na_2CO_3 as the natrifier, 800 °C as the roasting temperature and 90 min as the roasting time [9–11]. However, previous studies only started from the macro-level, analyzed the content of vanadium in the leached solution after sodium roasting by ICP-AES, and calculated the final conversion rate of vanadium slag under this condition. In the end, only the final result is obtained without in-depth exploration of the process of change and dynamic analysis of the whole roasting process. Therefore, on this basis, this study will further explore to provide theoretical support. From the perspective of dynamic and microscopic, this study explores in this condition, as the roasting time extension of vanadium slag phase change, so this study chose five points: 0 (no roasting), 5, 50, and 90 min of roasting vanadium slag at different times were observed, respectively. SEM/EDS technology and XPS scanning technology were used to explore the evolution rule of vanadium-containing phase in the interface reaction of vanadium slag roasting, and the microstructure, surface element distribution, and phase composition of the slag phase were obtained at different roasting times.

Experiment

Instruments and Reagents

All reagents were accurately weighed by FA2004 N model electronic scales. Samples were roasted and preserved using SRX2-5-12 model muffle furnace and WG-71 model draught drying cabinet. Samples were analyzed using TESCAN VEGA III LMHINCA model scanning electron microscope (SEM), INCA Energy 350 model X-ray spectrometer (EDS), and ESCALAB250Xi model X-ray photoelectron spectrometer (XPS).

The vanadium slag used in this experiment is high-chromium vanadium slag of converter, whose chemical composition is shown in Table 1. The sodium agent is Na_2CO_3 of analytical-reagent grade (Chengdu KeLong Chemical Reagent Factory).

Table. 1 The chemical compositions of vanadium slag (mass fraction, %)

Composition	FeO	SiO_2	MgO	Al_2O_3	V_2O_3	Cr_2O_3	CaO	MnO	TiO_2
Content	23.79	24.59	15.35	10.02	9.79	1.58	1.85	1.58	3.24

Experimental Methods

Na_2CO_3 was ground into a 200-mesh sieve (less than 74 μ) using an agate mortar, and then it is evenly sprinkled on the surface of the block vanadium slag (less than 0.1 g/cm^2), and sodium roasting process in the 800 °C muffle furnace, when the sodium roasting time reached 0 (noroasting), 5, 50, and 90 min sampling. The microstructure was observed by scanning electron microscopy (SEM) and the surface composition of block samples was analyzed by X-ray energy spectrometer (EDS).

Results and Discussion

Vanadium Slag Without Sodium Roasting

Figure 1a shows the morphology of unpolished vanadium slag on the surface. It can be seen that a large number of large-grain spinel with regular shape (three-pyramidal and four-pyramidal) grow out of the matrix. When the positions of several phases are close to each other, their growth will be restricted. The surface morphology of vanadium slag after polishing and grinding is shown in Fig. 1b1. Combined with the backscattered electron image of the polished surface in Fig. 1b2, it can be seen that each spinel has a regular and straight edge, with clear boundaries between each phase and large difference in composition contrast. EDS line scan was carried out on the red arrow through parts in Fig. 1b2. It can be found in Fig. 1b3 that the dark area at the tail and head of the red arrow is rich in Si, Mg, Ca, and Al elements, indicating that it is mainly rich area of peridot, while the light-colored area in the middle of the red arrow is rich in V, Ti, Cr, and Fe elements, indicating that it is mainly spinel phase, with regular and straight edges.

SEM/EDS analysis results of spinel phase after amplification are shown in Fig. 2. From EDS area-scan results, the atomic enrichment in the spinel phase was significantly different from that in the external matrix, mainly reflected in the differences of V, Cr, Fe, and Ti. There is almost no Si in spinel; Al and O are distributed in both spinel phase and matrix phase, with more distribution in the matrix.

As shown in Fig. 3, the elements on the surface of unroasted vanadium slag are O, Ca, Mg, Fe, V, Si, and Al; this is because their peak signal strength is greater through analysis. The signal of V and Fe on the surface of vanadium slag is weak, which is because vanadium is mainly existed in Fe_2VO_4 spinel. The results of XPS well confirm the conclusion of SEM/EDS.

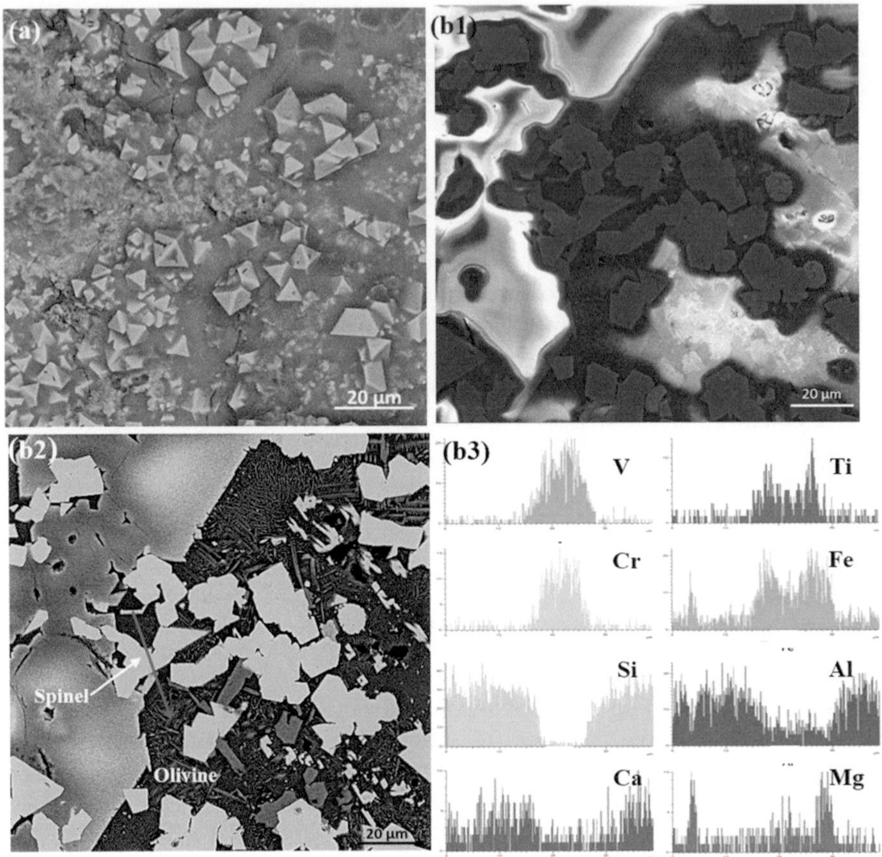

Fig. 1 SEM images of V-slag samples **a** surface morphology of unpolished V-slag; **b** surface morphology and line scanning images of polished surface

Vanadium Slag Was Sodium Roasted for 5 Min

When roasting for 5 min, it can be seen from Fig. 4 that the peridot phase has reacted with Na_2CO_3 to a certain extent, and the silicate matrix structure on the surface of a few areas has been destroyed and is fluffy; it provides a channel for the subsequent reaction of O_2 into vanadium slag spinel, and the main reaction between spinel and Na_2CO_3 is shown in Eq. (1). As can be seen from the composition contrast diagram, the boundary of spinel phase is still regular and smooth, Fe is concentrated on the surface of spinel and reacts with Na_2CO_3, while V and Cr still exist inside spinel, with less outer distribution than the central region.

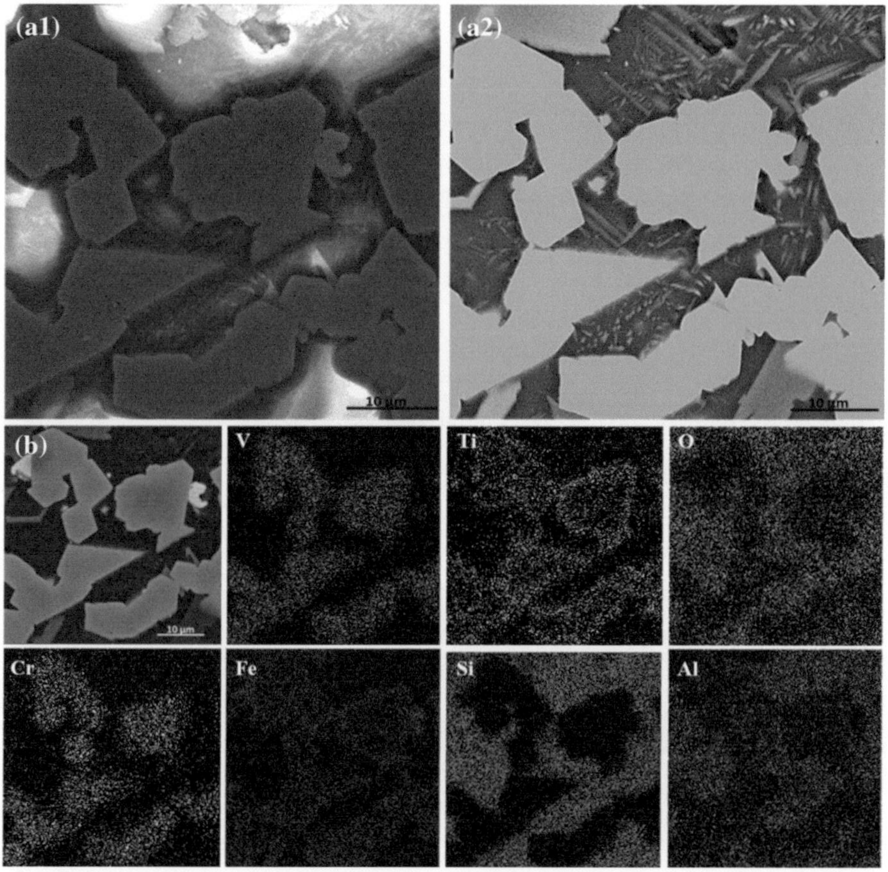

Fig. 2 SEM images of V-slag samples **a** surface morphology of the spinel **a2–b** backscattered electron image and elemental mapping images of polished surface

$$FeV_2O_4 + Na_2CO_3 + \frac{5}{4}O_2 = 2NaVO_3 + \frac{1}{2}Fe_2O_3 + 2CO_2 \quad (1)$$

Figure 5 shows the full-spectrum scanning results of element composition at the reaction interface of roasting for 5 min. As can be seen from it, the accumulated elements on the surface of vanadium slag are mainly O, Ca, Mg, Fe, V, and Si. The results of XPS well confirm the conclusion of SEM/EDS.

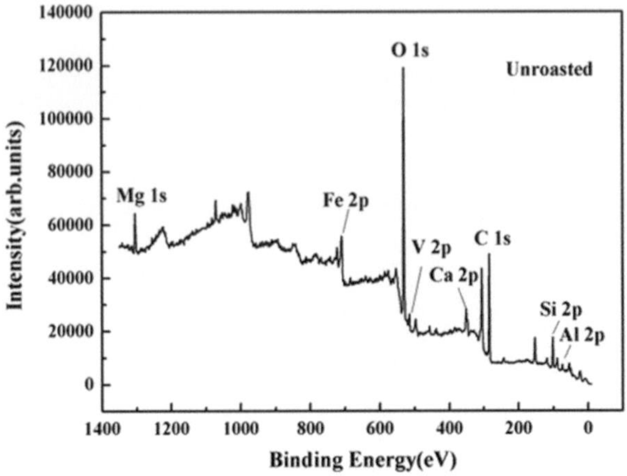

Fig. 3 XPS spectra of all elements in the unroasted V-slag

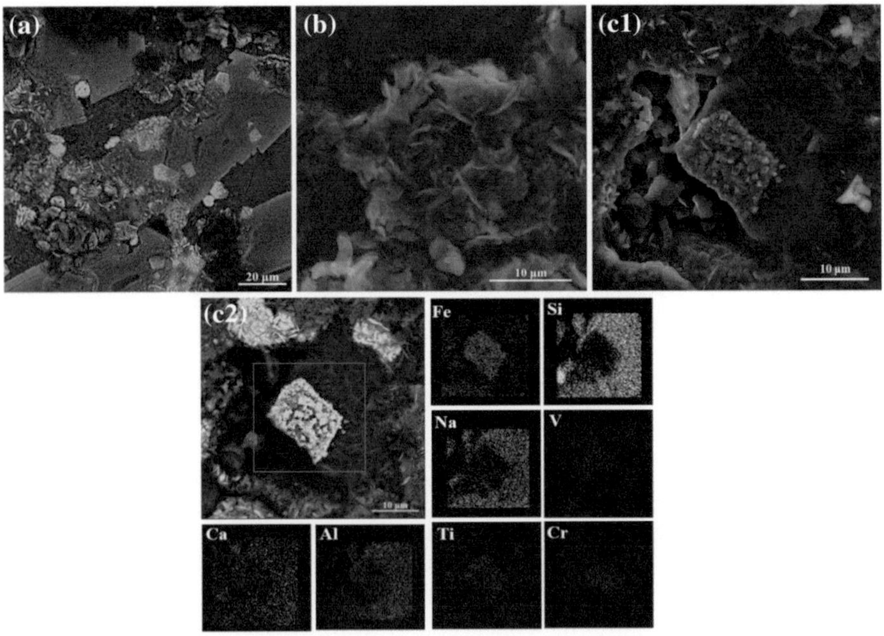

Fig. 4 SEM images of V-slag samples after roasting for 5 min at 800 °C **a–c1** secondary electron image; **c2** backscattered electron image and elemental mapping images

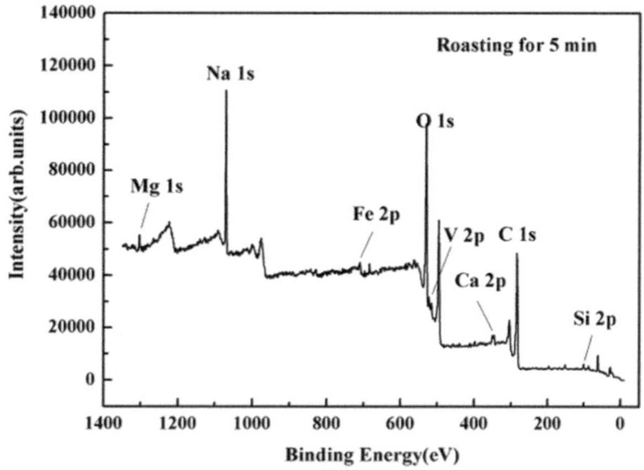

Fig. 5 XPS spectra of all elements in the V-slag after roasting for 5 min

Vanadium Slag Was Sodium Roasted for 50 Min

It can be seen from Fig. 6 that when sodium roasting for 50 min, the linear cluster of sodium vanadate crystals has been widely distributed on the surface of vanadium slag, and the size of most of them has exceeded 20 μ, further hindering the reaction between silicate matrix and Na_2CO_3. Backscattered electrons mainly reflect the composition characteristics of the sample surface, that is, the parts with large atomic number of the sample will generate strong backscattered electron signals and form brighter regions. Backscattered electrons mainly reflect the composition characteristics of the sample surface, that is, the parts with large atomic number of the sample will generate strong backscattered electron signals and form brighter

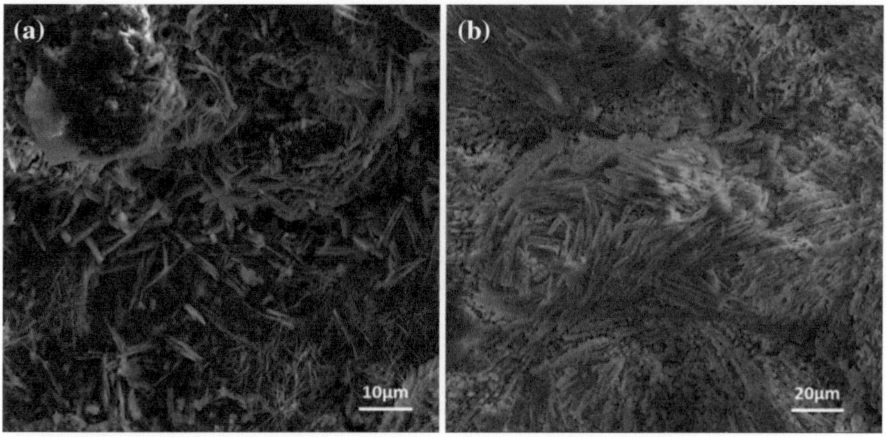

Fig. 6 SEM images of V-slag samples after roasting for 50 min at 800 °C **a** secondary electron image; **b** backscattered electron image

Fig. 7 XPS spectra of all elements in the V-slag after roasting for 50 min

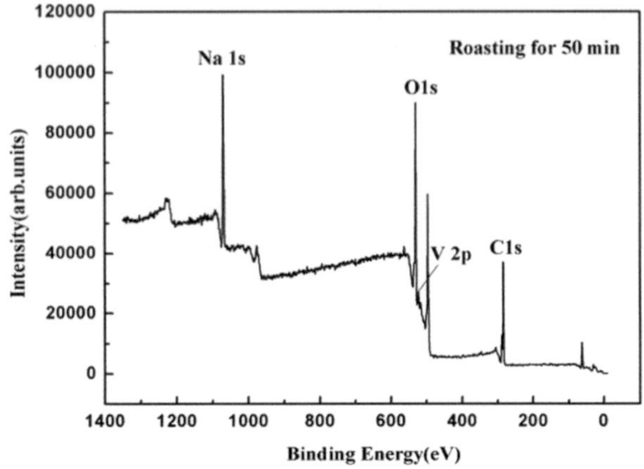

regions. As can be seen from Fig. 6b, the difference of distribution of elements on vanadium slag surface becomes smaller, and the distribution of elements becomes similar, so the difference of composition contrast decreases.

XPS results confirmed the conclusion of the SEM/EDS, as shown in Fig. 7. Reaction interface elements Na, C, O, and V show that sodium vanadate salts have covered on the surface of vanadium slag, burying the products of the reaction between silicate matrix and Na_2CO_3. That is to say, with the extension of roasting time, sodium vanadate crystal continuously diffuses and grows on the surface of vanadium slag, with obvious competitive advantages, and its inhibition effect on the reaction between silicate matrix and Na_2CO_3 is continuously enhanced.

Vanadium Slag Was Sodium Roasted for 90 Min

Figure 8 shows that a large number of sodium vanadate crystal particles have been generated and attached to the surface of vanadium slag when sodium roasting for 90 min. Figure 8b shows that dendrites grow out of the voids in the sodium vanadate particles, indicating that Na_2CO_3 has reacted with the deeper vanadate spinel to form vanadate.

The results of point-scan analysis of surface products are listed in Table 2; it can be seen that all the elements in the reaction interface (C_1,C_2,C_3, and C_4 points in Fig. 8 c1) are Na, V, and O, and the distribution proportion of these elements is basically the same; these points are formed brighter areas on the screen. So a large number of high-purity sodium vanadate salts have been generated. When the roasting time reaches 90 min, the reaction of sodium roasting is basically completed and the vanadium extraction stage is completed.

As can be seen from the XPS full-spectrum scanning results shown in Fig. 9, the elements at the reaction interface are Na, V, O, C, and Cr, and the relative content

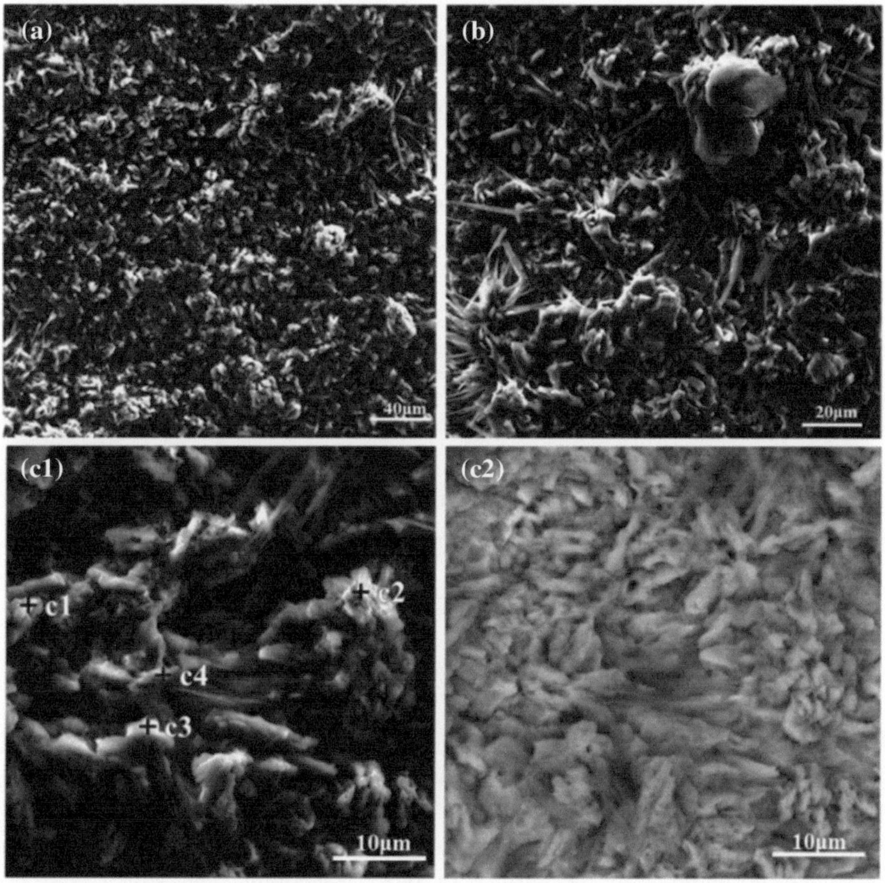

Fig. 8 SEM images of V-slag samples after roasting for 90 min at 800 °C **a–c1** secondary electron image; **c2** backscattered electron image

Table. 2 Element point scanning results of V-slag after roasting for 90 min, wt(%)

	O (%)	Na (%)	V (%)	Si
c1	44.21	22.82	32.97	/
c2	41.49	26.08	32.43	/
c3	40.13	26.39	33.48	/
c4	50.46	21.23	26.95	1.36%

of V element increases further with time. At the same time, Cr has begun to diffuse to the surface of vanadium slag to participate in the reaction. Cr atoms were exposed at the reaction interface to participate in the reaction; the reaction equation is shown in Eq. (2). The above results indicate that the vanadium reaction was

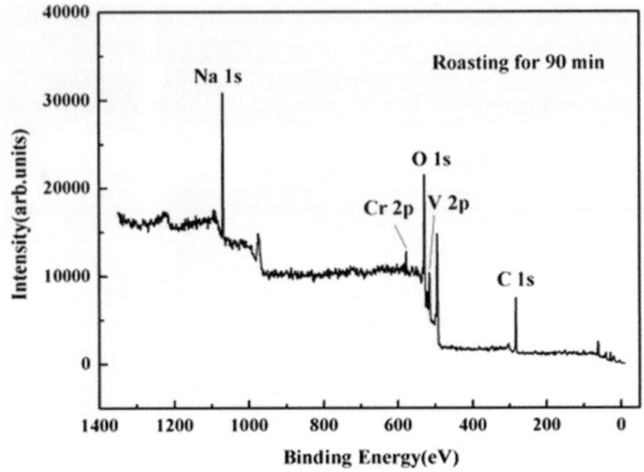

Fig. 9 XPS spectra of all elements in the V-slag after roasting for 90 min

basically completed when the roasting time reached 90 min, and the vanadium extraction stage of sodium roasting was completed.

$$Cr_2O_3 + 2Na_2CO_3 + \frac{3}{2}O_2 = 2Na_2CrO_4 + 2CO_2 \quad (2)$$

Conclusions

This paper has carried on the vanadium slag sodium roasting experiment in the 800 °C; SEM/EDS and XPS were used to analyze and characterize the composition and micromorphology of the surface of block vanadium slag, which were not roasted and roasted for different durations. The evolution law of vanadium phase in the interfacial reaction of sodium roasting was obtained:

1. The elements of unroasted vanadium slag are O, Ca, Mg, Fe, V, Si, and Al; at this time, the spinel is wrapped in silicate matrix phase in the shape of tri-pyramid and quad-pyramid. Polishing the vanadium slag to observe surface morphology, we can find that there is a regular, straight boundary between the spinel phase and the matrix phase. Because V, Cr, Fe, and other atoms in spinel phase have larger atomic radius than Si, Al, and other atoms in matrix phase, spinel phase is bright phase in the composition contrast diagram, while matrix is dark phase, which is contrary to the secondary electron morphology diagram.
2. In the initial stages of sodium roasting, silicate matrix first react with Na_2CO_3 on boundary, and spinel reaction has not been involved in so that structure did not change. With the increase of roasting time, the regular boundary between various phases on the surface of vanadium slag gradually becomes blurred in the reaction interface layer and the structure of each phase gradually decomposing. The smooth and compact surface of polished vanadium slag is gradually

destroyed and becomes rough, forming a sodium vanadate crystal there, and then spread to the surface of silicate and Na_2CO_3 reaction product. The contents of Ca, Mg, Fe, Si, Al, and other elements at the reaction interface continue to decrease, while V keeps accumulating at the interface and is exposed to the outermost layer of the interface reaction, which takes precedence over other elements in vanadium slag combining with Na^+ and O^{2-}.

3. In the intermediate stage of sodium roasting, the surface of vanadium slag is mainly enriched by V and Si elements, which are involved in the oxidation of sodium. V in spinel is gradually oxidized into new phase to form soluble sodium vanadate, and the linear cluster of sodium vanadate crystal is continuously coarsening. After that, the intermediate product produced by the reaction between silicate matrix and Na_2CO_3 will decompose, and Fe will be oxidized and decomposed, while the linear cluster sodium vanadate crystals are agglomerated into particles and continue to expand.
4. When roasted for 90 min, granular sodium vanadate crystals have been basically uniformly spread on the surface of vanadium slag and the phase containing vanadate at the reaction interface is soluble sodium vanadate, Cr also began to participate in the reaction at the interface, indicating that when the roasting time reaches 90 min, the vanadium reaction has been basically completed, and the vanadium extraction stage of sodium roasting has been completed, so there is no need to extend the roasting time.

Acknowledgements This work was financially supported by National Natural Science Foundation of China (grant no. 51474041 and no. 51674051), Chongqing Technological Innovation and Application Demonstration Project (cstc2018jscx-mszdX0066) and Fundamental Research Funds for the Central Universities of China (cqu2017hbrc1B08). We acknowledge Mr. Guocan Zheng in the analytical and testing center of Chongqing University for his kind help in MS characterization.

References

1. Moskalyk RR et al (2003) Processing of vanadium: a review. Miner Eng 16(9):793–805
2. Yao Z (2005) Review of vanadium processing in China. Eng Sci 3:58–62
3. Xu Y et al (2011) Study on recovery of vanadium from low-vanadium Slag by Calcified roasting methods. Adv Mater Res 284–286:2127–2130
4. Li HY, Fang HX, Wang K, Zhou W, Yang Z, Yan XM, Ge WS, Li QW, Xie B (2015) Asynchronous extraction of vanadium and chromium from vanadium slag by stepwise sodium roasting–water leaching[J]. Hydrometallurgy 156:124–135
5. Goddard JB, Fox JS (1981) Extractive metallurgy of refractory metals[M]. TMS-AIME, New York
6. Zhang JH, Zhang W, Zhang L (2015) Mechanism of vanadium slag roasting with calcium oxide[J]. Int J Mine Process 138:20–29
7. Farah H (2003) Oxidation-reduction equilibria of vanadium in CaO-SiO_2, CaO-Al_2O_3-SiO_2, and CaO-MgO-SiO_2 melts[J]. Journal of Mater Sci 38:1885–1894
8. Peng X, Zhong Z, Zheng P, et al. (2009) Aqua-Sim: an NS-2 based simulator for underwater sensor networks[C]

9. Lin M, Wang C, Xie B, et al. Study on the Roasting Mechanism of Vanadium–Chromium Slag with Sodium Hydroxide[J]
10. Liu H, Shi Y D, Yang G, et al. (2003) MECC determination of oleanolic acid and ursolic acid isomers in Ligustrum lucidum Ait.[J]. J Pharm Biomed Anal 32(3):479–485
11. Jena BC, Dresler W, Reilly IG (1995) Extraction of titanium, vanadium and iron from titanomagnetite deposits at pipestone lake, Manitoba, Canada[J]. Miner Eng 8(1–2):159–168

Indian Coal Ash: A Potential Alternative Resource for Rare Earth Metals (REMs)

Archana Kumari, Manis Kumar Jha, Sanchita Chakravarty and Devendra Deo Pathak

Abstract Huge scarcity of rare earth metals (REMs) globally, lack of good natural resources, and generation of tremendous coal ash containing REMs of power plant attracted the researchers to work in this area. The analysis of geologically distributed heterogeneous coal samples at CSIR-NML, India reports the presence of 0.5–1.5 kg/Ton REMs in particular seam of coal at Indian eastern part. In this regard, systematic leaching studies were made to recover REMs from Indian coal ash using hydrometallurgical technique. Maximum dissolution of REMs from coal ash take place using HCl of concentration ranging between 2 and 6 M at elevated temperature. From the obtained leach liquor, more than 90% REMs were recovered using oxalate precipitation. The process developed has tremendous potential to be commercialized after feasibility studies.

Keywords Rare earth metals (REMs) · Coal ash · Leaching · Precipitation

Introduction

Incessant development in the technological fields such as permanent magnet, lasers, automotive catalytic converters, fiber optics/superconductors, electronic devices as well as green energy sectors has increased the demand for rare earth metals (REMs) in the international market [1]. But the progressive incorporation of REMs in innovative technologies is mainly facing hindrance due to uncertain supply of REMs by China and absence of substitutes for individual REMs. These two factors

A. Kumari · M. K. Jha (✉) · S. Chakravarty
Metal Extraction and Recycling Division, CSIR-National Metallurgical Laboratory, Jamshedpur 831007, India
e-mail: mkjha@nmlindia.org

A. Kumari · D. D. Pathak
Department of Applied Chemistry, Indian Institute of Technology (ISM), Dhanbad 826004, India

G. Azimi et al. (eds.), *Rare Metal Technology 2020*, The Minerals, Metals & Materials Series, https://doi.org/10.1007/978-3-030-36758-9_25

have mainly raised a growing level of concern for constant supply of REMs for future use. This emphasized various researchers to identify new resources to ensure adequate supply of REMs for present and future use [2].

Nowadays, coal ash has been of great concern due to the presence of noteworthy amount of REMs in it. In developing countries, coal is a dependable and secure energy source. As a result, its consumption is continuously increasing which led to the production of huge amount of coal by-products. But major amount of coal by-products are disposed-of as waste [3]. Literature review reveals that limited lab-scale experiments were performed to check the possibilities for REMs recovery. Geochemistry, mineralogy, distribution, changes in organic structure, etc. of trace elements were mainly reported. Six samples of coal fly ash were collected from Europe for exploration studies and found that the sample of UK region was enriched in REMs and have potential to be exploited on commercial scale [2]. Determination and quantification of mineralogical association of REMs in coal fly ash of eastern US were carried out. Mechanical as well as chemical-based separation schemes were reported to be required for the separation and recovery of REMs from aluminosilicate glass [4]. To study the distribution of REMs and its partitioning in different coal ash, the Stanford-USGS SHRIMP-RG ion microprobe was used [5]. Laser-induced breakdown spectroscopy was reported to be used for analyzing the content of REMs in coal ash samples from Powder River Basin [6]. Studies were carried out to notice the distribution of mineral species in different coal seams of Talcher coalfield and their transformation at different temperatures [7]. Apart from mineralogical studies, the changes in the organic structures of coal during its combustion were also investigated [8].

In view of above, researchers worked and reported the presence of extractable amount of REMs in these coal by-products. Investigations revealed that the concentration of REMs in some coal samples is higher than that available in natural resources [9]. As a result, observation of REMs in coal by-products was confirmed. But still these are not habitually processed due to low content of REMs in these coal samples. Major focus is made on the behaviour of different elements or minerals present in coal/coal by-product and its geochemistry [10, 11]. Very limited work is reported for REMs extraction but the presence of significant amount of REMs in coal ash compared to primary resources attracted the researchers to explore possibility of REMs extraction [6, 12]. Recovery of rare earth elements from coal ash with a closed-loop leaching process was investigated [13]. Environmental Research and Educational Foundation also presented a process to recuperate REMs from coal fly ash [14].

In India, hardly any attention was found towards the examination of REMs content in coal ash. Therefore, emphasize on the development of a feasible process to extract REMs from coal by-products is required. Keeping these perspectives in view, work has been carried out at CSIR-NML, Jamshedpur to develop a process mainly focusing on the prospects carried out to check the possibilities for the extraction of REMs from coal ash, varying process parameters to get optimal conditions.

Materials and Methods

Materials

The coal samples utilized to carry out the whole work was from various Indian coal mines supplied by our Analytical Chemistry Department of CSIR-NML, Jamshedpur. The standard procedure of coning and quartering followed by crushing and grinding was opted to get fine particles of homogenized coal samples. Four different coal samples were collected, crushed, and put to combustion at elevated temperature for few hrs to get the coal ash required for further course of action (Fig. 1). The prepared coal ash were similar in composition with one generated from thermal power industries. The total REMs content present in the four coal ash sample is presented in Table 1. Data indicates that compared to heavy REMs, the light REMs are more abundant in coal ash samples. Along with REMs, coal ash also contains Al_2O_3, Fe_2O_3, CaO, MgO, SiO_2, etc. Different mineral acids used for the experimental studies were of analytical grade (AR) supplied by Rankem, India and used without further purification.

Methods

In order to determine REMs concentration in these coal ashes, they were dissolved using the standard laboratory procedure. After dissolution, leaching studies of these four coal ash samples were carried out using different lixiviants in a three-necked Pyrex reactor fitted with a condenser. The coal ash samples were put along with the lixiviant in a particular ratio and heated to a desired temperature with constant agitation. The samplings were made at regular time interval and analyzed using Inductively Coupled Plasma-Optical Emission Spectrometry (ICP-OES). After leaching, vacuum filtration was used to separate residue from the leach liquor

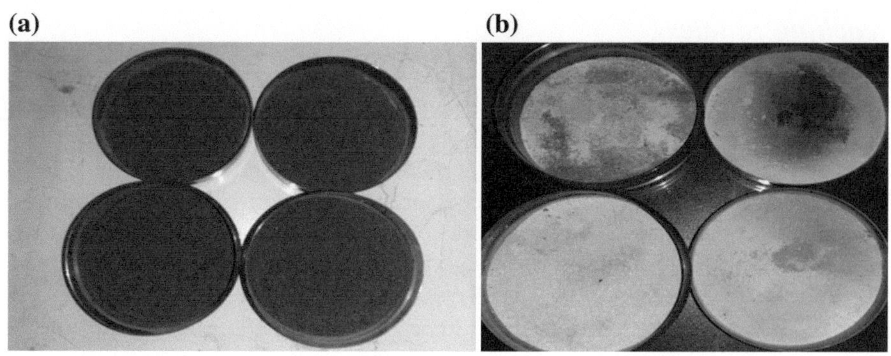

Fig. 1 **a** Coal sample **b** Coal ash sample

Table 1 Analysis of REMs in coal ash sample

Elements	REMs (%)			
	Sample A	Sample B	Sample C	Sample D
La	0.018	0.013	0.018	0.012
Ce	0.036	0.013	0.016	0.015
Nd	0.012	0.009	0.011	0.009
Pr	0.003	0.002	0.0029	0.004
Sm	0.014	0.027	0.038	0.016
Gd	0.0013	0.002	0.0037	0.001
Eu	0.0002	0.0002	0.0004	0.0003
Tb	0.0006	0.0017	0.0037	0.00004
Dy	0.0007	0.0004	0.0006	0.0005
Ho	0	0	0	0
Er	0.0022	0.001	0.0017	0.0016
Tm	0	0.000040	0.00004	0.001
Yb	0.0002	0.0004	0.005	0.0165
Lu	0	0.00016	0.002	0
Sc	0.0021	0.002	0.0034	0.002
Y	0.0031	0.004	0.0072	0.0043
Total (%)	*0.0934*	*0.0759*	*0.1112*	*0.0832*
Total (in 01 Ton)	*934 g*	*759 g*	*1112 g*	*832 g*

containing REMs. The obtained residue was washed, dried, and weighed. De-ionised water was used for making aqueous samples/solutions during the experiment. Satisfactory mass balance was obtained for each set of leaching experiment.

Results and Discussion

Leaching Studies

For effective recovery of REMs from the coal ash, selection of cheaper lixiviant and optimization of various process parameters are necessary. However, supercritical extraction and bioleaching methods are reported to be best but are not commercially feasible. The use of acid (instead of energy) is mostly recommended to treat tons of residues being generated each day in poor countries like India [15]. In order to carry out the leaching experiments, the coal ash containing maximum amount of REMs was selected for further studies, i.e. Sample C. Various process parameters viz. effect of acid concentration, leaching temperature, time, and pulp density on the behavior of different metals were optimized.

Leaching studies were carried out to study the dissolution behaviour of REMs from coal ash using sulfuric acid (H_2SO_4), hydrochloric acid (HCl), and nitric acid (HNO_3) for their selective dissolution leaving undesired components in the residue. From the studies carried out, H_2SO_4 does not present fruitful result for REMs dissolution from coal ash sample even at elevated temperature and high acid concentration. HNO_3 was rejected due to the evolution of toxic NOx into the environment. HCl was found to be the best lixiviant for leaching of REMs (especially Sm, La, Ce, Nd, and Pr) from coal ash (Fig. 2). 2 M acid concentration at 80 °C was used for 75 min for all the set of experiments maintaining solid-to-liquid ratio 100 g/L. At this condition, ~55% of total REMs was found to be leached from coal ash sample C using HCl. The analysis of light REMs along with some heavy REMs (those having significant amount in coal ash sample) is shown in the Fig. 2.

The dissolution percentage of REMs using HCl was noticed to increase up to ~80% at 80 °C with increase in acid concentration to 6 M in 75 min leaching time maintaining solid-to-liquid ratio 100 g/L (Fig. 3). Further increase in acid concentration showed negligible effect on REMs dissolution.

After optimizing the acid concentration, effect of different temperatures on REMs dissolution from coal ash was also studied. Experiments were carried out at room temperature and extended up to 90 °C. It was observed that the leaching percentage of REMs increases with the increase in temperature (Fig. 4). But further increase in temperature is not suggested due to evaporation. Similarly, variation in other parameters (i.e. time and solid-to-liquid ratio) was also carried out but minor increase in REMs dissolution was noticed. Thus, 6 M HCl at 90 °C for 90 min is sufficient to leach out maximum amount of REMs present in the coal ash sample keeping solid-to-liquid ratio of 100 g/L. Leaching studies reveal that after some systematic studies, Indian coal can be processed in a feasible manner to extract REMs from its ash.

Further, from the obtained leach liquor, pure solution containing REMs can be obtained using solvent extraction technique. From the purified solution, valuable products of REMs can also be prepared via precipitation/evaporation process.

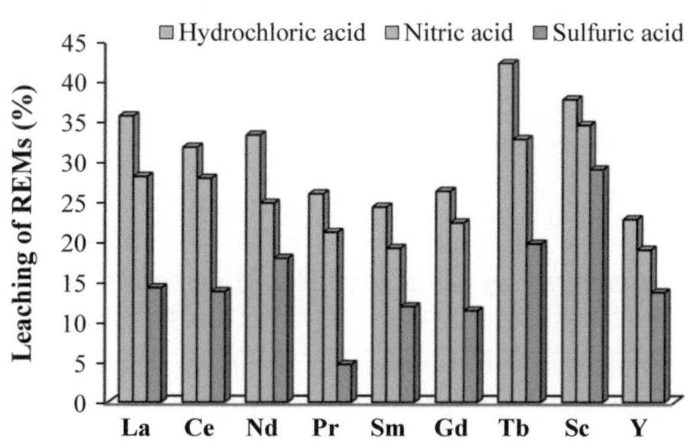

Fig. 2 Effect of different mineral acids on leaching of REMs from coal ash

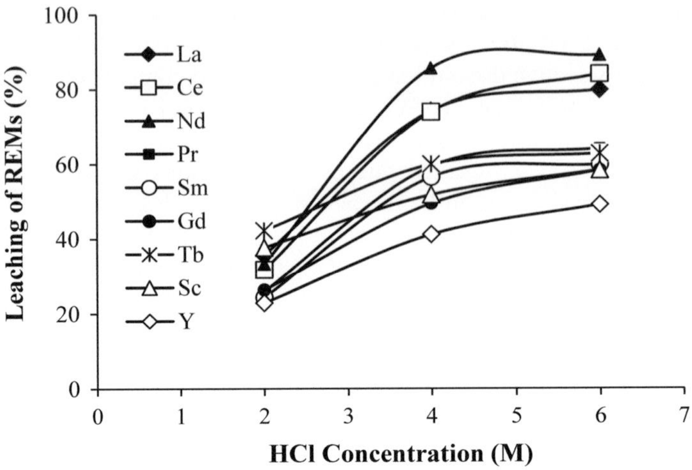

Fig. 3 Effect of HCl concentration on leaching of REMs from coal ash

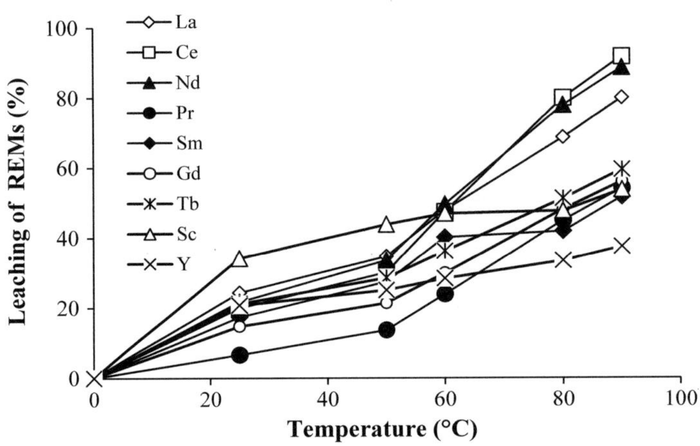

Fig. 4 Effect of temperature on leaching of REMs from coal ash

A complete hydrometallurgical process flowsheet to recover REMs from Indian coal ash is shown in Fig. 5.

Treatment of Leached Residue and Generated Effluent

Before disposing the leached residue to the environment, TCLP test is required to examine the leachability of the metals present in it, which should be under the

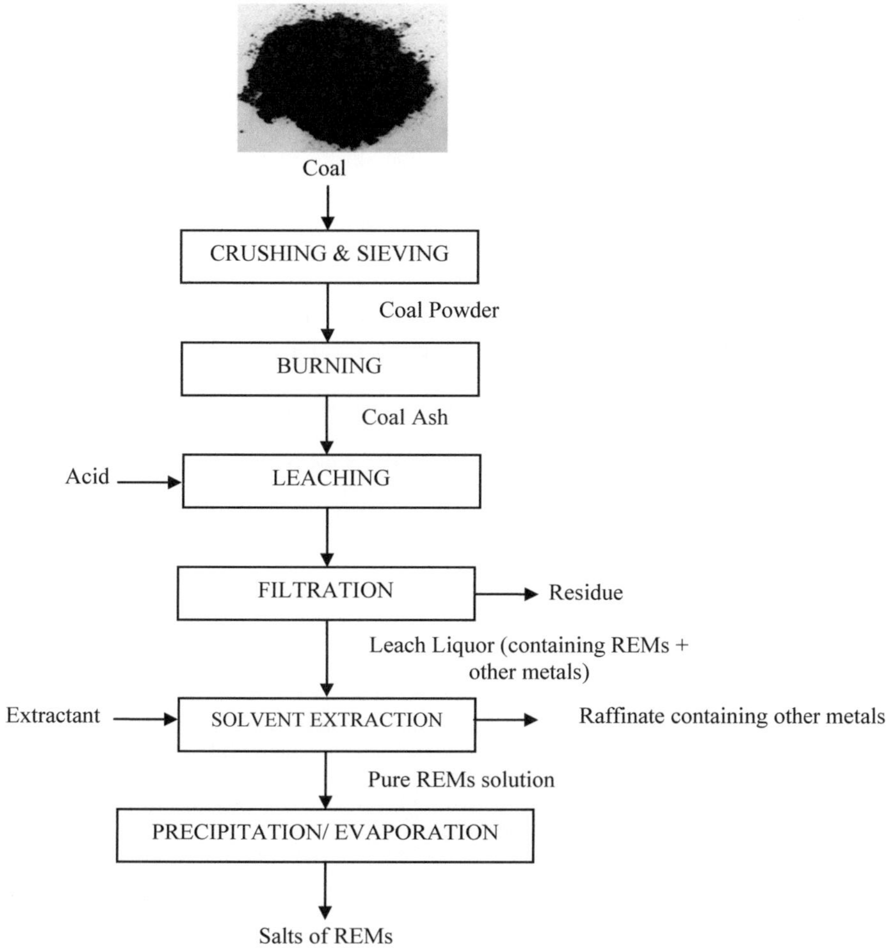

Fig. 5 A complete hydrometallurgical process flowsheet to recover REMs from Indian coal ash

permissible limit. Before disposal, treatment and required procedure for disposal of solid waste residue will be followed. The residue may be disposed of to the pit of required standard fixed in context of environmental regulations.

As far as generated effluent is concerned, they are often ignored and become the cause to increase ecological toxicity. Effluent also contain ample of metal ions which shows detrimental effect on the lives of human beings as well as animals, if not handled properly. To tackle this issue, Effluent Treatment Plant (ETP) should be established, where the effluent are neutralized before disposing to the environment.

Conclusions

From the above studies made, it can be concluded that Indian coal ash is a possible alternative resource of many elements with extractable amount of REMs. Laboratory-scale work carried out reveals that on an average ~500 to 1500 gm REMs are present in one ton of coal ash sample of Indian coalfield. For feasible recovery of these metals, HCl was observed to be a suitable lixiviant when treated at elevated temperature for few hrs. Possibilities to collect salts of REMs can be established using precipitation or evaporation. This set of work if carried out in a systematic manner will not only prove to be commercially viable but will also be a fruitful step to tackle REMs sustainability confrontations.

Acknowledgements Present paper is based on the research work related to REMs extraction from Indian coal ash carried out at CSIR-National Metallurgical Laboratory (CSIR-NML), Jamshedpur, India. Authors are thankful to the Director, CSIR-NML for the permission to publish this paper. One of the authors, Ms. Archana Kumari would like to extend her sincere gratitude to CSIR, New Delhi (Grant: 31/10(60)/2015-EMR-I) for providing Senior Research Fellowship to carry out this research work.

References

1. Kim JS, Lee CH, Han SH, Suh MY (1997) Studies on complexation and solvent extraction of lanthanide in the presence of diaza-18-crown-6-di-isopropionicacid. Talanta 45:437–444. https://doi.org/10.1016/s0039-9140(97)00151-3
2. Blissett RS, Smalley N, Rowson NA (2014) An investigation into six coal fly ashes from the united kingdom and Poland to evaluate rare earth elements content. Fuel 119:236–239. https://doi.org/10.1016/j.fuel.2013.11.053
3. Franus W, Wiatros-Motyka MM, Wdowin M (2015) Coal fly ash as a resource for rare earth elements. Environ Sci Pollut Res 22:9464–9474. https://doi.org/10.1007/s11356-015-4111-9
4. Thompson RL, Bank T, Montross S, Roth E, Howard B, Verba C, Granite E, (2018) Analysis of rare earth elements in coal fly ash using laser ablation inductively coupled plasma mass spectrometry and scanning electron microscopy. Spectrochim Acta B 143:1–11. https://doi.org/10.1016/j.sab.2018.02.009
5. Kolker A, Scott C, Hower JC, Vazquez JA, Lopano CL, Shifeng D, (2017) Distribution of rare earth elements in coal combustion fly ash, determined by SHRIMP-RG ion microprobe. Int J Coal Geol 184:1–10. https://doi.org/10.1016/j.coal.2017
6. Phuoc TX, Wang P, McIntyre D (2016) Detection of rare earth elements in powder river basin sub-bituminous coal ash using laser-induced breakdown spectroscopy (LIBS). Fuel 163:129–132. https://doi.org/10.1016/j.fuel.2015.09.034
7. Banerjee A, Mishra PR, Mohanty A, Chakravarty K, Biswas RD, Sahu R, Chakravarty S (2016) Distribution of mineral species in different coal seams of Talcher coalfield and its transformation behavior at varying temperatures. Int J Coal Sci Technol 3(2):97–103. https://doi.org/10.1007/s40789-016-0127-0
8. Mishra V, Sharma M, Chakravarty S, Banerjee A (2016) Changes in organic structure and mineral phases transformation of coal during heat treatment on laboratory scale. Int J Coal Sci Technol 3(4):418–428. https://doi.org/10.1007/s40789-016-0153-y
9. Seredin V, Finkelman RB (2008) Metalliferous coals: a review of the main genetic and geochemical types. Int J Coal Geol 76:253–289. https://doi.org/10.1016/j.coal.2008.07.016

10. Vassilev SV, Vassileva CG (1997) Geochemistry of coals, coal ashes and combustion wastes from coal fired power stations. Fuel Process Technol 51(1–2):19–45. https://doi.org/10.1016/s0378-3820(96)01082-x
11. Wang Y, Tang Y, Liu S, Wang Y, Finkelman RB, Wang B, Guo X (2018) Behavior of trace elements and mineral transformations in the super-high organic sulfur Ganhe coal during gasification. Fuel Process Technol 177:140–151. https://doi.org/10.1016/j.fuproc.2018.04.01
12. Lin R, Howard BH, Roth EA, Bank TL, Granite EJ, Soong Y (2017) Enrichment of rare earth elements from coal and coal by-products by physical separation. Fuel 200:506–520. https://doi.org/10.1016/j.fuel.2017.03.096
13. Peterson R, Heinrichs M, Glier J, Lane A, Taha R (2017) Recovery of rare earth elements from coal ash with a recycling acid leach process. In: Proceedings of the World of Coal Ash Conference, Lexington
14. Taggart R (2015) Recovering rare earth metals from coal fly ash. In: Proceedings of the World of Coal Ash Conference, Lexington
15. Kumari A, Parween R, Chakravarty S, Parmar K, Pathak DD, Lee JC, Jha MK (2019) Novel approach to recover rare earth metals (REMs) from Indian coal bottom ash. Hydrometallurgy 187:1–7. https://doi.org/10.1016/j.hydromet.2019.04.024

Recovery of Manganese and Cobalt from Discarded Batteries of Toys

Pankaj Kumar Choubey, Manis Kumar Jha and Devendra Deo Pathak

Abstract Compared to other electronic goods, life span of children's toys is very less, which resulted in the generation of huge amount of batteries and environmental pollution. Initially, the batteries are discharged, dismantled, crushed, and physically beneficiated to get black powder, metallic fraction, and plastics. Further, the black powder of batteries was processed for systematic leaching studies and found that 95.6% Mn and 86.05% Co were leached in 2 mol/L H_2SO_4 at 30 °C in 120 min using 10% H_2O_2 (v/v) as an oxidant, maintaining the pulp density 75 g/L. From the leach liquor, at pH 5–8 and above 12, the oxides of Co and Mn were obtained, respectively. The developed process has potential to be transferred in an industry after scale-up studies.

Keywords Toys batteries · Leaching · Precipitation · MnO_2

Introduction

Technological advancement and market expansion are continuously accelerating the replacement of old electrical and electronic equipment with latest one; as a result, huge amount of discarded electronic scrap are generated. The electronic devices like mobile phone, laptop, etc., contain rechargeable lithium–ion batteries as a source of power to store the energy. But batteries have certain life span for energy storage which becomes obsolete when it lost the capacity to store the energy. It was reported that 12 billion units of lithium–ion batteries were generated during 2017 which will be reached up to 25 billion units by the end of 2020 [1].

P. K. Choubey · M. K. Jha (✉)
Metal Extraction and Recycling Division, CSIR-National Metallurgical Laboratory, Jamshedpur 831007, India
e-mail: mkjha@nmlindia.org

P. K. Choubey · D. D. Pathak
Department of Applied Chemistry, Indian Institute of Technology (IIT), ISM, Dhanbad, India

G. Azimi et al. (eds.), *Rare Metal Technology 2020*, The Minerals, Metals & Materials Series, https://doi.org/10.1007/978-3-030-36758-9_26

Presently, only ~5% lithium–ion batteries are recycled by formal sectors; however, rest of the rechargeable LIBs treated by informal ways which may pollute the atmosphere as well as loss of valuable metals. Therefore, it needs to process the discarded batteries in a conventional ways for recovery of lithium, cobalt, manganese, etc., which will mitigate the demand of lithium, cobalt, and manganese in the near future [1, 2].

Various studies have been made to recycle the spent LIBs, nickel manganese hydride battery using the combine pyro- and hydrometallurgical routes. Usually, pre-treatment is required to enrich the metallic values prior to hydrometallurgical recovery of metals. The beneficiated products can be leached in minerals acid like hydrochloric acid, sulfuric acid, and nitric acid to get high extraction efficiency of Li, Co, Mn, etc., at elevated temperature in the presence of suitable oxidant. More than 99% Li and Co can be leached in hydrochloric acid under aggressive leaching conditions. But Cl_2, SO_3, and NO_x were also released during metals leaching which may cause air pollution. Apart from that, minerals acids are not selective in nature may dissolve other associated metals embedded in spent LIBs. Further, solvent extraction process is required to purify the leach liquor prior to recovery of metals by either precipitation or electro-winning process [2].

Alternatively, organic acids such as citric acid, oxalic acid, malic acid, ascorbic acid, etc. have been used to selectively leach the metals which may have advantages of being biodegradable and eco-friendly in nature [3, 4]. Li et al. [5] found that 98.5% Li and 94.8% Co leached in ascorbic acid at 70 °C in 20 min maintaining pulp density 25 g/L [5]. Interestingly, ascorbic acid also acts as a reductant to enhance the metals leaching efficiency. Further, oxalic acid is used to leach the lithium from spent LIBs. It was observed that 98% Li leached in 1.5 M oxalic acid at 80 °C in 120 min while cobalt gets precipitated as cobalt oxalate [6]. It suggests that lithium and cobalt could be selectively recovered in oxalic acid leaching process.

Besides the mobile phone batteries, discarded toys vehicle batteries are also one of the important sources of valuable metals which contain Co, Mn, Ni, etc. But there are no systematic studies for recovery of Co and Mn from spent toys batteries. The objective of this paper is to recover the Co and Mn from discarded batteries of toy vehicles. Various parameters such as acid concentration, oxidant concentration, temperature, and time were varied to optimize the conditions for leaching of Co and Mn from beneficiated toys batteries. Thereafter, precipitation studies were performed for recovery of Co and Mn from the obtained leach liquor of toys batteries. For this, different concentrations of sodium hydroxide were used to maintain the pH in certain range for selective recovery of Co and Mn in a form of their oxides as discussed in precipitation section.

Experiment

Materials and Sample Preparation

Spent toys LIBs were used for the experimental purpose. Initially, the spent toys batteries were crushed in the scutter crusher. Thereafter, crushed batteries (plastics, black powder) (Fig. 1), and metallic fractions were separated by flotation process. The obtained black powder contained 3.96% Co, 3.65% Li, 7.9% Mn, and 19.8% Ni, which was used for leaching studies. The hydrochloric acid and sulfuric acid supplied by Merck, India were used as lixiviant to dissolve the metals from beneficiated toys batteries. All the chemicals used were of analytical grade.

Method

Leaching experiments were conducted in a temperature-controlled Pyrex leaching reactor of capacity 250 mL having stirrer facility to maintain the constant stirring speed. Sulfuric acid solution was used as a leaching agent at pre-selected temperature (30–75 °C) to maintain the constant temperature throughout the leaching experiment. When the set temperature was reached, 18.75 g sample was charged into the leaching reactor under the stirring speed of 400 rpm, which was found sufficient to ignore the effect of mass transfer on leaching. Leach liquor was taken at different time intervals (15–120 min) and diluted in 5% hydrochloric acid solution to analyze the concentration of metal ions using atomic absorption spectrometer (AAS) (Perkin Elmer, analyst 200).

Results and Discussion

Effect of Oxidant Concentration

To effectively leach the metals from the black mass of discarded toys batteries, experiments were performed in 2 M H_2SO_4 solution at different oxidant (H_2O_2)

Fig. 1 Crushed batteries

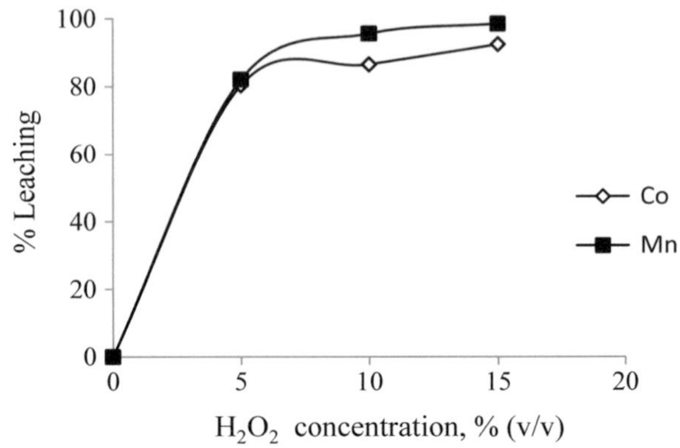

Fig. 2 Effect of oxidant concentration of Co and Mn leaching. *(Solid: black powder of toys batteries, Liquid: 2 M H_2SO_4, temperature: 30 °C, oxidant: 5 to 15% H_2O_2, Pulp density: 75 g/L)*

concentrations varying from 5 to 15% H_2O_2 (v/v) at 30 °C while maintaining the pulp density 75 g/L. Figure 2 shows that leaching of Co and Mn increases with increase of oxidant concentration. During cobalt leaching, the less soluble species of cobalt (Co^{3+}) ion transforms into more soluble of cobalt (Co^{2+}) ion resulting in the enhancement of cobalt leaching efficiency.

Effect of Acid Concentration

Experiments were also performed in a range of acid concentration varying from 1 M to 4 M H_2SO_4 for optimization of acid concentration under following leaching conditions: temp: 30 °C, time: 120 min, and pulp density: 75 g/L. Figure 3 shows that leaching of Co and Mn increases with increase in acid concentration from 1 to 2 M H_2SO_4, but after 2 M H_2SO_4 acid concentration, the leaching efficiency remains almost same with increase in acid concentration. It indicates that 2 M H_2SO_4 has sufficient acidity to leach the Co and Mn from spent toys batteries.

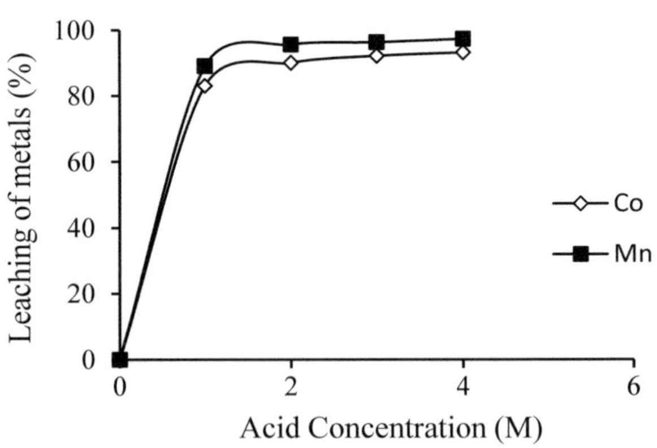

Fig. 3 Effect of acid concentration. *(Solid: black powder of toys batteries, liquid: 1 to 4 M H_2SO_4, temperature: 30 °C, oxidant: 10% H_2O_2, pulp density: 75 g/L)*

Effect of Temperature

Temperature is also one of the important parameters to enhance the cobalt and manganese leaching efficiency. Hence, experiments were performed in a range of temperature varying from 30 °C to 75 °C in 2 M H_2SO_4 in the presence of 5% H_2O_2 (v/v) in 120 min maintaining the pulp density 75 g/L. It was observed that cobalt leaching efficiency was decreased with the increase of temperature as shown in Fig. 4a. The decrease of cobalt leaching efficiency with rise in temperature is due to dissociation of hydrogen peroxide into H_2O and O_2; as a result, cobalt leaching was decreased. Similarly, Mn leaching was also decreased with the increase of temperature up to 75 °C but remain same in a temperature range of 30–50 °C (Fig. 4b). The obtained leach liquor was further processed by precipitation method for recovery of Mn and Co as discussed below.

Precipitation Studies

Effect of pH

Precipitation studies were performed in a range of pH varying from ~2 to ~12 at 30 °C to recover the Mn and Co in a form of their oxides. It was found that 99.21% cobalt were precipitated in a range of pH 8–9, while only ~25% Mn were precipitated under same experimental condition (Fig. 5a). Subsequently, solution pH was increased up to 12 by adding sodium hydroxide solution to completely precipitate the manganese as a manganese dioxide. Result shows that 99.2% Mn was precipitated at pH 11 as shown in Fig. 5b.

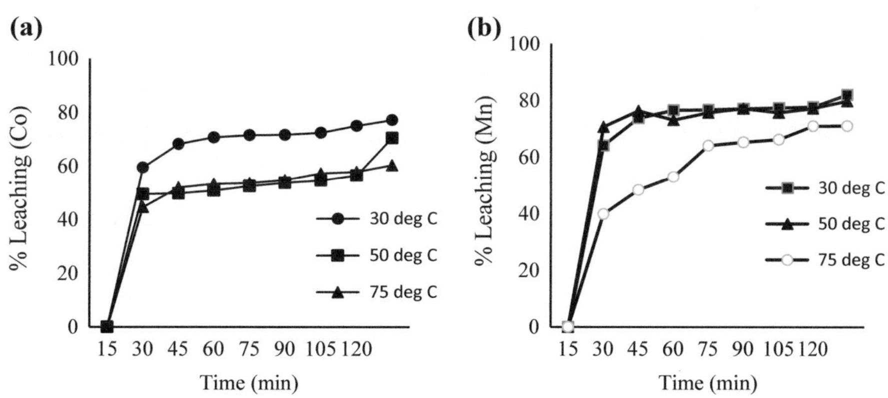

Fig. 4 **a** and **b** effect of temperature on leaching of cobalt and manganese. *(Solid: black powder of toys batteries, liquid: 2 M H_2SO_4, temperature: 30–75 °C, oxidant: 5% H_2O_2, pulp density: 75 g/L)*

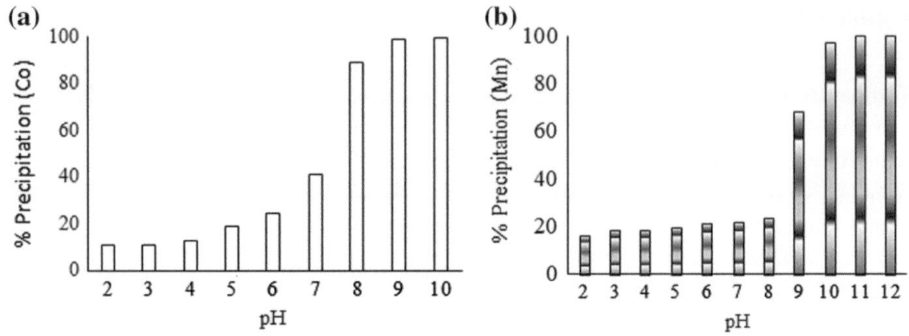

Fig. 5 **a** and **b** effect of pH on precipitation of cobalt and manganese. *(Liquid: Leach liquor of toys batteries, Precipitant: dilute sodium hydroxide solution)*

Conclusions

In order to recover the manganese and cobalt from spent toys batteries, at first batteries were pre-treated and beneficiated to get the black mass for hydrometallurgical extraction of metals. Thereafter, beneficiated sample was dissolved in sulfuric acid solution to leach the cobalt and manganese. 86.05% Co and 95.6% Mn were found to be dissolved in 2 M H_2SO_4 at 30 °C in 60 min in the presence of 10% H_2O_2 as an oxidant, while maintaining the pulp density 75 g/L. But leaching of Co and Mn was decreased when temperature reached 75 °C under the same experimental condition due to dissociation of hydrogen peroxide into water molecule.

Further, precipitation technique was used to selectively recover the cobalt and manganese in a form of their oxides by maintaining pH in a certain range using sodium hydroxide solution. It was observed that 99.21% cobalt were precipitated as cobalt hydroxide in a pH range of 8 to 9 at 30 °C. But manganese precipitation occurred in between the pH 11 and 12. The result is well matched with the Eh–pH diagram of cobalt and manganese precipitation.

Acknowledgements Authors are thankful to Director, CSIR-NML for giving permission to publish this paper.

References

1. International institute for sustainable development (2019) Sustainability and second life: the case of lithium and cobalt recycling
2. Choubey PK, Chung KW, Kim MS, Lee JC, Srivastava RR (2017) Advance review on the exploitation of the prominent energy storage element lithium. Part II: from sea water and spent lithium ion batteries (LIBs). Miner Engg 110:104–121

3. Li L, Ge J, Wu F, Chen R, Chen S, Wu B (2010) Recovery of cobalt and lithium from spent lithium ion batteries using organic citric acid as leachant. J Hazard Mater 176:288–293
4. Li L, Ge J, Chen R, Wu F, Chen S, Zhang X (2010) Environmental friendly leaching reagent for cobalt and lithium recovery from spent lithium-ion batteries. Waste Manage 30:2615–2621
5. Li L, Lu J, Ren Y, Zhang XX, Chen RJ, Wu F, Amine K (2012) Ascorbic-acid-assisted recovery of cobalt and lithium from spent Li-ion batteries. J Power Sources 218:21–27
6. Sun L, Qiu K (2012) Organic oxalate as leachant and precipitant for the recovery of valuable metals from spent lithium-ion batteries. Waste Manage 32(8):1575–1582

Zinc in Secondary Dust of Rotary Hearth Furnace Recovered by Water Leaching and Acid Leaching

Shuang Liang, Xiaoping Liang, Minghu Wu and Shilei Ren

Abstract In this paper, water leaching and acid leaching were combined to treat secondary dust of a rotary hearth furnace to achieve zinc enrichment. First, the solubles $ZnCl_2$, KCl, and NaCl in the dust were dissolved into the water solution through water leaching treatment, and separated from the other insoluble components in the dust. Then in order to recover zinc from slag, acid leaching was used to treat the residue. The acid leaching experiment showed that the leaching rate of zinc would increase with the increase of HCl concentration. Finally, in order to recover the zinc, NaOH precipitator was added to the aqueous solution and the acid solution, respectively. The experimental results showed that the content of $Zn_5(OH)_8Cl_2$ increased first and then decreased with the increase of the precipitator. The total recovery rate of zinc metal can reach 79.46% after the dust is treated by water leaching and acid leaching.

Keywords Secondary dust of rotary hearth furnace · Basic zinc chloride · Leaching

Introduction

With the rapid development of China's steel industry, the amount of scrap steel produced is increasing. Due to the high zinc content in scrap steel, the amount of zinc dust produced by steel enterprises in the process of using scrap steel is also increasing [1]. At present, the typical treatment processes of zinc dust in domestic iron and steel enterprises include rotary hearth furnace, rotary kiln, and OxyCup shaft furnace [2]. Among them, the rotary hearth furnace method has remarkable achievements in dust recovery and treatment of steelworks. However, as for the secondary dust generated by the rotary bottom furnace, the measures taken by

S. Liang · X. Liang (✉) · M. Wu · S. Ren
College of Materials Science and Engineering, Chongqing University,
Chongqing 400045, China
e-mail: xpliang@cqu.edu.cn

G. Azimi et al. (eds.), *Rare Metal Technology 2020*, The Minerals, Metals & Materials Series, https://doi.org/10.1007/978-3-030-36758-9_27

various steel enterprises are mainly to sell it at a low price as a Zn-rich raw material, which fails to be further effectively utilized.

Some scholars have also studied the method of recovering valuable metal from secondary dust of rotary hearth furnace. Tang et al. [3] studied the production of zinc oxide with secondary dust from the furnace as raw material, when the temperature was normal, the liquid–solid ratio was 10:1, and the time was 30 min, the water leaching rate of Zn reached over 95%. Wu [4] mainly studied the influence of dust leaching process parameters on zinc leaching rate by taking secondary dust of rotary hearth furnace as raw material and using sulfuric acid leaching at normal pressure. Under the conditions of particle size <140 mesh, sulfuric acid concentration 1.0 mol/L, leaching temperature 25 °C, solid–liquid ratio 1:8, and leaching time 0.5 h, the leaching rate of zinc in the furnace dust was up to 96%. Liu [5] took the secondary dust of rotary hearth furnace as raw material and treated the Zn-rich ash of the furnace by alkali leaching–electrolysis process. The obtained technological conditions for alkali leaching of dust from the furnace were as follows: bag ash particle size <140 mesh, NaOH concentration of 5 mol/L, leaching temperature of 60 °C, liquid–solid ratio of 11:1, and leaching time of 9 h. Under these conditions, the leaching rate of lead was more than 86% and that of zinc was more than 75%. Most of these methods are aimed at dust with high zinc content, and most of the zinc in dust exists in the form of ZnO, which is convenient to deal with. Moreover, most studies do not further deal with the residual residue of flooding.

This paper mainly studies the recovery process of zinc in secondary dust of rotary hearth furnace with low zinc content. In order to effectively recover zinc in secondary dust of the furnace, two technological routes are designed. One is to take the aqueous solution after water leaching and add NaOH to prepare $Zn_5(OH)_8Cl_2$ to recover the zinc in the secondary dust of the furnace. The other is to use hydrochloric acid leaching effluent to soak the slag, so that the zinc element in the slag into the acid solution, adding NaOH to the acid solution, the same preparation of $Zn_5(OH)_8Cl_2$, as shown in Fig. 1 is the secondary dust zinc recovery process flowchart.

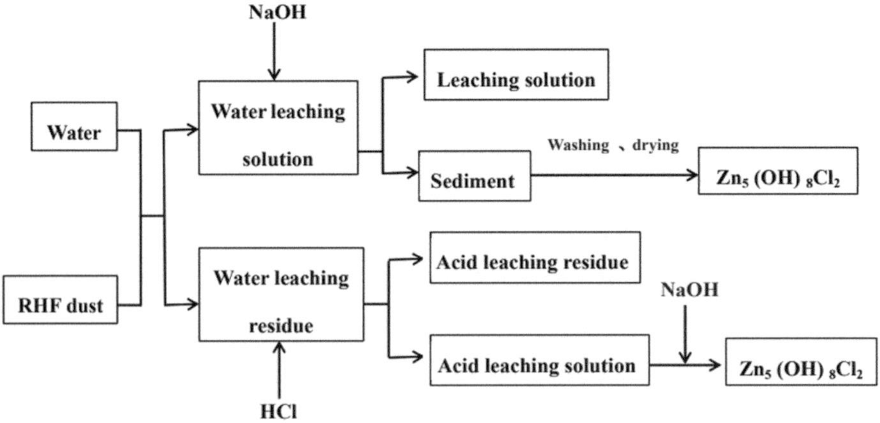

Fig. 1 Process flowchart of zinc recovery from secondary dust of a rotary hearth furnace

Experimental

Experimental Procedures

The experiment includes three parts: water leaching experiment, acid leaching experiment, and adding NaOH experiment. Under the water leaching experimental conditions shown in Table 1, the secondary dust of rotary hearth furnace is immersed in water. According to the first process route, NaOH was added to the water leaching solution to prepare $Zn_5(OH)_8Cl_2$. The effect of NaOH addition on the recovery of zinc from secondary dust of rotary hearth furnace was investigated. The second process uses hydrochloric acid to treat water leaching slag, mainly investigating the influence of different hydrochloric acid concentrations on the recovery of zinc from secondary dust of rotary hearth furnace. After each test, the leachate and leaching residues are analyzed. The research parameters of each process are shown in Table 1.

Analysis Method

In the process of zinc recovery of secondary dust of rotary hearth furnace, in order to detect the composition and phase of dust, leach liquid, and leach slag of the furnace, the main detection method is inductively coupled plasma emission spectrometer (ICP-OES) and X-ray diffractometer. The formula for calculating the distribution ratio of elements in dust in leaching solution (1) and the formula for calculating the distribution ratio of elements in leaching residue are as follows (2)–(3):

$$\eta = \frac{\mathrm{V_T} \times \mathrm{k_T}}{\mathrm{M_R} \times \mathrm{k_R}} \times 10^{-6} \times 100\% \tag{1}$$

$$\Psi = \frac{\mathrm{M_T} \times \mathrm{w_T}}{\mathrm{V_T} \times \mathrm{k_T} \times 10^{-6}} \times 100\% \tag{2}$$

$$R_w = \frac{M_T \times W_T}{M_R \times k_R} \times 100\% \tag{3}$$

Table 1 Main parameters investigated in the experiment

Experiment	Investigated parameters
Water leaching stage	Liquid–solid ratio(L/S):4:1; temperature (°C):25; stirring rate (r/min):350; mixing time (min):60
Acid leaching stage	Hydrochloric acid concentration(mol/L)1 ~ 3; liquid–solid ratio (L/S):4:1; temperature (°C):25
Add the NaOH stage	Theoretical consumption multiples of NaOH:0.75; 1.00; 1.125; 1.25

where η—Distribution ratio of elements in water, %; V_T—Water leaching volume, mL; k_T—Element concentration in solution, mg/L; M_R—Quality of secondary dust of rotary hearth furnace, g; K_R—Element content in secondary dust of rotary hearth furnace, %; Ψ—Distribution rate of elements in residue (Elemental precipitation rate), %; M_T—Quality of filter residue, g; W_T—Element content in filter residue, %; Rw—Element recovery rate of water leaching residue, %.

After adding a certain amount of NaOH to the water and acid leaching liquor, the precipitate and filtrate are obtained by filtration. The recovery of elements is shown in Eq. (4). The total recovery of elements is the sum of Rw and R_N.

$$R_N = \frac{M_N \times W_N}{M_R \times k_R} \times 100\% \tag{4}$$

where R_N—Element recovery rate in precipitate, %; M_N—Sediment quality, g; W_N—Element content in sediment, %.

Results and Discussion

Composition of Secondary Dust of Rotary Hearth Furnace

In order to get the element composition and phase composition of secondary dust of the rotary hearth furnace used in the experiment, the secondary dust of rotary hearth furnace was determined by inductively coupled plasma emission spectrometer (ICP-OES) and X-ray diffractometer. ICP-OES results show that the main elements in the secondary dust of rotary hearth furnace are Zn, K, Na, and Cl, and a small amount of impurity elements such as Ca, Mg, and Fe is also contained, as shown in Table 2. XRD results show that zinc in dust exists in the form of $K_2(ZnCl_4)$ and $ZnFe_2O_4$, and K and Na in dust in the form of $K_2(ZnCl_4)$, KCl, and NaCl, as shown in Fig. 2.

Water Leaching

Previous scholars have proved that water leaching of zinc in secondary dust of rotary hearth furnace is a good method [6]. As mentioned in the literature [7, 8], suitable water leaching conditions can create good conditions for later recovery of zinc, combined with the analysis results of secondary dust composition in rotary

Table 2 Element content of each element in the dust of the rotary bottom furnace (%)

Ca	Mg	Fe	Na	K	Zn	Cl
0.56	0.088	4.78	5.03	24.87	15.2	47.22

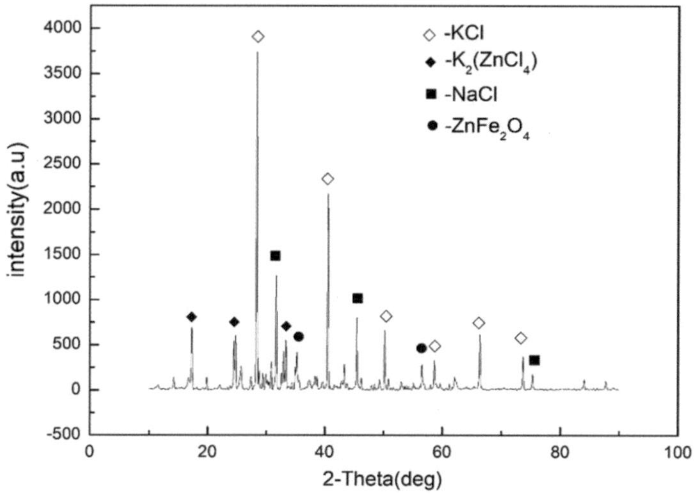

Fig. 2 XRD pattern of rotating bottom furnace dust

Table 3 Composition of aqueous extract (g/L)

Element	Zn	Fe	Cd	Pb	Cu
%	17.338	<0.001	0.598	1.965	<0.001

Table 4 Composition and content of leaching residue

Element	Zn	Fe	Cd	Pb
%	25.4	16.4	0.72	2.265

hearth furnace, under the conditions of liquid–solid ratio of 4:1, room temperature, leaching time of 60 min, and the water leaching experiment of secondary dust of rotary hearth furnace was carried out. Under these conditions, the main components and contents of the water leaching solution and the water leaching residue are shown in Tables 3 and 4, respectively. Distribution of elements in dust in water leaching solution and water leaching residue is shown in Table 5 and Fig. 3. The XRD phase detection results of water leaching residue are shown in Fig. 4.

From Table 5 and Fig. 3, it can be seen that 48.89% of zinc elements enter the water leaching residue, 45.97% of zinc elements enter the water leaching solution, and the distribution ratio of zinc elements between the residue and the liquid is 1:0.94. Therefore, in order to maximize the extraction of zinc from dust, it is better to adopt two technological routes.

Table 5 Distribution ratio of each element in the leaching residue and the leaching liquid (%)

Element	K	Na	Cl	Zn	Ca	Mg	Fe
Residue	8.00	8.84	5.12	48.89	84.64	93.09	100.51
Liquid	90.54	91.80	88.86	45.97	31.41	2.28	0.00

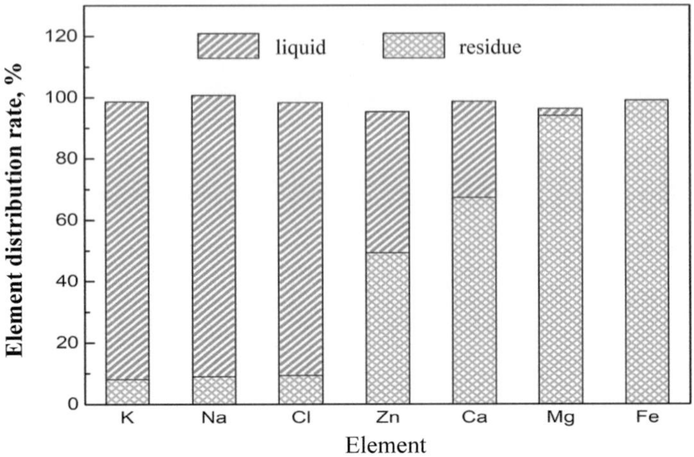

Fig. 3 The distribution ratio of elements in water leaching residue and water leaching liquid

Hydrochloric Acid Leaching

From Fig. 4 it can be seen that Zn and Fe mainly exist in the form of $ZnFeO_4$ in the water leaching residue. $ZnFeO_4$ belongs to the spinel shape, and the chemical grid of this compound is very stable. Because of the stability of $ZnFeO_4$, it is difficult to dissolve in dilute acid. The dissolution reaction of $ZnFeO_4$ is as follows:

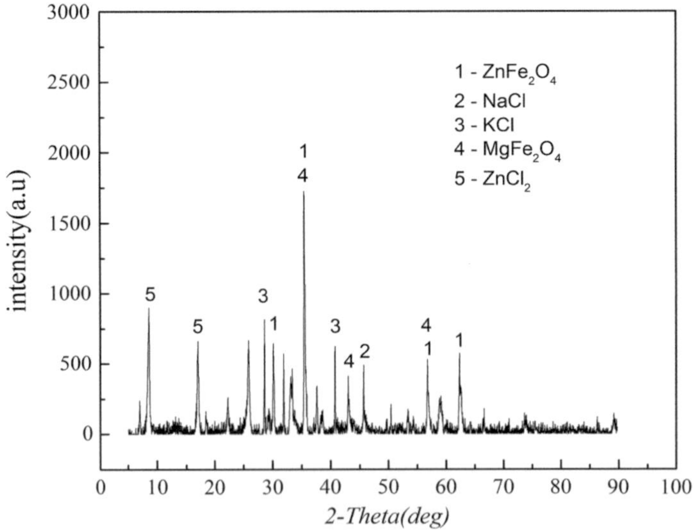

Fig. 4 The XRD pattern of aqueous leaching slag was used in the leaching experiment

Table 6 Composition of acid leaching solution at different HCL concentrations (g/L)

Concentration of HCL (mol/L)	Zn	Fe	Cd	Pb
1	14.22157	0.21546	0.04712	0.74209
2	15.19796	0.23675	0.05131	0.81707
3	15.42257	0.27956	0.05470	0.87624

$$ZnFeO_4 + 8H^+ = Zn^{2+} + 2Fe^{3+} + 4H_2O \quad (5)$$

The effect of hydrochloric acid concentration on the recovery of zinc in water leaching residue was studied under the conditions of reaction temperature 25 °C and liquid–solid ratio 4:1. The results are shown in Table 6 and Fig. 5.

From Table 6, it can be seen that when the concentration of hydrochloric acid increases from 1 mol/L to 2 mol/L, the leaching amount of zinc increases significantly, and the leaching rate of zinc increases from 85.35% to 91.21%. However, when the concentration of hydrochloric acid was further increased from 2 mol/L to 3 mol/L, the content of zinc extracted changed little, and the leaching rate of zinc was 91.21% to 91.76%. Under the experimental conditions, when the concentration of hydrochloric acid is higher than 2 mol/L, the change of zinc leaching rate is very small. Considering the environmental protection and operational safety, the concentration of hydrochloric acid of 2 mol/L is enough to leach zinc from the water leaching residue. Therefore, the maximum recovery rate of zinc in water leaching residue can reach 91.21% by hydrochloric acid.

Adding NaOH to the Water Leaching Solution

According to the results of water leaching experiment, 45.97% of zinc in the secondary dust of rotary hearth furnace enters the water leaching solution. After adding sodium hydroxide to the water leaching solution, the following reactions occur:

$$Zn^{2+} + 2NaOH = 2Na^+ + Zn(OH)_2 \downarrow \quad (6)$$

$$5Zn^{2+} + 8NaOH + 10Cl^- = 8NaCl + Zn_5(OH)_8Cl_2 \downarrow \quad (7)$$

It can be seen from (6) and (7) that there are two precipitation forms of zinc, and the precipitation products in water leaching solution will be different with the addition of NaOH. When the amount of NaOH added from 0 to 0.6 mol/L is calculated by using HYDRA/MEDUSA software, the effects of NaOH added on the morphology and coefficient of Zn^{2+} in the solution system are, respectively, shown in Fig. 6.

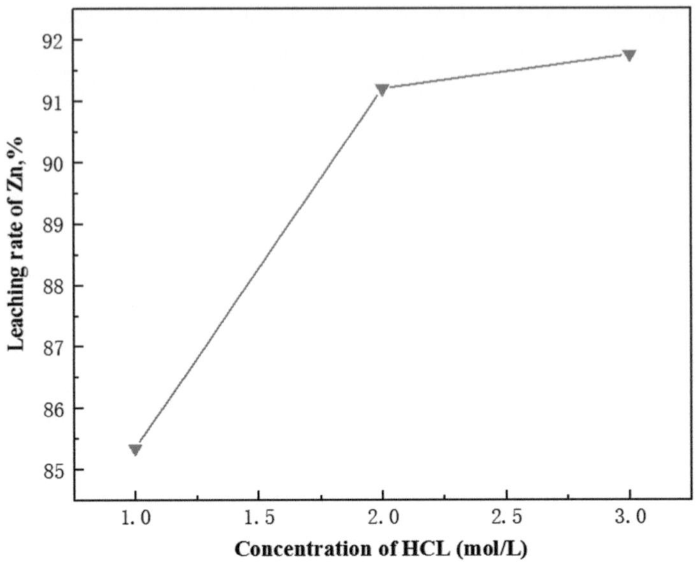

Fig. 5 Effects of different hydrochloric acid concentrations on the leaching rate of Zn

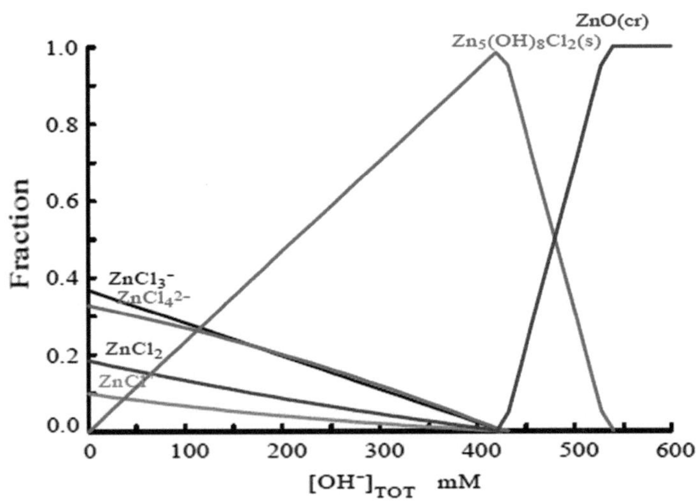

Fig. 6 Effect of NaOH on morphology and coefficient of aqueous solution Zn^{2+}

From Fig. 6, it can be seen that (6) will not occur in the range of 0–0.6 mol/L for NaOH addition in the water leaching system. When NaOH content is 1.6 times of zinc concentration in water leaching solution (0.42 mol/L), almost all of zinc is transformed into $Zn_5(OH)_8Cl_2(s)$.

Table 7 Experimental results of adding different amounts of NaOH to the water leaching liquid

Number of NaOH additions (relative to theoretical additions)	Dried sediment mass (g)	Solution pH
0.750	5.61	7.5
1.000	13.88	8
1.125	7.70	9
1.250	3.80	9.5

Under the experimental conditions shown in Table 1, the experimental results of adding sodium hydroxide to the water leaching solution are shown in Table 7 and Fig. 7.

As can be seen from Table 7 and Fig. 7, with the increase of NaOH addition, the net weight of sediment increases first and then decreases, and the decrease is obvious. This may be due to the fact that when NaOH content exceeds the theoretical value, the morphology of zinc may gradually change to amorphous solid ZnO (cr), which is unstable. Therefore, for the preparation of basic zinc chloride process, the addition of NaOH should be strictly controlled within 0.42 mol/L. The precipitate data and XRD diffraction patterns of 1.000 NaOH theoretical additions are shown in Table 8 and Fig. 8, respectively.

According to Table 8, the basic zinc chloride powder of 13.88 g can be obtained from 100 g dust after treatment, and the recovery of zinc is 43.86%. As can be seen from Fig. 8, the XRD diffraction patterns of the precipitates show that the phases in the precipitates are $Zn_5Cl_2(OH)_8$ and ZnO. The appearance of ZnO may be related

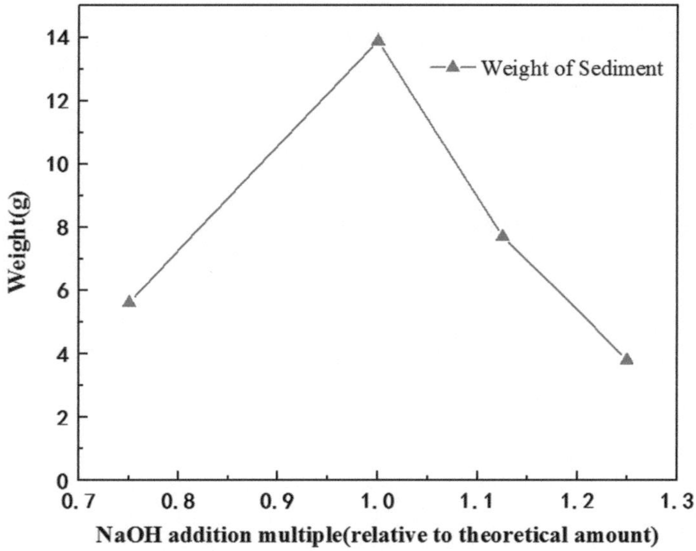

Fig. 7 Experimental results of adding different amounts of NaOH into water leaching

Table 8 Precipitate data record under optimal NaOH addition

Dried sediment mass (g)	Zn absolute mass (g)	Zn recovery rate (%)
13.88	6.667	43.86

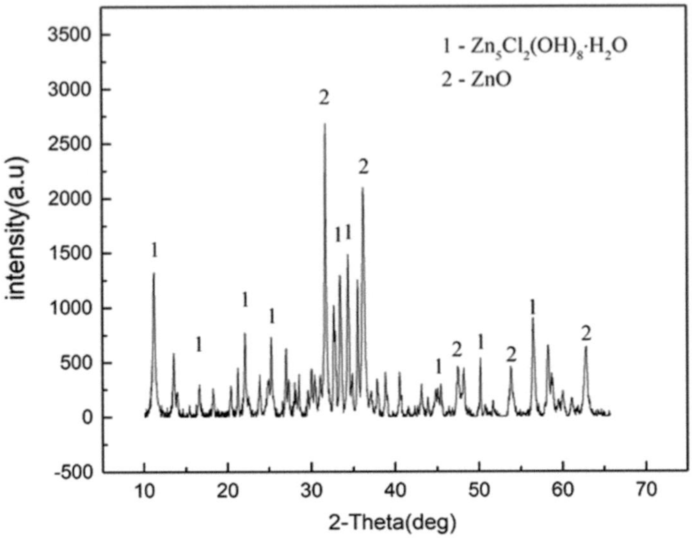

Fig. 8 XRD pattern of the precipitate with theoretical amount of NaOH added

to the way of adding NaOH and reaction conditions. If the added NaOH is too fast or stirred inadequately, the concentration of NaOH in the solution system is too high locally, and a small amount of non-equilibrium zinc oxide (cr) may be formed and filtered into the sediment.

Adding NaOH to the Acid Leaching Solution

When sodium hydroxide is added to the acid leaching solution, the reactions as shown in (6) and (7) will occur. Figure 9 shows the effect of HYDRA/MEDUSA software calculation on the morphology and coefficient of Zn^{2+} in acid solution system.

From Fig. 9 it can be seen that when the solution PH is 5.5, zinc begins to change into $Zn_5Cl_2(OH)_8$. When PH is 8.5, the production of $Zn_5Cl_2(OH)_8$ is the largest. After PH is greater than 8.5, zinc oxide (cr) begins to precipitate. This is consistent with the software calculation of adding NaOH to the aqueous solution. The precipitation quality obtained by adding NaOH with theoretical consumption of 1 time to the acid leaching solution is shown in Table 9.

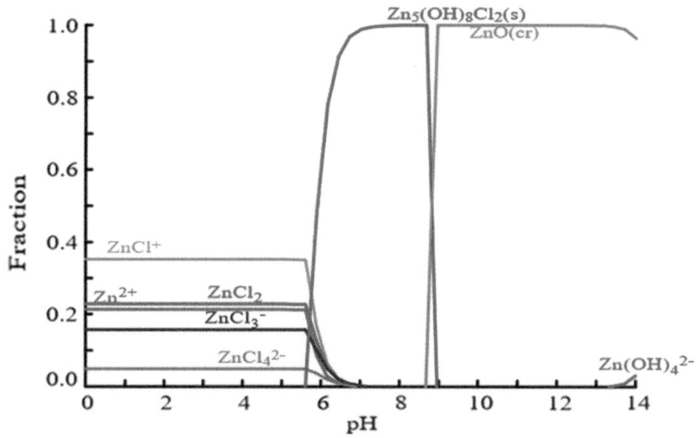

Fig. 9 Effect of NaOH on morphology and coefficient of acid solution Zn^{2+}

Table 9 Precipitate data record under optimal NaOH addition

Dried sediment mass (g)	Zn absolute mass (g)	Zn recovery rate (%)
7.490	6.5254	35.6

Adding NaOH precipitator to acid leaching solution, the quality of zinc extracted from secondary dust of 100 g converter is 6.5254 g, and the recovery of zinc is 35.6%. Therefore, 79.46% zinc can be recovered from secondary dust of rotary hearth furnace by water leaching and acid leaching.

Conclusions

This paper outlines two technological routes for recovering zinc from secondary dust of a converter by water leaching and acid leaching. After water leaching of the secondary dust of a rotary hearth furnace, 45.97 and 48.89% of zinc in the dust enter the water leaching solution and the water leaching residue, respectively, and the zinc in the water leaching residue is continuously extracted by hydrochloric acid. It is found that the maximum leaching occurs when the concentration of hydrochloric acid is 2 mol/L. The yield was 91.21%. The $Zn_5Cl_2(OH)_8$ was prepared by adding NaOH into the water leaching solution and acid leaching solution. The recovery of zinc in the water leaching solution was 43.86%, the recovery of zinc in the acid leaching solution was 35.6%, and the total recovery of zinc was 79.46%.

References

1. Liu P, Cao K (2013) Discussion on resource utilization of zinc and iron containing dust and sludge in steel works. World Steel 13(4): 20–26
2. DongRui L (2019) Status and prospect of zinc dust treatment technology in Chinese iron and steel enterprises. Ansteel Technol 03:7–10
3. Tang H, Zhang H, Fan L et al (2014) Zinc oxide preparation using rotary hearth furnace secondary dust. Paper presented at the 143rd TMS annual meeting, California, USA, 16–20 Feb 2014
4. Wu Z (2012) Technology research of acid treatment of rotary hearth furnace dust containing rich zinc and lead content. Master's thesis, Anhui University of Technology
5. Liu H (2012) Technology research of rotary hearth furnace rich lead-zinc ash by alkaline processing. Master's thesis, Anhui University of Technology
6. Ma R (2007) Principle of hydrometallurgy. Metallurgical Industry Press, Beijing
7. Tang Q (2018) Experimental study on extraction technology of valuable elements from secondary dust of rotary hearth furnace. Master's thesis, Chongqing University
8. Li A, Ge X, Xu A (2011) Dust removal of stainless steel and its comprehensive utilization. World Iron & Steel (6):32–37

Phosphate-Intensified Alkali Leaching to Recover Molybdenum from a Volatilizing Residue

Dapeng Shi, Guanghui Li, Hu Sun, Jun Luo and Tao Jiang

Abstract Volatilizing residues are commonly produced during the process of sublimating technical-grade molybdic oxide or oxidizing-volatilizing molybdenite concentrate, which contains a certain amount of Mo mainly existing in the form of molybdates. To achieve a deep Mo recovery from volatilizing residues, this work presented a new composite leaching reagent that consisted of sodium hydroxide (NaOH) and sodium phosphate (Na_3PO_4). It was found that the leaching of Mo in NaOH solution could be remarkably intensified with the addition of Na_3PO_4. Specifically, it showed that more than 90% of Mo was leached out from this material with 160 g/L NaOH and 20 g/L Na_3PO_4 at 95 °C for 120 min and liquid–solid ratio of 6 mL/g. Meanwhile, the rare earths were enriched in leach residue.

Keywords Molybdenum · Volatilizing residue · Alkali leaching · Phosphate

Introduction

Molybdenum is a silver-white refractory rare metal with high melting and boiling points, high strength, good corrosion resistance, excellent electrical conductivity, and small expansion coefficient [1, 2]. Its alloys and chemicals are widely used as steel additives, lubricants, catalysts, semiconductor materials, and high-temperature elements in aerospace industry [3]. Naturally, molybdenum occurs in several minerals including molybdenite (MoS_2), powellite ($CaMoO_4$), ferrimolybdite ($Fe_2(MoO_4)_3{\cdot}8H_2O$), wulfenite ($PbMoO_4$), etc. [4]. Among them, molybdenite has the highest industrial value and serves as the most common raw material of molybdenum industry.

Up till now, MoS_2 concentrates are mostly utilized through a combining process consisting of several pyrometallurgical and hydrometallurgical treatments. First,

D. Shi · G. Li (✉) · H. Sun · J. Luo · T. Jiang
School of Minerals Processing and Bioengineering, Central South University, Changsha 410083, Hunan, China
e-mail: liguangh@csu.edu.cn

G. Azimi et al. (eds.), *Rare Metal Technology 2020*, The Minerals, Metals & Materials Series, https://doi.org/10.1007/978-3-030-36758-9_28

MoS_2 concentrates are heated at 550–600 °C for 2–3 h to turn MoS_2 into MoO_3 and meantime release sulfur as SO_2 ($2MoS_2 + 7O_2 = 2MoO_3 + 4SO_2$). Since MoO_3 is soluble in alkali solutions, the subsequent operation is adding basic solvents, such as ammonia water, into the oxidized calcine, so that MoO_3 is transformed as ammonium molybdate and dissolved in solution. To obtain chemical pure product, the leachate is then purified, precipitated after adjusting pH value, and finally filtered [5–7]. The produced pure ammonium molybdates could be further processed into more valuable pure MoO_3 via thermal decomposing.

Recently, a shorter route of pure MoO_3 production from MoS_2 concentrate has been reported. Molybdenum concentrates mixed with additives of Al_2O_3–SiO_2 are made into pellets and roasted at 1050 °C under air atmosphere. With such high temperature, MoS_2 is rapidly oxidized as MoO_3, which is meantime evaporated and condensed as pure MoO_3. Although this new route remarkably improves the production efficiency of pure MoO_3 from MoS_2 concentrate, it also shares the same demerit with other sublimation methods: some MoO_3 would react with impurities to produce non-volatile molybdates, which were conserved in the volatilizing residue, decreasing the recovery of Mo to a low degree [8].

Our work has been focused on extracting molybdenum from the volatilizing residue. Elements and phases compositions of raw materials were first determined by instrumental analysis. Then, several leaching reagents including sodium hydroxide (NaOH), and mixtures of NaOH with sodium phosphate (Na_3PO_4) or sodium sulfate (Na_2SO_4) were used to leach the residue. It was found that the new composite reagent of NaOH and Na_3PO_4 worked best to dissolve Mo with the leaching rate up to 92.31%. The intensification effect of Na_3PO_4 on alkali leaching of Mo is a new finding, which might offer a new choice for Mo extraction from secondary resources.

Experimental

Preparation of Volatilizing Residues

MoS_2 concentrates were supplied from Huanglongpu, Shaanxi, China. Volatilization residues were prepared by oxidizing-volatilizing MoS_2 concentrate pellets at 1050 °C for 40 min with the addition of Al_2O_3 and SiO_2 (mass ratio of ore to additives = 2:1), which worked as the skeleton materials.

Leaching Tests

In each test, certain amounts of volatilizing residues and mixtures of water, NaOH, Na_3PO_4, or Na_2SO_4 were put into a beaker containing a rotor. Then, the beaker was placed in a water bath pot with the function of magnetic stirring and heating. After stirring for a desired duration, the solution was filtered and the residue obtained was washed with water and then dried for analysis.

The yield of leach residue is calculated as follows:

$$\mathrm{w} = \frac{m1 - m2}{m1} * 100\% \tag{1}$$

where w is the yield, m_2 is the mass of leached material, and m_1 is the mass of raw material.

The leaching rate of Mo is computed by

$$\mathrm{k} = \frac{c1 - c2}{c1} * 100\% \tag{2}$$

where k is the leaching rate of the element, c_2 is the element content in the leaching material, and c_1 is the element content in the leaching material.

Analytical Method

The chemical compositions of pulverized samples were examined using an X-ray fluorescence spectrometer (XRF, PANalytical, Axios mAX, the Netherlands). The mineral constituents were determined by using an X-ray diffraction (XRD, RIGAKU, D/Max 2500, Japan) under the conditions of radiation: Cu Kα tube current and voltage: 250 mA, 40 kV, scanning range: 10–80° 2θ, step size: 0.02° 2θ, and scanning speed: 8°/min.

The content of molybdenum in the solid material and aqueous solution was determined using a plasma emission spectrometer (ICP-AES, Thermo Fisher Scientific, Icap7400 Radial, USA).

The particle size distribution of the volatilizing residues is measured by a laser particle size analyzer (Master size 2000, USA).

Results and Discussion

Characterizations of Volatilizing Residue

The chemical composition of raw material was detected by XRF. Due to the volatilization of MoO_3 during roasting, impurities and additive components were greatly enriched. Specifically, it contained high contents of aluminum (Al_2O_3 = 63.16 wt%), silicon (SiO_2 = 23.45 wt%), molybdenum (Mo = 8.2 wt%), and rare earth (REO = 2.3 wt%). From the XRD pattern in Fig. 1, it was seen that

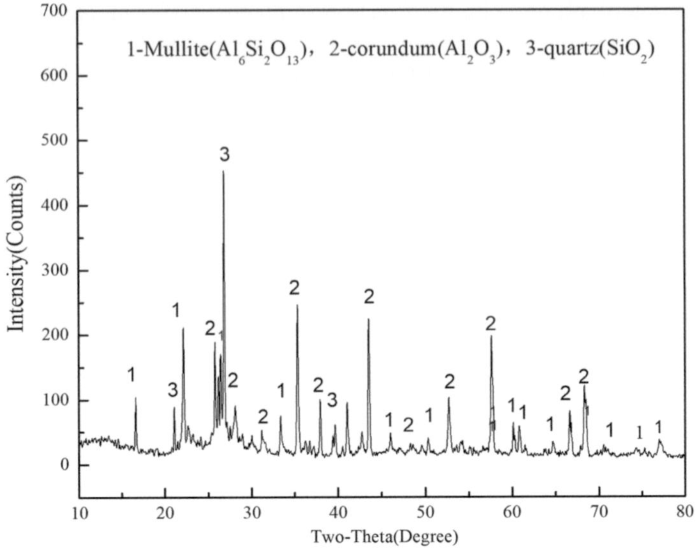

Fig. 1 X-ray diffraction pattern of the volatilizing residue

the main phases in the residue were corundum (PDF#78-2426), quartz (PDF#74-2419), and mullite (PDF#84-1205). Due to the low content, the rare earths contained in minerals were not detected. Further, the residue was observed by scanning electron microscopy-energy spectroscopy, which as shown in Fig. 2. It can be seen that the rare earth is mainly present in form of phosphate.

Particle size of raw materials were also determined as shown in Fig. 3, 10% of particles are smaller than 8.30 mm, 50% of the particles are smaller than 71.40 mm, and 90% of the particles were smaller than 162.62 mm.

Effect of Sodium Hydroxide Concentration on Mo Leaching Rate

The effect of different concentrations of sodium hydroxide solution (NaOH) on the leaching rate of molybdenum in volatilizing residue was investigated. The volatilizing residue was leached in different concentrations of sodium hydroxide for 120 min with liquid–solid ratio of 6 mL/g at temperature of 95 °C.

As shown in Fig. 4, the leaching rate of molybdenum increased with the increase of sodium hydroxide concentration while yield decreased with the increase of sodium hydroxide concentration. When the concentration of sodium hydroxide was up to 160 g/L, the leaching rate of molybdenum remained stable. Meanwhile, the yield of leaching tended to be stabilization. In the end, we choose the optimal

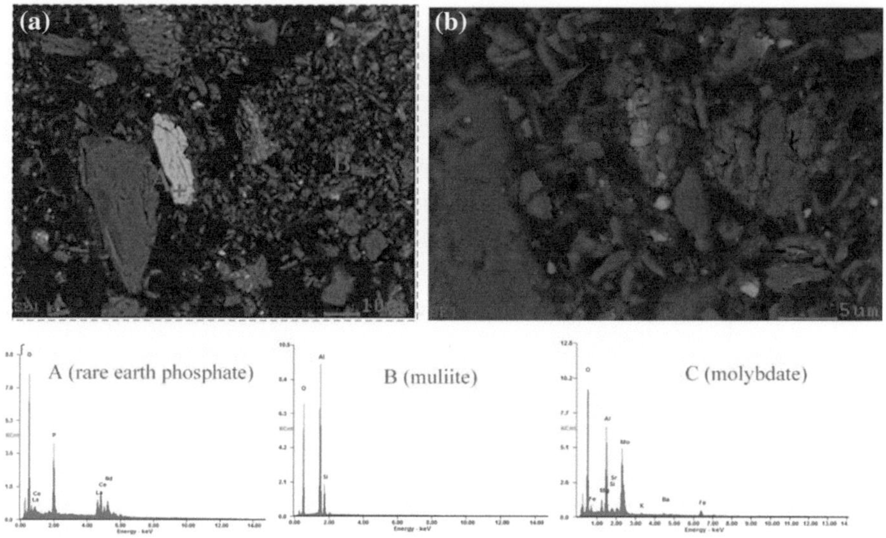

Fig. 2 Scanning electron microscope (SEM) images of the volatilizing residue (**a**, **b**)

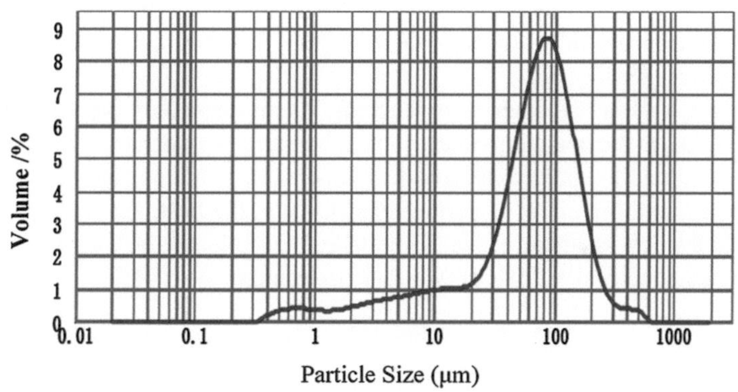

Fig. 3 Laser particle size analysis of volatilizing residue

sodium hydroxide concentration of 160 g/L. At the same time, the leaching rate of molybdenum was 86.28% and yield of leaching was 84.8%.

Effect of Combined Leaching Agent on Mo Leaching Rate

The effect of different kinds of leaching agents on the leaching rate of molybdenum in volatilizing residue was investigated. The volatilizing residue was leached in different kinds of leaching agents for 120 min with liquid–solid ratio of 6 mL/g at

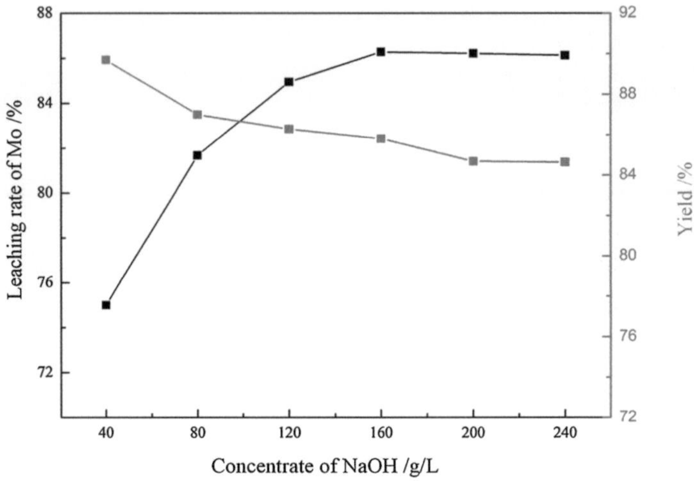

Fig. 4 Effect of sodium hydroxide concentration on molybdenum leaching

temperature of 95 °C. The combined leaching agent was an appropriate amount of sodium sulfate or sodium phosphate that was added to the sodium hydroxide solution. The concentration of Na_2SO_4 and Na_3PO_4 was both of 20 g/L in combined leaching agent.

As shown in Fig. 5, when the leaching agent was only sodium hydroxide solution (NaOH), the leaching rate of molybdenum was 85%, When Na_2SO_4 or Na_3PO_4 was added to sodium hydroxide solution, the leaching rate of molybdenum was improved, and the effect of adding Na_3PO_4 was the best. Therefore, we choose the combined leaching agent of NaOH and Na_3PO_4 to leach the molybdenum in the volatilizing residue and 92.3% of Mo can be leached.

Effect of Combined Leaching Agent Concentration on Mo Leaching Rate

The effect of adding different concentrations of sodium phosphate (Na_3PO_4) in sodium hydroxide solution (NaOH) on the leaching rate of molybdenum in volatilizing residue were investigated. The volatilizing residue was leached for 120 min with liquid–solid ratio of 6 mL/g at temperature of 95 °C, and the concentration of sodium hydroxide solution was 160 g/L.

As shown in Fig. 6, the leaching rate of molybdenum increased with increasing of sodium phosphate concentration. When the concentration of sodium phosphate was up to 20 g/L, the leaching rate of molybdenum remained stable. Finally, we choose the suitable sodium phosphate concentration of 20 g/L. At the same time, the leaching rate of molybdenum was 92.31%.

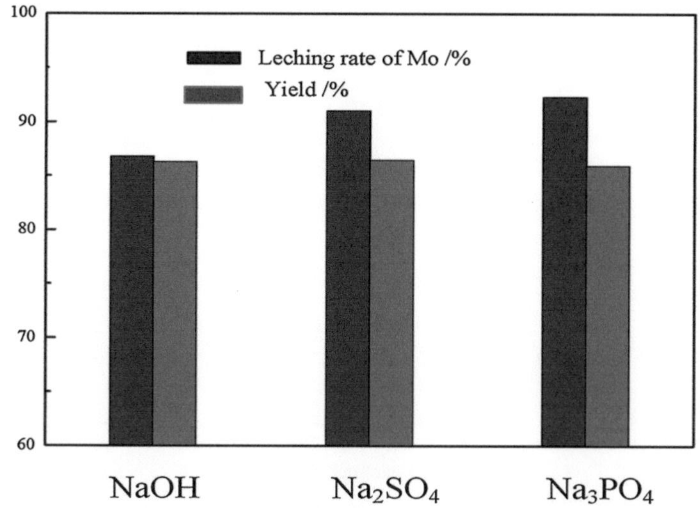

Fig. 5 Effect of combined leaching agent on molybdenum leaching rate

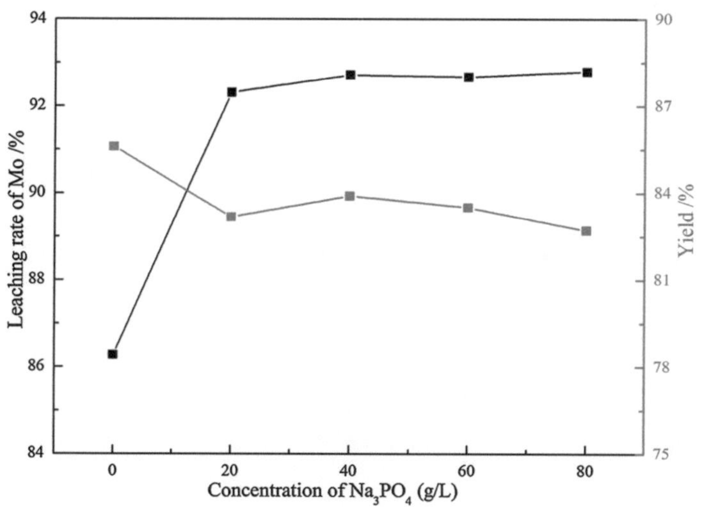

Fig. 6 Effect of combined leaching agent concentration on molybdenum leaching rate

Effect of Leaching Time on Mo Leaching Rate

The effect of different leaching times on the leaching rate of molybdenum in volatilizing residue was investigated. The volatilizing residue was leached with liquid–solid ratio of 6 mL/g at temperature of 95 °C. The concentration of sodium hydroxide was 160 g/L and the concentration of sodium phosphate was 20 g/L.

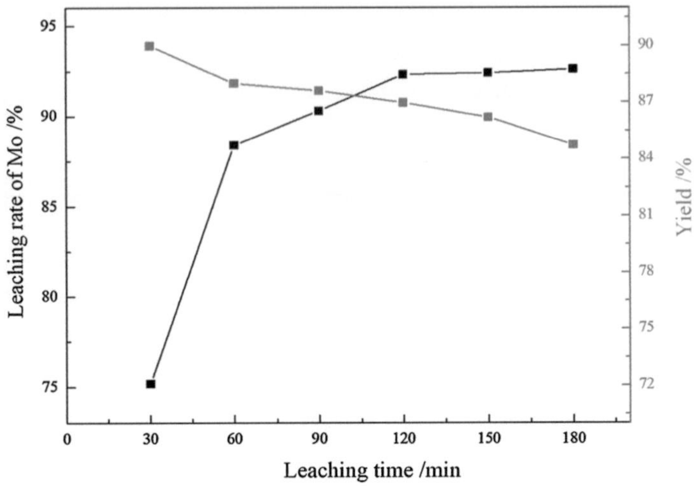

Fig. 7 Effect of leaching time on molybdenum leaching rate

As shown in Fig. 7, the leaching rate of molybdenum increased with the leaching time. When the leaching time was up to 120 min, the leaching rate of molybdenum remained stable. At the same time, the leaching rate of molybdenum was 92.31%.

As shown in Figs. 8 and 9, with molybdenum being selectively leached out, rare earths were enriched in leach residues, which was favorable for subsequent treatment.

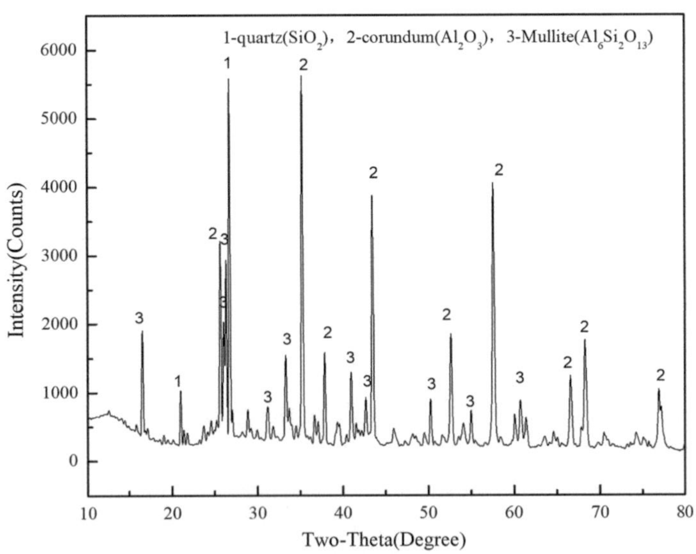

Fig. 8 X-ray diffraction pattern of leaching slag

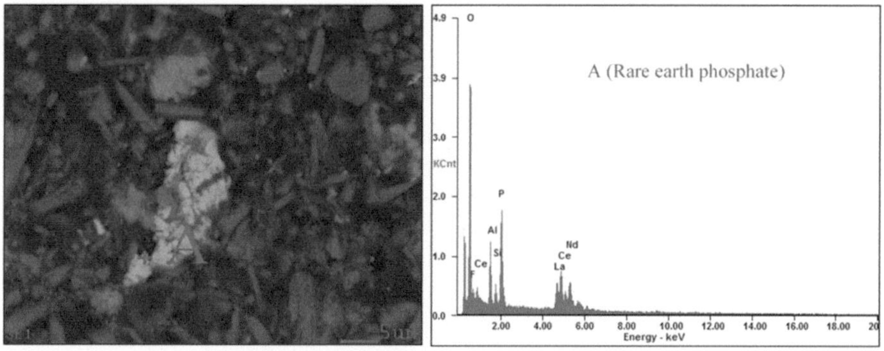

Fig. 9 Scanning electron microscope (SEM) images of the leaching slag

Conclusion

The composite leaching reagent consisting of NaOH and Na_3PO_4 has been proven better to recover Mo from a rare earth contained volatilizing residue than only NaOH or mixtures of NaOH and Na_2SO_4. A high leaching rate of 92.31% was achieved under conditions of NaOH concentration of 160 g/L, Na_3PO_4 concentration of 20 g/L with leaching time 120 min at temperature of 95 °C and liquid-to-solid ratio of 6 mL/g.

Acknowledgements This work was supported by the National Natural Science Foundation of China (No. 51874355) and the Co-Innovation Center for Clean and Efficient Utilization of Strategic Metal Mineral Resources.

References

1. Wang MY, Wang XW, Liu WL (2009) A novel technology of molybdenum extraction from low grade Ni-Moore. Hydrometallurgy 97(1):126–130
2. Ronzio RA, Davis PK, Ziegler RC (1978) Process for producing ammonium heptamolybdate and/or ammonium dimolybdate
3. Yonglin J, Bingguo L, Peng L, Jinhui P, Libo Z (2017) Dielectric properties and oxidation roasting of molybdenite concentrate by using microwave energy at 2.45 GHz frequency. Metall Mat Trans B
4. Khomoksonova DP et al (2018) Improvement of oxidizing roasting of molybdenite concentrate by addition of magnesite. J Min Sci 54(2):300–305
5. Wang L, Zhang GH, Dang J et al (2015) Oxidation roasting of molybdenite concentrate. Trans Nonferrous Met Soc China 25(12):4167–4174
6. Wang P, Pan Y, Sun X, Zhang Y (2019) Leaching molybdenum from a low-grade roasted molybdenite concentrate. SN Appl Sci 1(4)

7. Zhou QS, Yun WT, Xi JT, Li XB (2017) Molybdenite-limestone oxidizing roasting followed by calcine leaching with ammonium carbonate solution. Trans Nonferrous Met Soc China 27 (7):1618–1626
8. Sun H et al (2019) A novel simultaneous oxidizing-volatilizing process for efficient separation of pure MoO3 from structure self-sustained molybdenite concentrate pellets. Powder Technol 345:338–345

Extraction of Rare and High-Valued Metals from Blast Furnace Dust

Xiong Xiao, Shengfu Zhang, Hua Zhang, Guibao Qiu, Yuntao Xin and Jintao Wang

Abstract The treatments of blast furnace dust such as returning to blast furnace sintering, their use as flocculants and cement raw materials and adsorbents, recycling carbon and iron, as well as recovering rare and high-valued metals were discussed. The current processing status of blast furnace dust in the iron and steel industry and the research trend of rare precious metals represented by indium were investigated. Finally, some advanced research directions on dealing with blast furnace dust were analysed in the context of solid waste recycling; meanwhile, the advantages and disadvantages of pyrometallurgy and hydrometallurgy in the metallurgical industry were analysed by means of the process of indium extraction.

Keywords Blast furnace dust · High-valued metals · Pyrometallurgy
Hydrometallurgy

Introduction

Blast furnace iron-making is the main output source of the iron and steel industry, and plays an important role in the development of the world industrial economy especially in China; however, the steel industry produces a lot of solid waste like blast furnace dust [1]. As we all know, China's steel production is far ahead of other countries globally. In 2012, China's steel outputs exceeded 700 million tons while blast furnace dust reached 10 million tons. China's steel outputs exceeded 800 million tons in 2014 and maintained sustained growth. Following the policy of reducing production capacity in 2015, steel production continued to rebound and exceeded 900 million tons in 2018 with more than 18 million tons of dust produced. According to statistics, a ton of steel produced produces about 20 kg of blast furnace dust [2].

X. Xiao · S. Zhang (✉) · H. Zhang · G. Qiu · Y. Xin · J. Wang
College of Materials Science and Engineering, Chongqing University,
Chongqing 400044, China
e-mail: zhangsf@cqu.edu.cn

G. Azimi et al. (eds.), *Rare Metal Technology 2020*, The Minerals,
Metals & Materials Series, https://doi.org/10.1007/978-3-030-36758-9_29

Blast furnace dust is composed of blast furnace burden powders and particles generated by intense reaction in high-temperature area [3]. It mainly contains iron, carbon, oxides of Si, Al, Ca, and Mg, but also contains Pb, Zn, and alkali metal oxides with low boiling points. It is also a kind of material with light-weight and small particle size with the average particle size being 20–25 μm with non-adhesive properties [4, 5]. Nonferrous metals such as zinc, in the blast furnace dust, are mainly distributed under 200 mesh, and the moisture content fluctuates greatly and has a strong hydrophobicity, so it is easy to form harmful floating dust in the air, among which alkali metals are corrosive and heavy metals are toxic [6, 7]. According to relevant national standards, the pollution discharge of metallurgical industries has not fully met the requirements of clean production, and the comprehensive recovery rate of dust is very inadequate; therefore, these resources and environmental problems need to be continuously improved all over the world.

In addition, with the rapid development of the global economy, the demand for metals continues to expand, and this increase in consumption has caused serious problems in resources, energy, and environment, which becomes an important factor restricting the sustainable development of society and the economy. Therefore, the research on recycling secondary metal resources and the development of efficient separation and recycling technology of secondary resources can make full use of precious resources and reduce environmental pollution [8, 14].

In recent years, the consumption of indium has been increasing promptly with the rapid development of the information industry [9]. Dissipated metal indium is widely used in high-tech fields such as computer, military, nuclear industry, and modern information industry, and has an important strategic position [10]. The indium market is often in short supply, meanwhile, the limited indium resources are also decreasing day by day [11]. Therefore, the recycling of rare precious metals like indium in blast furnace dust is of great significance to the sustainable development of metallurgical industry and the promotion of circular economy even protection of the global environment [12]. This paper aims to provide a comparative analysis of the recovery and utilization of rare and precious metal resources at present, and to analyse the combined development of indium resource and metallurgical industry in the future.

Research Work

In the process of blast furnace smelting, the dust is discharged from the top of the furnace along with the blast furnace dust with the coarse particles being removed in the gravity dust collector through the descending pipe before the fine particles are collected by the cloth bag filter. The blast furnace dust is the general name of gravity and cloth bag dust [13]. According to the results of various researches and analyses, the general chemical composition of blast furnace dust is shown in Table 1. These data can accord basically with the composition analysis results of most blast furnace dust.

Table 1 Content distribution of elements in blast furnace dust (wt%)

Fe	C	Zn	Pb	Cd/Sn	Bi	In	Si/Al/Mg/Ca/F/Cl
30–45	15–20	5–15	1–10	10^{-2}–10^{-1}	10^{-4}–10^{-2}	10^{-3}–10^{-1}	20–30

In recent years, the treatments of blast furnace dust such as returning to blast furnace sintering, using as flocculants and cement raw materials and adsorbents, recycling carbon and iron as well as recovering rare and high-valued metals were studied [8, 13, 14]. Many process technologies have been more mature in enterprise application; however, the rare precious metals such as indium and bismuth are not comprehensively utilized, or only one of the valuable metals are recycled, so the comprehensive efficiency is low [15]. There are three main problems as follows [8]:

(1) Low added value of products. Blast furnace dust is heavily applied in the primary stage of construction like building and paving, and some high value-added metals are not used effectively.
(2) Low comprehensive utilization. At present, it is limited to the extraction application of some average-valued metals, and the comprehensive value is not fully utilized; therefore, its economic and social benefits are not obvious.
(3) Lower industrialization. Some research results about rare and precious metals are still in the laboratory stage and have not been applied to large-scale industry. Weak systematics are attached by the way.

Returning to Blast Furnace Sintering

Practice has proved that blast furnace dust can directly be used as sintering burden, and research has showed that solid fuel consumption of sinter decreased by 4.75 kg per ton of steel produced when 1% of the blast furnace dust was added [4]. After 10% magnetic separation of the dust was used to replace the non-magnetic separation in sintering process, the indices of the sintering process and iron-making process were improved to different degrees, among which the utilization coefficient of blast furnace increased by more than 0.2 t/(m^3 · d), and the focal ratio decreased by more than 10 kg/t [16]. However, nonferrous metals in dust are corrosive with long-term use of dust for sintering causing corrosion to the blast furnace. Studies showed that when the dust containing more than 5% of zinc or lead was used in sintering, it would easily damage the furnace lining and reduce the furnace life [6].

In order to avoid the risk of corrosion, the blast furnace dust, iron concentrate, charcoal powder, and binder were mixed for pelletizing, which was placed in the rotary hearth furnace after drying, and then zinc and lead were reduced, evaporated, and discharged together with flue gas deposition in the dust collector through external heating. At the same time, iron could be directly added to blast furnace iron-making after reduced to metallized pellets by carbon. However, the blast furnace industry requires that the pellet contains more than 60% iron in the iron-making process while the content of impurities in the metallized pellet is high, and the heavy metals such as cadmium and tin will also pollute and corrode the blast furnace due to enrichment in this process, so the dust cannot be directly used alone [17].

Recycling Carbon and Iron

The blast furnace dust is so rich in carbon and iron that researchers started to focus on it, and the commonly used methods of mineral separation came to be magnetic separation, gravity separation, and flotation for recovering iron resources. Magnetic separation is the separation under the action of magnetic field with the different permeabilities of each component where the magnetic force on the minerals should be greater than the sum of the reverse mechanical forces. Gravity separation is based on the density difference of minerals, in addition to the gravity separation equipment, there must also be a medium such as water or air. This equipment is simple in structure and low in cost, so it can be used for appropriate conditions. Flotation is the most widely used method to separate mineral particles through different physical and chemical properties of the mineral surface. Flotation is suitable for the treatment of fine particles and fine materials especially those less than 10 μm which are difficult to recover by other methods. The factors of flotation process must be determined and selected according to the characteristics of ore properties and experimental research to obtain the optimal technical and economic conditions [18, 19].

Scholars also studied the combination of the above three methods in order to find better production conditions, Hu et al. [20] tested the recovery of iron by using conventional ore dressing methods and four technologies contain single shaker, fine sieve-shaker, magnetic separation-shaker, and fine sieve-magnetic separation-shaker, and concluded that single magnetic separation and flotation are difficult to obtain high-grade iron concentrate, but easy to discard low-grade tailings. Xu et al. [21] adopted the combined floatation and gravity separation technology, and results showed that the process was simple, reliable, and stable, and the iron concentrate with the total iron content of 61.13%, a recovery rate of 56.12% and the carbon content of 80.09% with a recovery rate of 88.04% could be obtained. The obtained iron concentrate and carbon could be used as sintering raw materials, and tailings could be used as slag brick raw materials.

Using as Flocculants and Cement Raw Materials and Adsorbents

Blast furnace dust can be used as flocculants, raw materials of cement production and adsorbents, which is conducive to the integration and efficient utilization of resources.

(1) The main function of flocculants is to make groups with positive or negative electricity and some particles with negative or positive electricity in water close to each other and reduce their potential to an unstable state, and concentrate particles with the aggregation property, and separate particles by physical or chemical methods. Therefore, flocculants are mainly used in sewage treatment [22].
In recent years, the development of compound inorganic polymer flocculants has become a hot spot. It is a new kind of flocculant, which has gradually become mainstream due to its wide application. Li and Zhang [23] researched the preparation of the compound inorganic polymer flocculants such as poly-aluminum ferric chloride (PAFC). The optimal dissolution effect of hydrochloric acid was achieved with temperature of 100 °C, time of 3 h, dust/acid ratio of 1/3, and the iron and aluminum dissolution rates were 67.61% and 11.35%, respectively. On the basis of the optimal dissolution conditions, $Al(OH)_3$ sol was added into the dissolution solution and stirred for 4 h at 60–70 °C, and the PAFC obtained was stable and the turbidity removal rate of wastewater was 99.36%. Taking blast furnace dust and aluminum smelting dust as raw materials, Yang [24] added neodymium compounds in the process of synthesizing PAFC and rare earth neodymium modified PAFC was developed, and the removal effect was 4.47% higher than self-made PAFC.
(2) The blast furnace dust can also be used as adsorbents to adsorb heavy metal ions. The dust has small particle size, large specific surface area, and high activity, and contains a large number of activated carbon particles, silicate, and other oxides, which can effectively adsorb organic compounds in sewage. The study showed that the adsorption of Pb^{2+} by dust mainly occurred in hematite phase. The adsorption of Pb^{2+} increased with the increase of temperature and reached 79.89 mg/g at 80 °C. The adsorption capacity of dust increased with the increase of Fe_2O_3/C ratio, and the adsorption capacity of different heavy metal ions was Pb > Cu > Cr > Cd > Zn [25, 26].
(3) The blast furnace dust can be used as part of raw materials in cement production such as limestone, clay, iron ore powder, or other auxiliary additives. Zou [27] used the dust as raw materials for iron correction instead of copper slag in Huangshi cement plant of Guangzhou military region. He studied the effect of blast furnace dust on raw cement preparation, shaft kiln calcination, clinker, and cement quality, and obtained better economic benefits when it was used in replacing some ingredients of cement raw material mixing.

Recovering Rare and High-Valued Metals

It is an important means for comprehensive utilization of blast furnace dust to recover valuable metals such as zinc, lead, tin, bismuth, and indium and also a secondary utilization of rare and precious metal resources. Zinc is the most abundant nonferrous metal in the blast furnace dust and the extraction cost is relatively low with research showing that zinc can be easily extracted and recovered by both pyrometallurgy and hydrometallurgy. It has therefore achieved a mature application in industry. The recovery of lead and tin is often accompanied by the recovery of indium and bismuth is also easier to separate because of its physical and chemical properties.

The classic method is to enrich nonferrous metals in dust by pyrometallurgy. Jianhui Liu carried on the fuming treatment in the rotary kiln, which took the blast furnace dust as the main raw material, and added the appropriate amount of reducing agent like coal and coke and the solvent containing silicon and calcium for the mixture containing about 10% zinc. Results showed that Fe mainly remained in kiln slag and accounted for 40%, and by controlling the fuming process, we could make the iron exist mainly in the form of magnetite in slag as raw material of beneficiation while valuable metals were enriched 5, 4, 5, and 3 times in oxidation of zinc powder, respectively [28].

There is much research on rare and precious metals like indium. Mao et al. [29, 30] recovered indium by acid leaching with 0.4 g gelatin added to every 20 g blast furnace dust, and the leaching rate of indium under high-speed agitation was 15.31% with 1 mol/L H_2SO_4, solid–liquid ratio 1/5 for 2 h at 80 °C. The indium in acid leaching solution was enriched and extracted by W/O liquid membrane extraction. The W/O emulsion liquid membrane was prepared by P507 of 5%, SPAN80 of 5%, industrial kerosene of 85% and liquid paraffin of 5% as film-forming components, and 99% of the extraction rate was achieved with indium concentration of 1.18×10^{-6} g/ml, pH 2.96, emulsion water ratio 1/3 for 6 min at room temperature.

In order to improve extraction efficiency, Li et al. [31] used the process of neutral leaching and high concentration acid leaching with $MnCO_3$ to recover indium from dusts. The optimum process conditions were achieved by using the organic phase composed of 30% P204 and 70% sulfonated kerosene with level 3 extraction, O/A ratio of 3/1 for 5 min at 20 °C, and results showed that the extraction rate of indium was greater than 99.4%, and the total recovery rate was greater than 95.0%. In the study of different leaching solutions, Gupta [32] studied the extraction of cyanex923 in HCl, H_2SO_4 and HNO_3 solutions, respectively, and the recovery rate of In^{3+} reached 95% and purification rate reached 99% when the ratio of indium ion and extractant agent were 1/2. It was pointed out that the extraction rate of In^{3+} did not change significantly with the concentration of the extractant, but decreased with the increasing polarity of the extraction solution. We can realize that the leaching solution of indium is mostly sulfuric acid with extractant P204, as Bina Gupta studied, under the condition of high acidity, the order of effect of different acid solutions on extraction rate is HNO_3 > H_2SO_4 > HCl, while the reverse is true at low acid concentration [32].

Therefore, the combination of pyrometallurgy and hydrometallurgy has become a more mature extraction process. Zhao et al. [33] used 80% of raw dust and 20% of coke powder and was first enriched by reducing volatilization in rotary kiln with the temperature of 1100–1200 °C. Indium ingot was obtained through neutral leaching, second-stage acid leaching, extraction with P204, stripping with HCl and $ZnCl_2$, replacing with aluminum 4–5 h with 50–60 °C, melting casting, electrolysis and other processes. The total recovery rate was more than 70%, which has been successfully applied to industrial practice and has generated respectable economic benefits.

Tech-Development Trend

In recent years, with the development of organic chemistry, hydrometallurgy is becoming more and more diverse with more and more related research and results on hydrometallurgy. Li et al. [34] studied the direct solvent extraction of indium from a zinc residue reductive leach solution by D2EHPA, which effectively avoided the traditional steps of intermediate precipitation, solid–liquid separation, and re-leaching. Indium was effectively extracted from zinc sulfate solution by 20% D2EHPA dissolved in kerosene at an initial pH of 0.5 and aqueous-to-organic phase ratio (A/O) of 6/1 using three countercurrent stages. The Fe^{3+} was reduced to the ferrous state to avoid the possibility of its co-extraction with indium. Finally, the extraction rate was 96.1%.

The production of indium mainly comes from the residue of zinc and lead smelting, and Yuhui Zhang studied leaching-solvent extraction to separate and recover indium from lead smelting hazardous dust (LSHD), which has higher extraction percentage and less environment pollution. The novel technological keys included in the pressure sulfuric acid leaching of LSHD, precipitation and enrichment of indium, atmosphere sulfuric acid leaching of indium-enriched residues, and solvent extraction of indium. More than 95% indium was leached by two-stage countercurrent pressure sulfuric acid leaching using the initial concentrate of H_2SO_4 of the first stage 30–50 g/L and the second stage 120–150 g/L, temperature of 413–433 K, total pressure of 0.6–1.0 MPa, time of 1–1.5 h, liquid/solid (L/S) ratio of 3–5/1 and agitation speed of 400 r/min, respectively [35, 36].

Industrial indium is mainly contained in a mixture of other metals such as gallium and tin; therefore, Lee et al. [37] studied the separation of indium and gallium from sulfuric acid leaching solution with D2EHPA. In the extraction process, the separation factor between indium and gallium decreased with pH, and it was found that aliphatic kerosene was superior to benzene as a diluent in respect of loading capacity. In the stripping process, the stripping percentage of indium decreased from 95% to 5% with pH in the range of 0.2–1.4, while separation factors increased with the pH of the stripping solutions. The results showed that 99% of indium was extracted into 0.05 M D2EHPA and nearly complete separation of indium could be accomplished by a two-stage extraction from simulation experiments. In order to improve the leaching efficiency, Zhang et al. [38] reduced and leached indium-bearing zinc slag with sphalerite concentrate as reducing agent in sulfuric acid. The optimal leaching condition was determined as 150 g/L H_2SO_4, 0.95 times of theoretic amount of sphalerite concentrate for 4 h at 90 °C while using particles in the range of 74–58 μm. The leaching efficiencies were 94.8% of indium, 96.1% of zinc and 92.8% of iron, respectively, and a Fe^{2+}/Fe^{3+} molar ratio of 7.5 in the leach solution was also obtained. This process could effectively extract indium and converted Fe^{3+} into Fe^{2+} at the same time to prevent its synergistic extraction with indium ions.

To sum up, in the process of hydrometallurgy, the process of leaching, extraction, stripping, and electrolysis are mostly a system for metal refining, and based on the research of hydrometallurgy recovery of nonferrous metals, we can find that the key to the advantages and disadvantages of the technology lies in the two processes of leaching and extraction. However, we should not only optimize the process and conditions but also recover valuable metals from resources such as LCD waste, which can be regarded as the comprehensive utilization of rare metal resources. For example, Silveira et al. [39] studied the recovery of indium from discarded LCD screens of mobile phones. First, the screens were manually removed from cell phones, and then a pre-treatment was developed for removal of the polarizing film from the glass of LCD panels. The leaching rate of 96.4% was obtained with 1.0 M H_2SO_4 and solid/liquid ratio of 1/50 for 1 h at 90 °C, and agitation speed of 500 rpm. Results showed that recovery of indium from H_2SO_4 solution by precipitation with NH_4OH was most satisfactory at pH 7.4, resulting in precipitation of 99.8 wt% of the indium. Finally, the overall mass balance of the procedure showed that it was possible to recover 0.2986 kg of indium for every 1000 kg of LCD screens, which corresponded to 0.0299 wt% of the LCD screens.

Meanwhile, Rocchetti et al. [40] studied the technology of indium recycling from LCD screens by cementing with zinc powder in sulfuric acid leaching solution, and the effect of zinc concentration, pH values, and cementing time on cementing efficiency and purity of solid products were analysed. It was found that almost all the indium presenting in the leaching solution passed to the solid phase when cementation was performed with a low (2–5 g/L), a medium (15–20 g/L) and a high (100 g/L) concentration of zinc at pH of 3. Results showed us that indium cementation took place also with low concentration of zinc at pH of 3, allowing important reagent saving, and the highest purity of 62% of the indium product (calculated in the four-metal system of indium–aluminum–calcium–iron) was achieved after a cementation of 10 min, whereas the content of impurities increased with time.

As we can see, each method has its special purpose, and we need to be specific and also comprehensive in our practical application. And Table 2 shows some extraction ideas of indium presently; it is mainly for the two processes of leaching and extraction.

Table 2 Extraction methods and thoughts of indium

Ideas	Process	Purpose
Oxidation leaching [41]	Oxidant as O_3 and H_2O_2	Oxidize lower ions
Reduction leaching [38]	Zn or Cu powder	Prevent Fe^{3+} co-extraction
Pressure leaching [35]	0.6–1.0 MPa pressure leaching	Improve leaching rate
T-type microreactor [42]	Microfluidic extraction	Improve separation factor
Specific extraction [29]	W/O emulsion liquid membrane	Exchange In^{3+} precisely

Conclusion

(1) Blast furnace dust has important value as it is rare and valuable. The comprehensive utilization cannot only relieve environmental pressure and realize clean production but also save metallic resources and create economic and social value.
(2) Pyrometallurgy is characterized by high productivity, short process, low investment in equipment, high energy consumption, serious pollution, and high ore grade requirements. Hydrometallurgy is characterized by low productivity, long process, serious corrosion in equipment, low operating temperature, and no special requirements for ore grade.
(3) Leaching and extraction are the most important processes in hydrometallurgy, and so many research methods are aimed to improve this process.
(4) The utilization of rare precious metals such as indium will be more and more developed in the direction of ultra-purity, and selective purification technology will be developed and applied to the high-tech fields.

Acknowledgements This work was supported by the Natural Science Foundation of China (Project No. 51474042 and 51774061) and the Fundamental Research Funds for the Central Universities (Project No. 106112017CDJQJ138801).

References

1. Tong XZ (2016) Technology of recovery and harmless treatment of valuable metals in blast furnace gas ash. China Met Bull 04:35–37
2. Jia JH, Han HL, Duan DP et al (2013) Study on comprehensive recovery and recycling of blast furnace gas ash. Iron Steel Vanadium Titan 34(5):33–37
3. Jia YZ (2018) Investigations on the fundamental characteristics and efficient dezincification of zinc-bearing blast furnace dust. Master's thesis, Taiyuan University of Technology
4. Gu WS, Deng MZ (1997) Harmless treatment and comprehensive utilization of blast furnace gas ash. Fly Ash Compr Util 3:54–56
5. Jia GL, Zhang BH (2007) Study on the mixed injection of blast furnace gas ash and pulverized coal. China Metall 17(5):18
6. Wang SK (2009) Recovery and recycling of non-ferrous metal from the ash gas. Resour Recycl 10:48–50

7. Liu Y, Ma L, Zhao HB (2014) The process of removal zinc from the blast furnace dust of Bx Steel. Bengang Technol 5:12–16
8. Liu BG, Peng JH, Zhang LB et al (2007) Present condition of research on recycling utilization of blast furnace sludge or flue dust. Express Inf Min Ind 23(5):14–19
9. Nong JT (2002) Present situation and development of China's indium industry. China Nonferrous Metall 31:12–14
10. Alfantazia AM, Moskalyk RM (2003) Processing of indium: a review. Miner Eng 16(8):687–694
11. Lupi C, Pilone D (2014) In(III) hydrometallurgical recovery from secondary materials by solvent extraction. J Environ Chem Eng 2(1):100–104
12. Wang SK (2008) Progress of extraction and application of indium. Eng Sci 10:85–94
13. Deng YC (2014) Research situation of the blast furnace gas ash comprehensive utilization. Hunan Nonferrous Met 30:25–29
14. Li C, Cong SC (2011) The recovery utilizing method of gas ash resources. Environ Prot Circ Econ 31:15–17
15. Zhang ZB (2015) A recycling process of recovering indium, germanium and bismuth from zinc-containing waste in smelting industry. Eng Constr 47:53–56
16. Sun BY (2006) Production practice of sintering mixed with magnetic separation gas ash. Iron Mak Technol J 7:16
17. Wang SK (2010) Industrial practice of recovering non-ferrous metals from blast furnace dust. Res Iron Steel 38:45–49
18. Pan GT (2012) Brief discussion on carbon and iron extraction technology from blast furnace gas ash. Sci Technol View 20:79–80
19. Wang WS, Feng LJ, Pan XF et al (2004) Test research on comprehensive recovery of carban and iron from blast furnace flue dust. Met Mine 8:498–500
20. Hu XH, Zhang ZF, Ni YH (2004) Research on comprehensive utilization of blast furnace gas mud. Express Inf Min Ind 8:14–16
21. Xu BH, Wang EJ, Yang JW (2007) Research on recovery of iron and carbon from blast furnace gas ash. Conserv Util Miner Resour 3:51–54
22. Li SP, Fan SS, Zhao YX et al (2008) Study on the treatment of brewing wastewater by using the composite flocculant. Ind Water Treat 04:19–22
23. Li SP, Zhang YL (2004) Study on the preparation of composite inorganic macromolecule flocculant PAFC from steel plant waste. In: Published in May 2004 of conference and symposium on water treatment chemistry of china chemical society
24. Yang W (2012) Research of neodymium modified flocculant made of blast furnace dust and aluminum scrap material. Master's thesis, Shandong University
25. Lopez-Delgado A (1996) The influence of carbon content of blast furnace sludge and coke on the adsorption of lead ions from aqueous solution. Carbon 34(3):423–431
26. Lopez-Delgado A (1998) Sorption of heavy metals on blast furnace sludge. Water Res 32 (4):989–996
27. Zhou JP (2000) Application of blast furnace gas ash in cement production. Jiangxi Buil Mater 2:34–37
28. Liu JH, Wang ZR, Luo BH et al (2008) Decontamination treatment and comprehensive utilization of zinc materials with waelz technology in production practice. Hunan Nonferrous Met 24(6):16–18
29. Mao L, Wan J, Xu LJ et al (2009) Recovery of indium from acid leachate of blast furnace dust with emulsified liquid membrane. Nonferrous Met (Extractive Metallurgy) 4:23–25
30. Mao L, Gao Q, Yan JB et al (2009) Leaching indium from blast furnace flue dust with sulphuric acid. China nonferrous metallurgy, volume b: research and development paper. Exp Res 1:68–70
31. Li SQ, Liu WF, Tang MT et al (2004) Extracting indium from reduced solution with adding manganese in the hydrometallurgy of zinc without producing iron residue. J Jishou Univ (Natural Science Edition) 25(4):14–18

32. Gupta B (2004) Liquid-liquid extraction and recovery of indium using Cyanex 923. Anal Chim Acta 513:463–471
33. Zhao T, Zheng ZE, Dong P (2014) Process design for scattered metal indium recovery from blast furnace dust. Rare Met Cem Carbides 42(4):14–16
34. Li XB, Deng ZG, Li CX et al (2015) Direct solvent extraction of indium from a zinc residue reductive leach solution by D2EHPA. Hydrometallurgy 156(1):1–5
35. Zhang YH, Jin BJ, Ma BZ (2017) Separation of indium from lead smelting hazardous dust via leaching and solvent extraction. J Environ Chem Eng 5(3):2182–2188
36. Sawai H, Ismail MMR, Yoshinori T et al (2015) Selective recovery of indium from lead-smelting dust. Chem Eng J 277
37. Lee MS, Ahn JG, Lee EC (2002) Solvent extraction separation of indium and gallium from sulphate solutions using D2EHPA. Hydrometallurgy 63(3):269–276
38. Zhang F, Wei C, Deng ZG et al (2016) Reductive leaching of indium-bearing zinc residue in sulfuric acid using sphalerite concentrate as reductant. Hydrometallurgy 161:102–106
39. Silveira AVM, Fuchs MS, Pinheiro DK et al (2015) Recovery of indium from LCD screens of discarded cell phones. Waste Manag 45:334–342
40. Rocchetti L, Amato A, Beolchini F (2016) Recovery of indium from liquid crystal displays. J Clean Prod 116:299–305
41. Tian QH, Xin YT, Wang HL et al (2015) Research status on application of clean oxidants in oxidizing leaching of metals in hydrometallurgy. Hydrometall China 34(03):167–172
42. Le T, Xiao BQ, Ju SH et al (2019) Separation of indium from impurities in T-type microreactor with D2EHPA. Hydrometallurgy 183:79–86

Mechanism of Extraction of Vanadium from Vanadium Slag with MgO

Chengjie Wang, Yiheng Yuan, Bing Xie and Hong-Yi Li

Abstract The widely applied methods to extract vanadium are sodium roasting and calcified roasting. However, these roasting methods have difficulties in recycling of wastewater and efficient utilization of leaching residue, which contains high concentration of sodium or calcium sulfate, due to the variety of roasting additives radically. Therefore, a novelty extraction medium by roasting with MgO and leaching with sulfuric acid was introduced in this study. The oxidation and roasting mechanism were systematically studied, providing a theoretical basis for industrial extraction of vanadium. The result shows that under the optimum conditions, the roasting temperature of 900 °C, the Mg/V molar ratio of 0.6, and the roasting time of 1.5 h, the leaching efficiency of vanadium is above 90%. In the processes, vanadium spinel is converted into $Mg_2V_2O_7$ and $Mn_2V_2O_7$.

Keywords Vanadium slag · MgO · Roasting mechanism

Introduction

Vanadium is a very important material widely applied in the iron and steel industry, aviation and aerospace, chemical, and pharmaceutical industry due to its special chemical and physical properties [1]. Almost 88% of vanadium worldwide is extracted from vanadium-titanium magnetite ore, and the remaining is extracted from stone coal, spent catalyst, and so on [2–4]. The converter vanadium slag, a by-product of vanadium-titanium magnetite smelting process, is the major resource for vanadium extraction. Vanadium mainly exists in the form of FeV_2O_4 spinel.

C. Wang · Y. Yuan · B. Xie · H.-Y. Li (✉)
College of Materials Science and Engineering, Chongqing University,
Chongqing 400044, China
e-mail: hongyi.li@cqu.edu.cn

C. Wang · Y. Yuan · B. Xie · H.-Y. Li
Chongqing Key Laboratory of Vanadium-Titanium Metallurgy and New Materials,
Chongqing 400044, China

G. Azimi et al. (eds.), *Rare Metal Technology 2020*, The Minerals, Metals & Materials Series, https://doi.org/10.1007/978-3-030-36758-9_30

There are several methods to extract vanadium from vanadium slag, molten NaOH roasting method, and roasting–leaching methods [5, 6]. Molten NaOH roasting method was researched systematically by Yi Zhang from the Institute of process engineering, Chinese academy of sciences [7]. The sub-molten salts method can decrease the cost of energy due to the lower production temperature. But it consumes large amount of expensive NaOH and has the problem of refractory solid wastes and wastewater which leads to high production cost [8]. And the requirement of reactors with high corrosion resistance and the large consumption of alkaline limit its wide application.

The roasting–leaching methods are the most economical and cost-effective methods at present, which are widely applied in the vanadium production industry. It contains sodium roasting–water leaching method and calcified roasting–acid leaching method. The sodium roasting method, transforming vanadium spinels into water-soluble sodium vanadates to gain production of vanadium oxides, is mature and operationally steady [9]. However, it seriously pollutes the environment and corrodes equipment due to the corrosive gases, including HCl, Cl_2, and SO_2, which are generated in the roasting process [10–12]. And it has difficulties in comprehensive utilization of the leaching residue and in the recycling of wastewater, which contains high concentration of sodium. The calcified roasting-acid leaching, converting vanadium spinels into acid-soluble calcium vanadates to gain production of vanadium oxides, is efficient and clean [13]. However, a large amount of calcium sulfates are produced to enwrap calcium vanadates partially to lead a low leaching efficiency of vanadium. And the leaching residues with a mass of calcium sulfates are difficult to utilize synthetically, leading to a serious wasting of resources [14, 15]. By analyzing the actual methods of vanadium extraction, it can be observed that the corrosive gases and the difficulties in comprehensive utilization of the leaching residue are caused by the roasting additives. Hence, in order to solve the wasting problems and the pollution problems of vanadium extraction, we should start with the roasting additive.

In this paper, the novel method of roasting with MgO and acid leaching is proposed to extract vanadium efficiently. The method translates vanadium spinels into acid-soluble magnesium vanadates. In the roasting and leaching process, it produces no corrosive gases and the leaching residues can as the raw material for converter steelmaking. Hence, the method of roasting with MgO and acid leaching is believed to be an environment-friendly method. This work is focused on roasting mechanism and optimization of the roasting conditions, including the amount of MgO additive, roasting temperature, and roasting time. This work has laid the foundation for the industrial application of roasting with MgO and acid leaching and provided new insights into leaching of vanadium in vanadium slag.

Experimental

Instruments and Reagents

The original material, vanadium slag was from Pan Steel, China. The chemical compositions of vanadium slag are shown in Table 1. All chemical reagents used are of analytical grade. MgO as additive was dried at 105 °C for hours before using. All aqueous solutions were prepared with deionized water produced by Millipore Aquelix 5.

Chemical compositions of vanadium slag was determined by X-ray Fluorescence (XRF, Shimadzu XRF-1800). Phase compositions of solid samples were identified by X-ray diffraction analysis (XRD, Rigaku D/MAX 2500PC) using Cu Kα radiation. Concentrations of vanadium element in the leach liquor were determined by Inductively Coupled Plasma-Atomic Emission Spectroscopy (ICP-AES, Thermo Scientific iCAP 6300 Duo).

Experimental Methods

In this study, vanadium slag was dry sieved and gave a particle size fraction 45–74 μm. During the roasting process, MgO was mixed with vanadium slag in Mg/V molar ratio of 0.4–1.5 in a ceramic evaporation dish. The mixture was heated in a muffle furnace at 600–1000 °C for 1–2.5 h and the door of the muffle furnace was not closed completely to ensure the complete oxidation. The vanadium slag roasting with MgO was cooled down to room temperature by air.

The roasted vanadium slag was milled and screened to size portion (45–74 μm). The roasted slags were leached by dilute sulfuric acid at solid/liquid ratio of 1:5. The pH of dilute sulfuric acid is 0.5. Leaching experiments were performed in a commercial magnetic stirring water bath pot with a stirring speed of 300 r/min at 70 °C for 1.5 h. After leaching, leaching liquor and leaching residue were collected separately by filtration. Leaching liquor was used to calculate the extraction efficiency of vanadium with Eq. (1):

$$\%\text{Extraction} = \mathrm{M_L}/\mathrm{M_V} \times 100\% \tag{1}$$

where M denotes vanadium, M_v is the mass of the roasted vanadium slags, and M_L is the mass of the leaching liquor.

Table 1 Chemical compositions of vanadium slag (wt%)

Compositions	FeO	V_2O_3	TiO_2	MnO	SiO_2	MgO	CaO	Al_2O_3
	38.53	15.37	12.60	9.47	13.51	4.01	3.02	3.47

Note All the Fe oxides were counted in the form of FeO for content calculation

Results and Discussion

The Effect of MgO Additive Amount on the Vanadium Extraction

The effect of Mg/V molar ratio on the extraction of vanadium was conducted from 0.4 to 1.5 while keeping fixed roasting temperature of 900 °C and roasting time of 1.5 h. The results are shown in Fig. 1. It can be observed that the leaching efficiency of vanadium increased gradually and reached the maximal of 92.19% at Mg/V molar ratio of 0.6, then decreased in the range of 0.6–1.0. And in the range of 1.0–1.5, the leaching efficiency of vanadium has slightly increased. This indicates that MgO added functions as the critical material to convert vanadium spinels into vanadates during roasting. The vanadium spinels and olivine-related reactions during roasting process were expressed as below:

$$FeV_2O_4 + MgO + 1.25O_2 = MgV_2O_6 + 0.5Fe_2O_3 \tag{2}$$

$$FeV_2O_4 + 2MgO + 1.25O_2 = Mg_2V_2O_7 + 0.5Fe_2O_3 \tag{3}$$

$$FeV_2O_4 + 3MgO + 1.25O_2 = Mg_3V_2O_8 + 0.5Fe_2O_3 \tag{4}$$

$$Fe_2SiO_4 + MgO + 0.5O_2 = MgSiO_3 + Fe_2O_3 \tag{5}$$

In the roasting process, different molar ratios of MgO and vanadium spinels will generate different productions, magnesium metavanadate $Mg(VO_3)_2$, magnesium pyrovanadate $Mg_2V_2O_7$, and magnesium orthovanadate $Mg_3(VO_4)_2$. According to

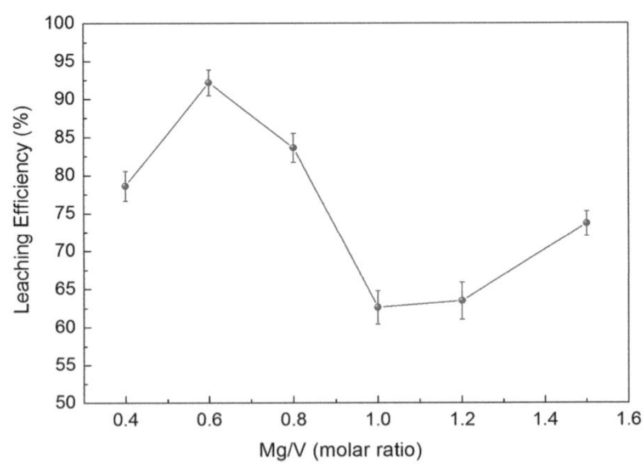

Fig. 1 Effect of Mg/V molar ratio on vanadium extraction

the solubility of different magnesium vanadates, $Mg(VO_3)_2$ has obviously higher solubility in sulfuric acid. At Mg/V molar ratio of 0.6, the production of magnesium metavanadate $Mg(VO_3)_2$ is facilitated. With the increase of the Mg/V molar ratio, reduction in the relative amount of $Mg(VO_3)_2$ and increasing in production of $MgSiO_3$ wrapping the magnesium vanadates give rise to the decreasing of the leaching efficiency. In contrast, the formation of $Mg_2V_2O_7$ and $Mg_3(VO_4)_2$ are facilitated at Mg/V molar ratio range of 1.0–1.5, which results in the slightly increasing of the leaching efficiency. Therefore, the optimal Mg/V molar ratio is considered to be 0.6.

The Effect of Roasting Temperature on the Vanadium Extraction

The roasting temperature is the key parameter for the oxidation of vanadium with MgO. The effect of roasting temperature on the extraction of vanadium was conducted from 600 to 1000 °C, while keeping fixed Mg/V molar ratio of 0.6 and roasting time of 1.5 h. The results shown in Fig. 2. It can be observed that the leaching efficiency of vanadium increased dramatically at the temperature range from 600 to 850 °C and reached the maximal of 92.19% at the temperature of 900 °C. This shows that the increase in the roasting temperature contributes to destruction of olivine phases which wrap the spinels phases and the formation of magnesium vanadates. However, with a further increase in roasting temperature from 900 to 1000 °C, the leaching efficiency of vanadium decreased. The possible reason is that magnesium vanadates are enwrapped partially by sintered phases due to excessive temperature. Therefore, the optimal roasting temperature is considered to be 900 °C.

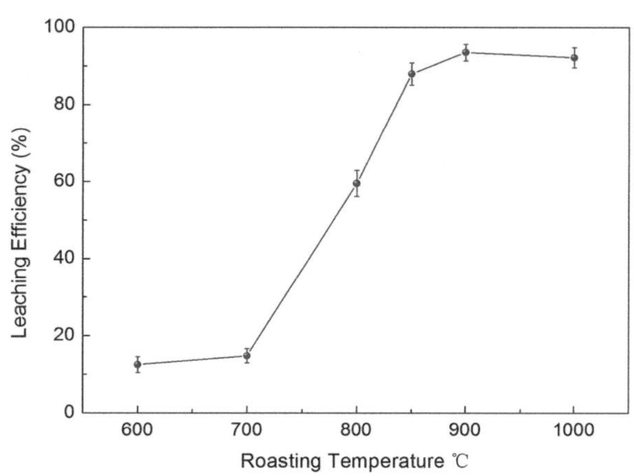

Fig. 2 Effect of roasting temperature on vanadium extraction

The Effect of Roasting Time on the Vanadium Extraction

The effect of roasting time on the extraction of vanadium was conducted from 1 to 2.5 h, while keeping fixed Mg/V molar ratio of 0.6 and roasting temperature of 900 °C. The results are shown in Fig. 3. The results indicate that the leaching efficiency of vanadium reached a peak value of 92.19% at the roasting time of 1.5 h. The oxidation of vanadium proceeds rapidly, leading to leaching efficiency of 82.85% after roasting for 1 h. With the increasing of roasting time, more vanadium spinels transform into acid-soluble magnesium vanadates to enhance the leaching efficiency. However, with a further increase in roasting time from 1.5 to 2.5 h, the leaching efficiency of vanadium decreased. More side reactions occur and the by-products wrap the vanadium phase to defer the leaching process because of the overlong roasting time, which causes lower vanadium efficiency. Therefore, the optimal roasting time is considered to be 1.5 h.

Characterization of Roasting Mechanism

The XRD pattern of original vanadium slag was presented in Fig. 4. The vanadium slag mainly consists of spinel phases ((Fe, Mg)V_2O_4, $MnFe_2O_4$, and Mg_2TiO_4), olivine phase (Fe_2SiO_4), and pyroxene phase ((Ca(Ti, Mg, Al)(Si, Al)$_2$$O_6$). After roasting of optimal roasting conditions, Mg/V molar ratio of 0.6, roasting temperature of 900 °C and roasting time of 1.5 h, the XRD pattern of the sample roasted was presented in Fig. 5. Compared with original vanadium slag, vanadium spinels (Fe, Mg)V_2O_4 are completely oxidative decomposed to $Mg_2V_2O_7$ and $Mn_2V_2O_7$. The Fe_2SiO_4 olivine is converted to Fe_2O_3 and $MgSiO_3$. In the XRD spectrogram, the diffraction peak of MnV_2O_6 can't be observed. This is major owing to the melting point of MnV_2O_6 is 742 °C, which is below the roasting

Fig. 3 Effect of roasting time on vanadium extraction

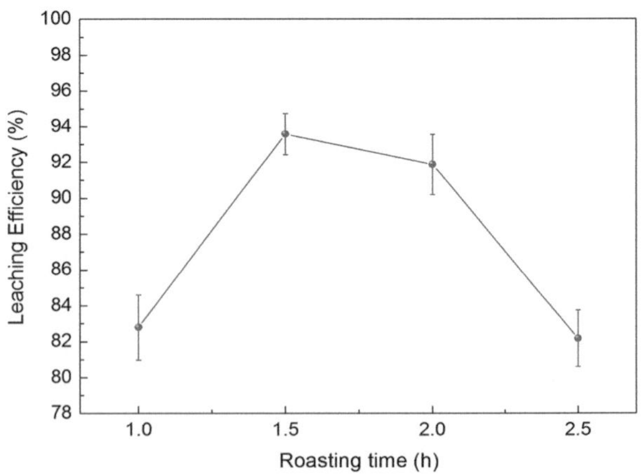

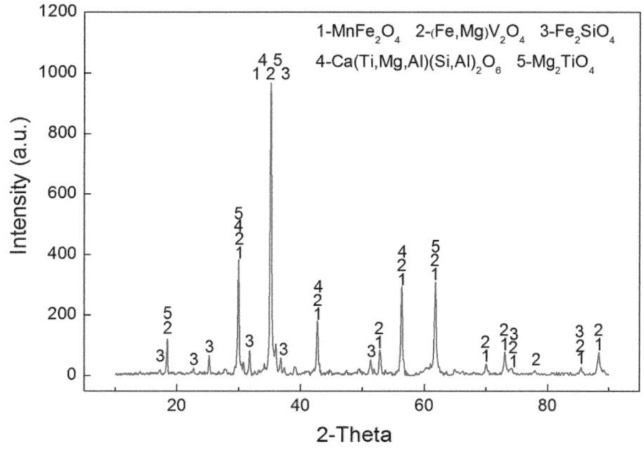

Fig. 4 XRD pattern of vanadium slag

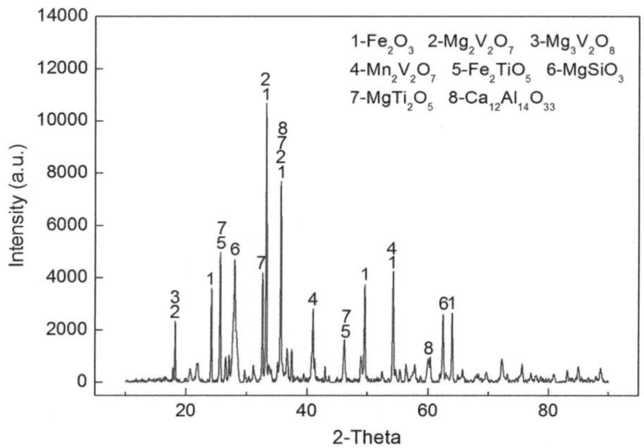

Fig. 5 XRD pattern of sample roasted

temperature. This results in the low crystallinity of $Mg(VO_3)_2$ due to no conditions for good crystal nucleus growth in the cooling process of sample roasted. Titanium spinels are decomposed to $MgTi_2O_5$ and Fe_2TiO_5 and pyroxenes are decomposed to $Ca_{12}Al_{14}O_{33}$. The transformation of phases and the leaching efficiency of vanadium under the optimal roasting conditions indicate that the method of roasting with MgO and acid leaching can perpetuate efficient use of vanadium slag.

Conclusions

In this work, a novelty extraction method by roasting with MgO and leaching with sulfuric acid was introduced. The oxidation and roasting mechanism were systematically studied, providing a theoretical basis for industrial extraction of vanadium. During the roasting process, vanadium spinels in the vanadium slag are converted to acid-soluble vanadates, including MgV_2O_6, $Mg_2V_2O_7$, and $Mn_2V_2O_7$. The result shows that under the optimum conditions, the roasting temperature of 900 °C, the Mg/V molar ratio of 0.6, and the roasting time of 1.5 h, the leaching efficiency of vanadium is above 90%. This experimental has provided a new strategy to extract vanadium efficiency which is environment-friendly and cost-effective.

Acknowledgements This work was financially supported by National Natural Science Foundation of China [grant numbers 51474041, 51674051], Chongqing Science and Technology Bureau [grant number cstc2018jscx-mszdX0066], and Ministry of Education of China [grant numbers cqu2017hbrc1B08, 2019CDXYCL0031].

References

1. Yao Z et al (2005) Review of vanadium processing in China. Eng Sci 3:58–62
2. Bradbury D et al (2002) The production of vanadium pentoxide. In: Vanadium—geology, processing and applications, the 41st annual conference of metallurgists of CIM, pp 115–130
3. Bale C et al (2002) Chartrand FactSage thermochemical software and databases. Calphad 26 (2):189–228
4. Tavakoli M et al (2014) The kinetics of oxidative leaching of vanadium trioxide. Hydrometallurgy 147:83–89
5. Chen D et al (2013) A novel process for recovery of iron, titanium, and vanadium from titanomagnetite concentrates: NaOH molten salt roasting and water leaching processes. J Hazard Mater 244:588–595
6. Liu H et al (2013) Kinetics analysis of decomposition of vanadium slag by KOH sub-molten salt method. Trans Nonferrous Metals Soc, China 23(5):1489–1500
7. Wang Z-H et al (2014) Research and prospect on extraction of vanadium from vanadium slag by liquid oxidation technologies. Trans Nonferrous Metals Soc China 24(5):1273–1288
8. Wang Z et al (2015) Electrochemical decomposition of vanadium slag in concentrated NaOH solution. Hydrometallurgy 151:51–55
9. Li X-S et al (2012) Extraction of vanadium from high calcium vanadium slag using direct roasting and soda leaching. Int J Miner Metall Mater 19(7):595–601
10. He D et al (2017) An environmentally-friendly technology of vanadium extraction from stone coal. Miner Eng 20(12):1184–1186
11. Tavakoli M et al (2014) Separation of vanadium from iron by solvent extraction using acidic and neutral organophosporus extractants. Hydrometallurgy 141:17–23
12. Zhang Y et al (2011) The technology of extracting vanadium from stone coal in China: history, current status and future prospects. Hydrometallurgy 109(1):116–124
13. Li HY et al (2012) Vanadium extraction from high calcium-content vanadium slag by calcification roasting. In: TMS 2012 annual meeting, T.T. Chen honorary symposium on hydrometallurgy, electrometallurgy and materials characterization, pp 621–627
14. Zhang JH et al (2017) Oxidation kinetics of vanadium slag roasting in the presence of calcium oxide. Min Proc Ext Met Rev 38:265–273
15. Xiang JY et al (2018) Extraction of vanadium from converter slag by two-step sulfuric acid leaching process. J Clean Prod 170:1089–1101

Effect of Sulfuric Acid Concentration on Marmatite Dissolution in the Presence of Cupric Ions

Xiaoyu Meng, Hongbo Zhao, Yisheng Zhang, Yanjun Zhang, Xin Lv and Shuai Wang

Abstract Marmatite often associates with copper sulfide ore with the role of Cu^{2+} in marmatite dissolution being important in hydrometallurgy. In this work, we found that cupric ions (Cu^{2}) can accelerate marmatite dissolution significantly at high temperature, regardless of sulfuric acid concentration (pH 0.5–2.5). Marmatite dissolved faster at lower acid concentrations and compared with cupric-free conditions, and the concentration of zinc dissolved in the presence of Cu^{2+} ions increased by 5500 mg/L. When the acid concentration was high (pH above 2.5), the acceleration effect of high Cu^{2+} concentration was more obvious. In addition, the consumption of Cu^{2+} was low at high acid concentration (pH below 1.5), but Cu^{2+} consumption increased significantly at low acid concentration (pH above 2.5). X-ray diffraction (XRD) of the leaching residues proved that no copper-containing mineralogical phase was produced at high acid concentration, but copper-containing products were formed at low acid concentration. Kinetic analysis showed that marmatite dissolution was mainly controlled by surface reaction. The reaction between Cu^{2+} and marmatite should be different at different concentrations of sulfuric acid.

Keywords Marmatite · Cupric ions · Sulfuric acid concentration · Kinetic analysis

---

X. Meng · H. Zhao (✉) · Y. Zhang · Y. Zhang · X. Lv · S. Wang
Central South University, No. 932 Lushan South Road, Yuelu District, Changsha 410083, Hunan, China
e-mail: alexandercsu@126.com; zhbalexander@csu.edu.cn

X. Meng
e-mail: 369784790@qq.com

H. Zhao
Key Laboratory of Biohydrometallurgy of Ministry of Education, No. 932 Lushan South Road, Yuelu District, Changsha 410083, Hunan, China

G. Azimi et al. (eds.), *Rare Metal Technology 2020*, The Minerals, Metals & Materials Series, https://doi.org/10.1007/978-3-030-36758-9_31

Introduction

Zinc is an important nonferrous metal and can be widely used in various fields. Sulfides (sphalerite, marmatite) are the most abundant zinc mineral and are widely present in nature. The sphalerite containing no impurities is a colorless crystal but sphalerite and iron usually appear in a form of marmatite in nature [1]. The main process for extracting metallic zinc from sulfides is Roast–Leach–Electrowinning Process (RLE) and more than 85% of the zinc production is obtained by this process [2]. Although the REL process is very mature, the route has many restrictions to treated low-grade zinc ore which contain high silica, copper, and iron [3]. In addition, the technology leads to extensive environmental problems due to SO_2 emission control, which makes challenging the future of this technology [3]. Recently, hydrometallurgical techniques has taken more attention, because it can treat low-grade ores at comparatively lower cost and sulfur in minerals is converted to elemental sulfur rather than SO_2 [4]. Nevertheless, the complexities of the minerals, low solubility, and low dissolution kinetics pose challenges during hydrometallurgical technology, which obstruct the processes entering commercial production. What is more, sphalerite is rarely present in nature as a pure mineral, usually appear in a form of sulphide complex minerals, which may contain chalcopyrite, galena, and pyrite. These minerals are generally separated from each other by flotation. However, since Cu^{2+} ions have an activating effect on sphalerite in flotation, they will float up with the sphalerite, making it difficult to completely separate, so the doping of Cu^{2+} ions in sphalerite hydrometallurgy is difficult to avoid.

Much effort has been made to enhance the dissolution kinetics of leaching. A few researchers have point out that adding metal ions can significantly accelerate the dissolution of sulfide minerals [5]. Investigation of chalcopyrite leaching with Ag^+ has been conducted by many authors [6, 7]. A few authors estimated that elemental sulfur produced during leaching is the main component of the passivation film which is a bad electric conductor [8]. However, in the presence of Ag^+, silver ions participated in the reaction to form Ag_2S, and the passivation layer becomes porous and conductivity increases due to the formation of Ag_2S [9]. In addition, the acceleration effect of Cu^{2+} ions on marmatite and realgar dissolution has also been reported [10–12]. Song et al. [11] considered that the acceleration the leaching of marmatite by Cu^{2+} ions was an indirect effect, and his paper was not an in-depth discussion of the mechanism of the reaction. Therefore, a number of authors have provided different reaction mechanisms for the reaction between Cu^{2+} ions and marmatite, and the intermediation products continue to remain unclear and controversial. In previous work, we found that the dissolution process of marmatite in presence of Cu^{2+} was affected by many factors, and the sulfuric acid concentration in the leaching process will also be an important factor [13]. This work would provide reference for the research filed of marmatite hydrometallurgy.

Experiment Materials and Methods

Materials

In this paper, the nature marmatite mineral was obtained from Anhui Province of China. The marmatite used in the test was ground until the particle size was less than 74 μm. X-ray diffraction (XRD) analysis of the mineral showed that the chemical composition of sample was 51.45% Zn, 12.28% Fe, and 32.82% S.

Dissolution Experiments

The marmatite sample was leached in sulfuric acid at different acid concentrations (pH 0.5, 1.5, 2.5), certain temperature (65 °C), and time (336 h). The leaching experiments were performed at 250 mL Erlenmeyer flasks and Shaking incubator. For each leaching flasks, 100 or 2500 mg/L Cu^{2+} ions, 2 g marmatite, and 100 mL solution of predetermined acid concentration (pulp density 2% w/v) were charged into the reactor and the marmatite powder and leaching agent was well mixed at 170 rpm. Sulfuric acid was used to maintain the acid concentration every day.

At selected time intervals (every 2 days), about 100 μl of supernatant solution sample was taken for analysis. The chemical analysis of solution samples was determined by an inductively coupled plasma optical emission solution spectrometry (ICP-OES) (ICAP 7400, Thermo Fisher Scientific Co.). The redox potentials and pH values of the solution were measured by using an ORP meter (BPP-922) and a pH meter (PHSJ-4A). Deionized water was added to avoid evaporation losses at high temperatures. After the leaching experiment, the leaching residues were washed with deionized water and dried and then to be chemically analyzed.

Characterization of the Residues

The mineralogical components of marmatite samples and leaching residues were analyzed by XRD. The tests were conducted using an X-ray diffractometer (Advance D8/Bruker) with Cu Kα radiation (λ = 1.54056 Å) in the range of 2-theta from 15° to 80°.

Results and Discussion

Dissolution Experiment

A plot of total extracted Zn and Fe concentration against time is presented in Fig. 1a–c using different sulfuric acid concentrations in the range of pH 0.5–2.5 at a constant temperature of 65 °C. It was clear that Cu^{2+} ions accelerated marmatite dissolution obviously at different sulfuric acid concentrations. Marmatite dissolved faster at lower acid concentrations and compared with cupric-free conditions, the concentration of zinc dissolved in the presence of Cu^{2+} ions increased by 5500 mg/L. In addition, when the sulfuric acid concentration of the solution was high (pH below 1.5), the acceleration effect of high concentration Cu^{2+} ions was basically the same as that of low concentration Cu^{2+} ions. However, with the decreasing of sulfuric acid concentration (pH above 2.5), the acceleration effect of high concentration Cu^{2+} ions were significantly stronger than that of low concentration Cu^{2+} ions.

Figure 1d, e showed the influence of sulfuric acid concentration against Cu^{2+} ions consumption. The results showed that Cu^{2+} ions were hardly consumed when the sulfuric acid concentration of the solution was high (pH below 0.5). However, the consumption of Cu^{2+} ions increased with decreasing sulfuric acid concentration, and Cu^{2+} ions were completely consumed when the solution pH was higher than 2.5. As shown in Fig. 1f–h, the solution redox potential significantly increased after adding Cu^{2+} ions. Actually, a few authors considered that marmatite was easily dissolved at higher redox potentials. Therefore, the results were in good agreement with the other researchers [14]. Nevertheless, the situation changed when the sulfuric acid concentration decreased (pH above 1.5). After the addition of low concentration Cu^{2+} ions, the redox potential did not increase significantly. Despite adding high concentration Cu^{2+} ions increased the redox potential in the initial stage, but it gradually decreased with increasing of leaching time (Fig. 1h). We considered that the reaction between Cu^{2+} and marmatite should be different at different concentrations of sulfuric acid, and the solution oxidation environment had little effect on accelerating marmatite dissolution at low sulfuric acid concentration (pH above 1.5). We considered that the reaction between Cu^{2+} ions and marmatite was surface reaction when the sulfuric acid concentration was high (pH below 1.5), and only a small part of Cu^{2+} ions participated in the reaction. In previous work, we have demonstrated that this surface interfacial reaction product can improve the electrochemical activity of marmatite and then accelerated dissolution [13]. The main reaction may follow Eq. (1):

$$ZnS + 2H^{+} = Zn^{2+} + H_2S \quad (1)$$

However, the reaction between Cu^{2+} ions and marmatite may change with the decrease of sulfuric acid concentration, combined with the change Cu^{2+}

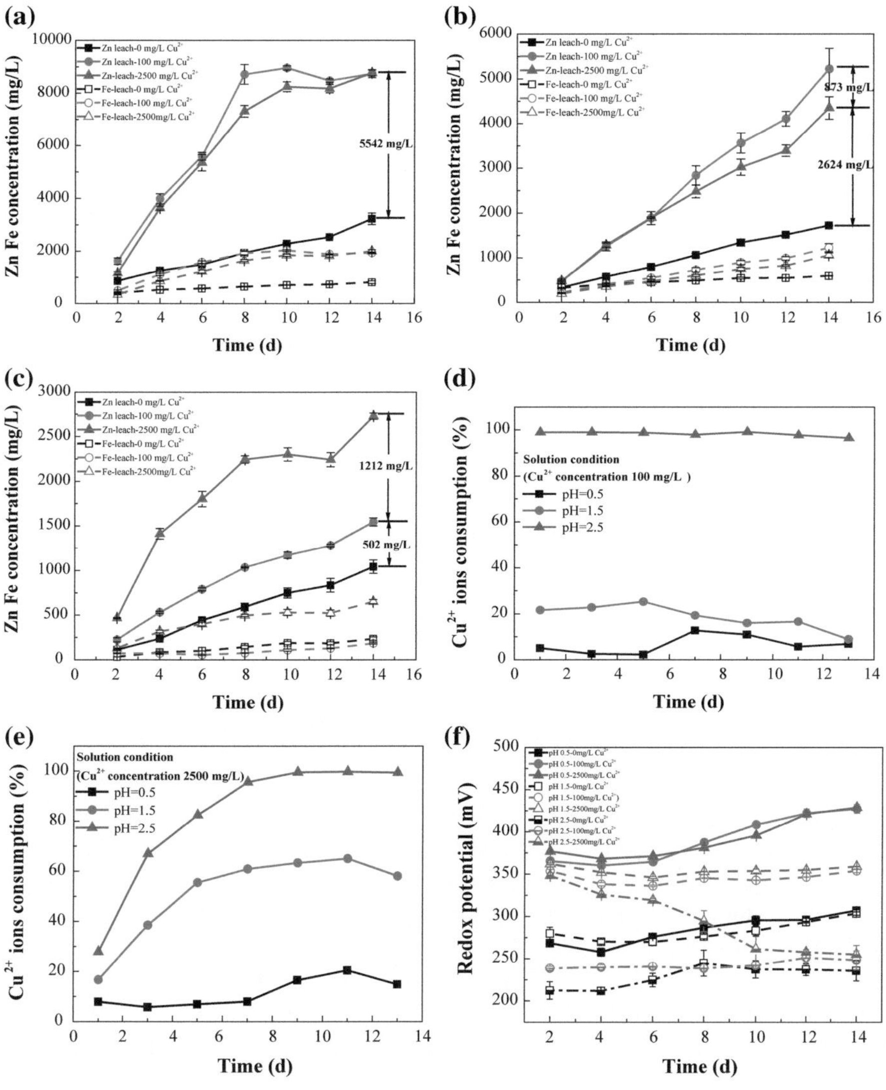

Fig. 1 Dissolution behaviors of marmatite under different conditions: **a** leached at pH 0.5; **b** leached at pH 1.5; **c** leached at pH 2.5; **d** 100 mg/L Cu^{2+} consumption; **e** 2500 mg/L Cu^{2+} consumption; and **f** redox potential

consumption, we considered that the reaction may transform from surface reaction to substitution reaction which presented in Eq. (2) [15].

$$ZnS_{(s)} + Cu^{2+}_{(aq)} \rightarrow CuS_{(s)} + Zn^{2+}_{(aq)} \tag{2}$$

Kinetic Study

This paper has done a fitting model on the influence of sulfuric acid concentration on marmatite dissolution kinetics. The fitting kinetic model was the shrinking-core model (SCM), which was the most common model considered for sulfide minerals dissolution in acid leaching agents. The unreacted shrinking-core model mathematic model has been widely used in describing the kinetics in heterogeneous reactions. A few researchers considered that in the solid–liquid reaction like acid leaching, the reaction rate was controlled either by the liquid film diffusion, or through the product layer diffusion, or by the surface reaction and chemical reaction [16, 17]. If the control step of the reaction rate was dominated by product layer diffusion, the reaction rate expression can be summarized as Eq. (3) [16]:

$$\left[1-\frac{2}{3}\alpha-(1-2/3\alpha)^{2/3}\right] = \frac{2\mathrm{M_B DC_A}}{\rho_\mathrm{B}\mathrm{ar}_0^2}t = k_\mathrm{d}t \tag{3}$$

If the reaction rate was mainly controlled by surface chemical reactions, the rate equation is expressed as Eq. (4) [17]:

$$\left[1-(1-\alpha)^{1/3}\right] = \frac{k_c\mathrm{M_B C_A}}{\rho_\mathrm{B}\mathrm{ar}_0}t = k_\mathrm{r}t \tag{4}$$

In this paper, the experimental data were fitted with two equations, respectively, and the results of the fitting are shown in Fig. 2. The rate constants values, k_d and k_r, and their correlation coefficients for each sulfuric acid concentration and Cu^{2+} concentration are given in Table 1. Through the calculated parameters we found that the kinetics of reaction was proportional to sulfuric acid concentration and in the same acid concentration, the presence of Cu^{2+} ions had a positive effect on the dissolution kinetics of marmatite. Furthermore, the correlation coefficient of surface reaction control was higher than the correlation coefficient of product layer diffusion control equation, indicating that leaching of marmatite with Cu^{2+} ions in sulfuric acid can be clearly controlled by surface chemical reactions as described by Eq. (4).

Residue Analysis

A typical XRD pattern of the leaching residue at high temperature in presence of high Cu^{2+} concentration was given in Fig. 3. It can be seen clearly in Fig. 3 that constituent phases of the residue which leaching at low sulfuric acid concentration (pH above 2.5) changed obviously a1s compared with that leaching at high sulfuric acid concentration (pH below 1.5). The main phase in the residue was sulfur, and no copper-containing compounds were detected at high acid concentration. As the

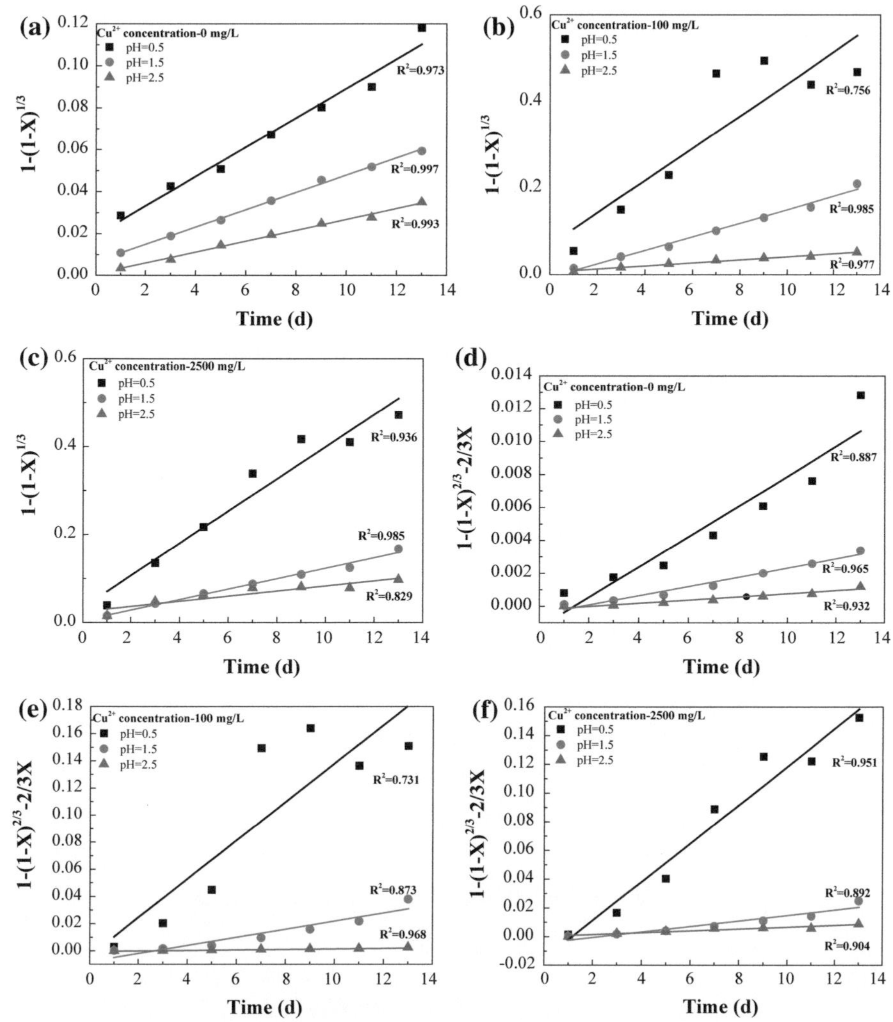

Fig. 2 Plot of $1 - (1 - X)^{1/3}$ versus t (**a–c**); $1 - (1 - X)^{2/3} - 2/3X$ versus *t* (**d–f**) for different sulfuric acid concentrations and Cu^{2+} concentrations

concentration of sulfuric acid decreased, undissolved marmatite can be detected and additional peaks which can be assigned to copper–sulfur compound appeared. The analysis results of the leaching residues were consistent with the laws obtained from the leaching experiment. The marmatite dissolved faster and more thoroughly when the acid concentration was high, and the effect of Cu^{2+} ions and marmatite only occurred on the mineral surface. With the decrease of sulfuric acid concentration, the peak of copper–sulfur compounds was more obvious, indicating that the

Table 1 The k_r, $k_{d,}$ and correlation coefficients values for different sulfuric acid concentrations and Cu^{2+} concentrations

Cu^{2+} concentration	Solution (pH)	Apparent rate constants $(10^{-3}d^{-1}) \times 10^{-3}$		Correlation coefficient (R^2)	
		k_r	k_d	$[1 - (1 - \alpha)^{1/3}] = kt$	$[1 - 2/3\alpha - (1 - 2/3\alpha)^{2/3}] = kt$
0 mg/L	0.5	7	0.918	0.973	0.887
	1.5	4.41	0.278	0.997	0.965
	2.5	2.59	0.095	0.992	0.932
100 mg/L	0.5	37.2	14.2	0.756	0.730
	1.5	15.7	2.95	0.985	0.873
	2.5	3.6	0.21	0.977	0.968
2500 mg/L	0.5	36.5	13.4	0.936	0.951
	1.5	11.8	1.89	0.985	0.892
	2.5	5.8	0.64	0.829	0.903

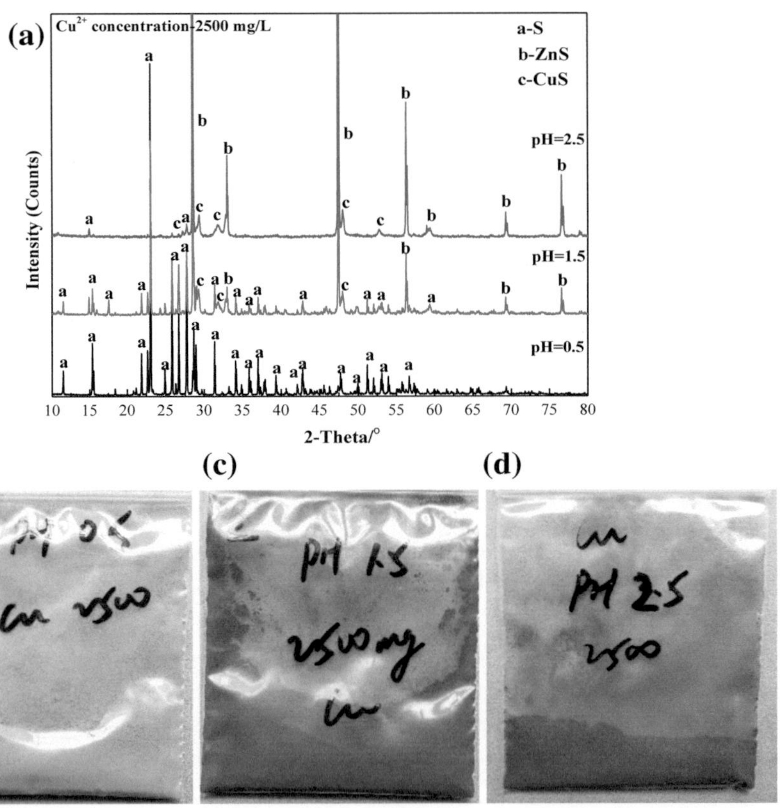

Fig. 3 Analysis of residue: **a** 15–80° XRD; **b** actual residue at pH 0.5; **c** actual residue at pH 1.5; and **d** actual residue at pH 2.5

reaction between Cu^{2+} ions and marmatite conversed from surface reaction to bulk substitution, and Cu^{2+} ions participated the reaction in a large amount.

In addition, we observed a significant change in the color of the residue under different sulfuric acid concentrations. When the acid concentration was high, the color of the leaching residue was pale yellow, which was similar to the color of elemental sulfur. The color deepened with the decrease of sulfuric acid concentration. This may be due to the presence of incompletely dissolved marmatite. In addition, the substitution reaction of Cu^{2+} ions at low acid concentration, which enriched the copper–sulfur compounds on the mineral surface also contributes to the deepening of color. Combined with the changes in Zn^{2+} ions dissolution and Cu^{2+} ions consumption in the leaching experiment, we considered that the reaction between Cu^{2+} and marmatite was different at different concentrations of sulfuric acid.

Conclusion

The change of sulfuric acid concentration in the solution has a significant effect on the dissolution process of marmatite in the presence of cupric ions (Cu^2) at high temperature. Marmatite dissolves faster and more thoroughly when the sulfuric acid concentration is high. Cupric ions (Cu^2) can accelerate marmatite dissolution significantly, regardless of sulfuric acid concentration (pH 0.5–2.5) but the mechanism of action of cupric ions (Cu^2) changes with decreasing of acid concentration. When the sulfuric acid concentration is low (pH below 1.5), the Cu^{2+} ions consumption is small, the reaction between Cu^{2+} ions and marmatite is a surface reaction, and the dissolution concentration of Zn^{2+} increased about 5500 mg/L compared with the case of no Cu^{2+} ions. The concentration of Cu^{2+} ions has little effect. However, Cu^{2+} ions consumption increases as the acid concentration decreases (pH above 2.5), and the reaction changes from surface reaction to bulk substitution reaction, the substitution effect increases the Zn^{2+} dissolution concentration by 1700 mg/L. The accelerating effect of high Cu^{2+} concentration is more obvious. Kinetic studies show that the dissolution process of marmatite in the presence of Cu^{2+} ions conforms to the shrinkage core model and the control step of the process is reaction-controlled.

Acknowledgements This work was supported by the National Natural Science Foundation of China (51704331), Young Elite Scientists Sponsorship Program by CAST (2017QNRC001), and Innovation-Driven Project of Central South University (2018CX019).

References

1. Schaefer S, Gokcen N (1982) Electrochemical determination of the thermodynamic properties of sphalerite, ZnS (beta). High Temper Sci 15(2–3):225–237
2. De Souza AD, Pina PS, Leão VA (2007) Bioleac hing and chemical leaching as an integrated process in the zinc industry. Min Eng 20(6):591–599
3. Deller G (2005) World zinc supply and demand-heading for a late-decade price spike. Lead & Zinc 5:17–25
4. Dehghan R, Noaparast M, Kolahdoozan M (2008) Statistical evaluation and optimization of factors affecting the leaching performance of a sphalerite concentrate. Int J Min Process 89 (1):9–16
5. Córdoba EM, Muñoz JA, Blázquez ML, González F, Ballester A (2008) Leaching of chalcopyrite with ferric ion. Part III: effect of redox potential on the silver-catalyzed process. Hydrometallurgy 93(3):97–105
6. Yuehua H, Guanzhou Q, Jun W, Dianzuo W (2002) The effect of silver-bearing catalysts on bioleaching of chalcopyrite. Hydrometallurgy 64(2):81–88
7. Zhao H, Zhang Y (2019) Catalytic mechanism of silver in the oxidative dissolution process of chalcopyrite: experiment and DFT calculation. Hydrometallurgy 187:18–29
8. Miller JD, Portillo HQ (1981) Electrochemistry in silver catalysed ferric sulfate leaching of chalcopyrite. J Macromol Sci Part B Phys
9. Price D, Warren G (1986) The influence of silver ion on the electrochemical response of chalcopyrite and other mineral sulfide electrodes in sulfuric acid. Hydrometallurgy 15(3): 303–324
10. Liu ZX, Yin ZL, Hu HP, Chen QY (2012) Catalytic-oxidative leaching of low-grade complex Zinc Ore by Cu(II) Ions produced from Copper Ore in Ammonia-Ammonium Sulfate solution. Metall Mater Trans B 43(5):1019–1026
11. Chen S, Qin W-Q, Qiu G-Z (2008) Effect of Cu^{2+} ions on bioleaching of marmatite. Trans Nonferrous Metals Soc China 18(6):1518–1522
12. Pei G (2011) Catalytic effect of Ag^{+} and Cu^{2+} on leaching realgar (As_2S_2). Hydrometallurgy 106(1):99–103
13. Meng X (2019) The role of cupric ions in the oxidative dissolution process of marmatite: a dependence on Cu^{2+} concentration. Sci Total Environ 675:213–223
14. Zhao H (2017) Stepwise bioleaching of Cu–Zn mixed ores with comprehensive utilization of silver-bearing solid waste through a new technique process. Hydrometallurgy 171:374–386
15. Popov SR, Vučinić DR (1990) The ethylxanthate adsorption on copper-activated sphalerite under flotation-related conditions in alkaline media. Int J Min Process 30(3):229–244
16. Aydogan S, Aras A, Canbazoglu M (2005) Dissolution kinetics of sphalerite in acidic ferric chloride leaching. Chem Eng J 114(1):67–72
17. Dehghan R, Noaparast M, Kolahdoozan M (2009) Leaching and kinetic modelling of low-grade calcareous sphalerite in acidic ferric chloride solution. Hydrometallurgy 96(4): 275–282

Recovery of Rare Earth Elements from Waste Permanent Magnets Leach Liquors

Rajesh Kumar Jyothi, Kyeong Woo Chung, Chul-Joo Kim and Ho-Sung Yoon

Abstract The scrap magnet contains significant amounts of rare earth metals such as neodymium, praseodymium, dysprosium, and terbium. In the twenty-first century, faster growing rate of hi-tech industrialization leads to the generation of the secondary wastes across the globe, which is becoming an environmental concern. To resolve the environmental issue vis-à-vis to meet the demand for rare earth metals, this investigation is focused on the development of a suitable process for extraction of Nd, Dy, Pr, and Tb from the scrap permanent magnet using leaching followed by solvent extraction. During leaching, oxidation roasting process was adopted to minimize the dissolution of iron. The resulted leach liquor was subjected to liquid–liquid extraction of Nd, Pr, Dy, and Tb using organophosphorus derivatives reagents. Process optimization followed by extraction isotherm study was carried out to obtain quantitative extraction of these metals from the aqueous media. The crowding effect was further investigated to improve the separation factor of the RE metals and reported.

Keyword Rare earths · Recovery · Waste permanent magnets · Hydrometallurgy

Introduction

South Korea faces a major issue of landfilling in its peninsula region pertaining to the industrial waste management. Particularly, the wastes/scraps generated from the electrical, electronics, and automobile industries in South Korean peninsula are becoming a challenge because of their direct disposal in land fields. On the other hand, these scarp/wastes are certainly considered a critical source because they

R. K. Jyothi (✉) · K. W. Chung · C.-J. Kim · H.-S. Yoon
Korea Institute of Geoscience and Mineral Resources (KIGAM), Daejeon 34132, Korea
e-mail: rkumarphd@kigam.re.kr

G. Azimi et al. (eds.), *Rare Metal Technology 2020*, The Minerals, Metals & Materials Series, https://doi.org/10.1007/978-3-030-36758-9_32

contain considerable amounts of Nd, Tb, Dy, and Pr. Thus, a suitable recycling technology is required to address the environmental issues as well as to exploit the RE metals out of the scrap sources. Hydrometallurgical processes based on leaching and solvent extraction processes are gaining greater attention for the extraction of metal values from these secondary sources. Based on the reports from DOE and USA, in magnet scrap, neodymium, praseodymium, dysprosium, and terbium are recognized as critical rare metals based on their limited reserves, supply, and demand [1–4]. Moreover, these four REEs play a key role in clean energy technologies [1–4].

In recent years, rare earth-based permanent magnets utilization in energy conversion devices has been growing fast [1–4]. Permanent magnets are categorized to two types; (i) NdFeB type, which is stronger, and (ii) CoSm type, which is more useful for high-temperature applications [5]. Recycling of these magnet scraps can offer economic benefits while offering a new resource for these critical REEs. In REE processing industry, recycling and reuse is important since they enable the circular economy and sustainability [3]. This study is focused on recycling of REEs from waste permanent magnets, through hydrometallurgical routes.

In a previous study for the recovery of REEs, after the physical separations and oxidation and roasting process, acid leaching was performed for waste permanent magnets and leaching parameters were optimized and applied the Taguchi method. This method was performed to recognize the significant process parameters for higher rate of recovery as well as increased leaching efficiency of rare earths from waste permanent magnets [6]. This study reported that following experimental parameters were optimized for rare earths leaching from waste permanent magnets: acid of 5 mol/L hydrochloric acid, temperature of 368 K, S/L ratio of 2%, reaction time of 24 h, particle size of 0.250 mm, and stirring speed of 800 rpm [6]. With these optimized conditions, 78% of neodymium, 83% of praseodymium, and 76% of dysprosium were leached [6]. In another study, spent NdFeB magnet of wind turbine was leached to recover rare earths [7]. To minimize the iron interference, oxidation roasting process was utilized (at 1123 K temperature, 360 min time). Selective leaching was performed with following experimental conditions: 0.5 mol/L HCl for 100 g/L of pulp density at 368 K temperature with 300 min time. After leaching oxalic acid precipitation and roasting process was performed to obtain the final mixed rare earth oxide product [7]. To separate the RE metals, solvent extraction and/or membrane separation methods are extensively employed.

In another study, membrane-based solvent extraction was investigated and reported for selective extraction of REEs from permanent magnet scraps (PMs) [8]. From the PMs Nd, Pr, and Dy were extracted with 6 mol/L HNO_3. lower nitric acid was able to strip the extracted rare earths. The strip solution was treated with oxalic acid to produce rare earth oxalates which were then converted to rare earth oxide through calcination [8]. The above studies demonstrate the feasibility of the recovery of REE metals from scrap magnets.

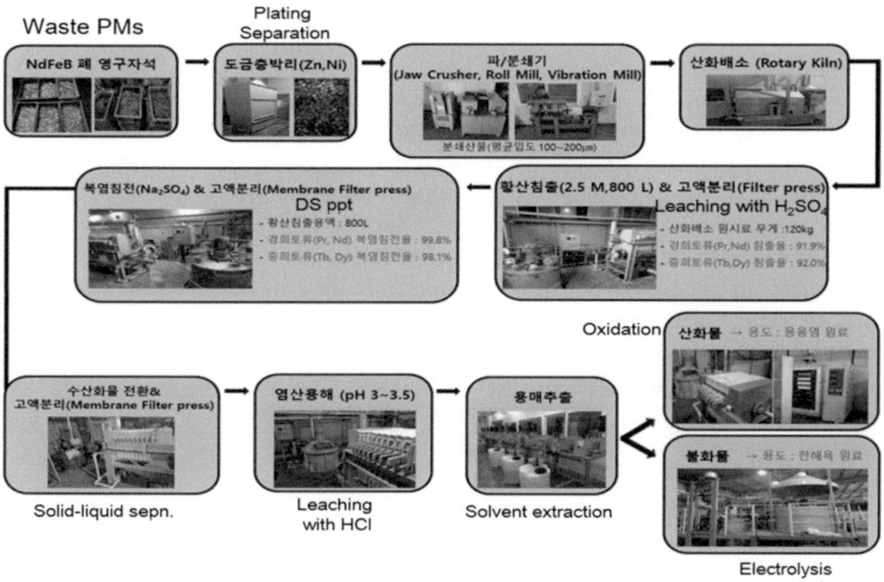

Fig. 1 Schematic diagram of hydrometallurgical processing of waste permanent magnets

The current study focuses on the recovery of REEs from waste permanent magnets (WPMs) using hydrometallurgical methods. Figure 1 presents a schematic diagram of the process.

Experimental

Apparatus and Reagents

Compositional analysis of Dy, Tb, Pr, Nd, and Fe was carried out using by inductively coupled plasma optical emission spectroscopy (ICP-OES) (Thermo Scientific, USA,. model iCAP 6000 Series ICP Spectrometer). The pH meter (Thermo Scientific) was used for the pH measurements in aqueous samples. Various commercial extractants such as Cyanex 272 (bis(2,4,4-trimethylpentyl) phosphinic acid supplied by Cytec), D2EHPA (di-(2-ethylhexyl)phosphoric acid supplied by Aldrich), PC 88A (2-ethylhexylphosphonic acid mono-2-ethylhexyl ester, supplied by Daihachi Chemical Industry, Japan), and TBP (tributyl phosphate supplied by Aldrich) were used. All other reagents used were of analytical grade (Ar). Except Cyanex 272, all the other extractants used in this study were prepared on 1 M basis with Exxol D-80 diluent. Cyanex 272 extractant was prepared with Exxol D-80 (D80) diluent combined with 3.5% (v/v) TBP to avoid the turbidity.

Solvent Extraction Procedure

Equal volumes (30 mL each, except for extraction isotherm experiments) of the aqueous phase containing rare earths and organic phase containing the 'P'-based extractant such as Cyanex 272 (or) D2EHPA (or) PC 88A (or) TBP were equilibrated for 30 min in glass stopped separating funnels using a shaking incubator for proper shaking at 25 °C temperature and 250 rev min^{-1} speed of shaker. After phase disengagement, the aqueous phase was separated and the metal concentrations were analyzed by ICP-OES. The concentration of the metal in the organic phase was calculated based on the mass balance. All experiments were conducted at a temperature of 25 ± 1 °C and the obtained general agreement with distribution ratio (D) values was within ± 3% (extraction procedure was adopted from our own previous reports) [9, 10]. After experiment, metals in aqueous solution were analyzed to determine distribution coefficient (D_E) and percentage extraction (%E) values.

For extraction, $D_E = [\text{Metal}]_O/[\text{Metal}]_A$

$[\text{Metal}]_O$ = Metal concentration in organic phase

$[\text{Metal}]_A$ = Metal concentration in aqueous phase

% Extraction ($\%E$) = $D_E \times 100/(D_E + 1)$ phase ratio.

Results and Discussion

Oxidation and Roasting Process

In a first step, scrap magnet samples were grinded to obtain a suitable size fraction for roasting. Oxidation roasting process in rotary kiln was used to prevent the iron leaching. The optimum conditions for roasting oxidation process were determined to be roasting temperature of 600 °C and particle size of 50–100 micron (Fig. 2).

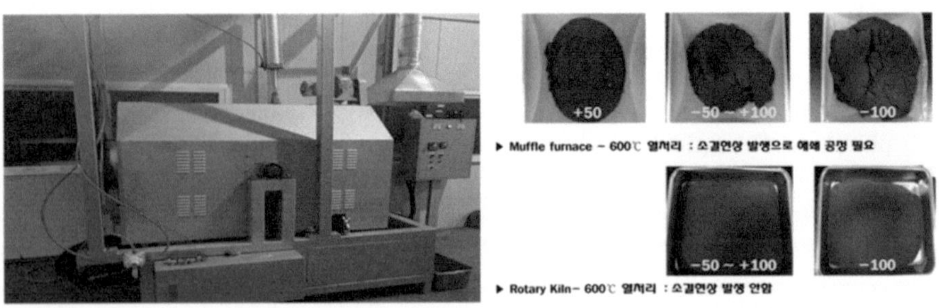

Fig. 2 Oxidation and roasting process of waste permanent magnets by rotary kiln

Sulfuric Acid Leaching of the Waste Permanent Magnets

After oxidation and roasting of waste permanent magnets, they were treated with sulfuric acid to selectively leach rare earths from other elements. In this experiment, 120 kg of WPMs was leached with 800 L of concentrated sulfuric acid and 91.9% extraction of RE metals was achieved with about 50.5% co-extraction of iron. The overall leaching of RE metals is as follows:

$$NdFeO_3 + 6H^+ = Nd^{3+} + Fe^{3+} + 3H_2O \quad (1)$$

Double Salt Precipitation of Sulfuric Acid Leach Liquor

Due to high iron concentration in sulfuric acid leach liquors, the double salt precipitation (DSP) process was applied using Na_2SO_4. This process resulted in $98 \pm 1\%$ rare earths precipitation with negligible co-precipitation of iron. Furthermore, solid–liquid separation process was carried out using membrane filter press.

Hydrochloric Acid Leaching of Solid Product Gained from DSP Process

The precipitated total REEs was subjected to hydrochloric acid leaching. At the optimum leaching condition, the final leach liquor contained 2,498 mg/L of Dy, 10,680 mg/L of Tb, 13 mg/L of Nd, 540 mg/L of Pr, and 22 mg/L of Fe.

Solvent Extraction Processing of WPMs Leach Liquor

Selection of Extractant

Various organophosphorous-based extractants such as Cyanex 272, D2EHPA, PC 88A, and TBP were tested to find their suitability for REE separation. Solution pH was varied from −0.5 to 6 and then equilibrated with 1.0 mol/L of either of the extractants dissolved in kerosene at room temperature (25 °C), A/O = 1. The results are presented in Figs. 3, 4, 5 and 6.

Based on the results of the above experiments, it was found that phosphorus-based extractants, i.e., Cyanex 272, D2EHPA, PC 88A are better for good extraction of REEs compared with TBP. The lower pH range of the solution (−0.5) is not suitable for REE extraction independent of the type of P-based

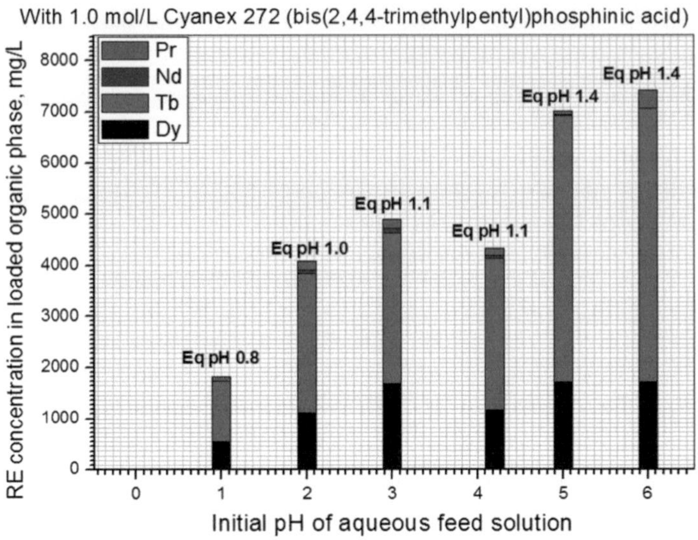

Fig. 3 Cyanex 272 effect on REE extraction from waste permanent magnets leach liquors

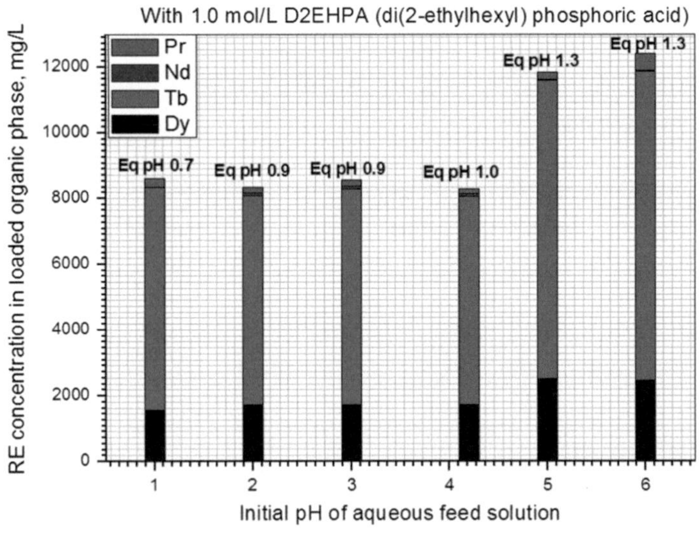

Fig. 4 D2EHPA effect on REE extraction from waste permanent magnets leach liquors

extractant. Using 1.0 mol/L of D2EHPA (or) PC 88A, ~99% of Dy and Tb could be extracted at a pH range of 1 to 4 (corresponding to equilibrium pH of 0.6 to 1.0). In the contrary, when 1.0 mol/L of Cyanex 272 was used, only 35–69% of Dy and

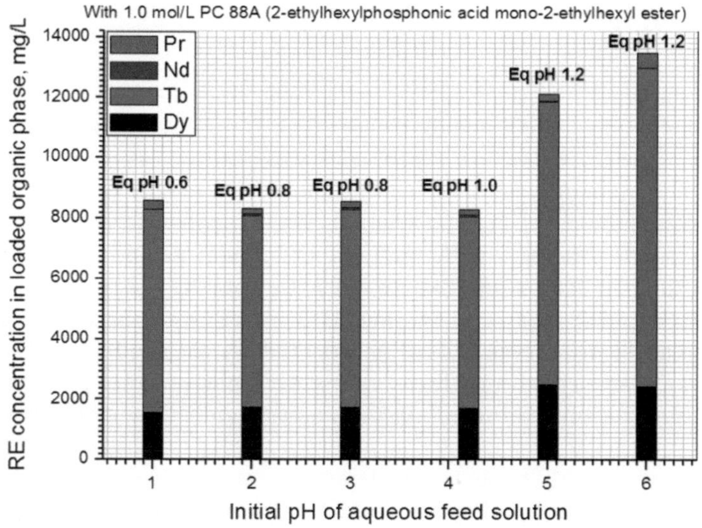

Fig. 5 PC 88A effect on REE extraction from waste permanent magnets leach liquors

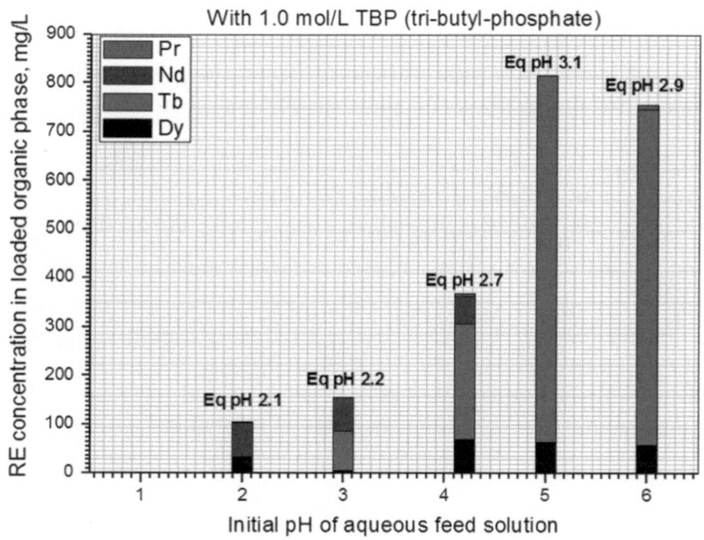

Fig. 6 TBP effect on REE extraction from waste permanent magnets leach liquors

17–58% of Tb was extracted at the entire pH range studied, i.e., 1–6 (corresponding to equilibrium pH of 0.8–1.4).

From this investigation, it is clear that Cyanex 272 is preferentially more suitable than the other two P-based extractants (D2EHPA (or) PC 88A).

Crowding Effect

Based on the results of screening of extractants, three P-based extractants were selected for crowding effect studies: Cyanex 272, D2EHPA, and PC 88A. To study the scrubbing effect, extraction experiments were carried out at pH 5.0 and 6.0, followed by scrubbing with a Dy-rich solution (6,089 mg/L of Dy). The results obtained are presented in Figs. 7, 8 and 9. The experimental conditions were as follows: Dy-rich solution = 6089 mg/L of Dy, Initial pH of 5 (or) Eq. pH of 1.3 ± 0.1, time of 30 min, temperature of 25 °C, and A/O = 1. In Figs. 7, 8 and 9, LO indicates loaded organic phase and SLO indicates scrubbed loaded organic phase. As can be seen, both D2EHPA and PC 88A are good for extraction, but loadings of Dy and Tb are very high because the replacement of Dy from LO was difficult with the scrubbed reagent. In the contrary, in the case of Cyanex 272, the loading of both metals was less and scrubbing was easy to replace Dy with Tb and Dy concentration in SLO was higher than that when other extractants were used. From this study, it is concluded that Cyanex 272 is the most suitable extractant for further development of crowding studies.

Cyanex 272 as an Extractant

After choosing Cyanex 272 as the extractant, the next step was to fix the extraction time and hence for this experiment, the experimental conditions were selected as

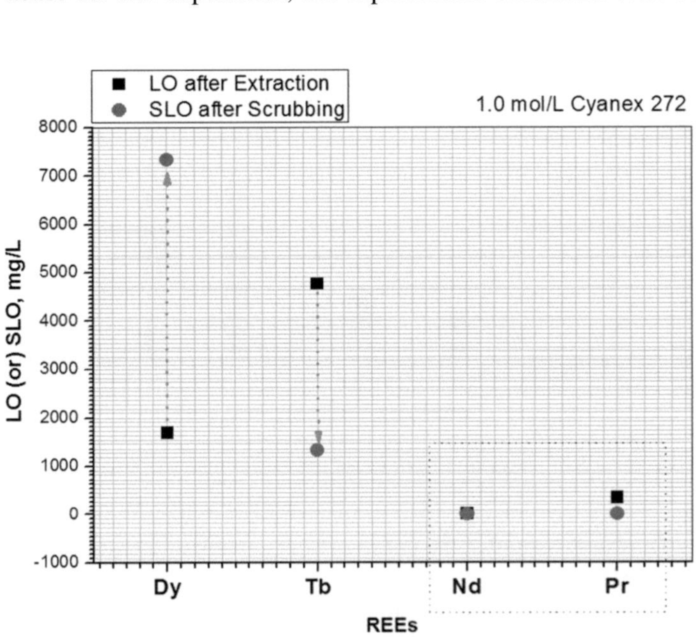

Fig. 7 Crowding effect on REE loading to 1.0 mol/L Cyanex 272-D80 system

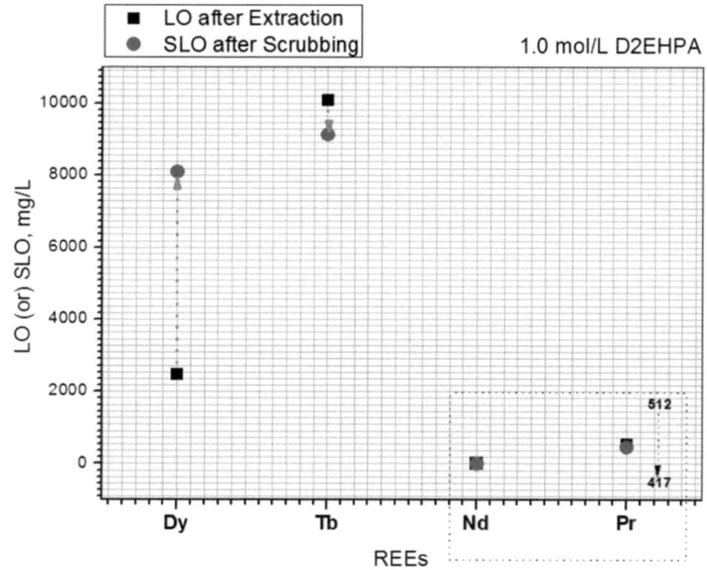

Fig. 8 Crowding effect on REE loading to 1.0 mol/L D2EHPA-D80 system

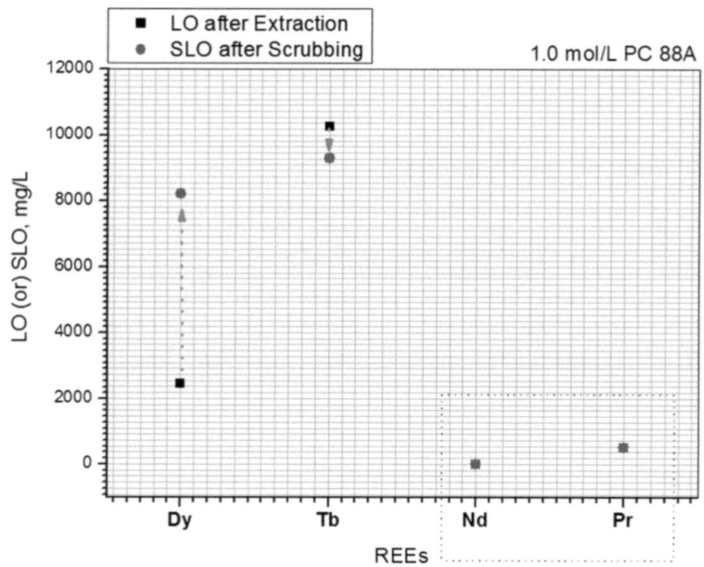

Fig. 9 Crowding effect on REE loading to 1.0 mol/L PC 88A-D80 system

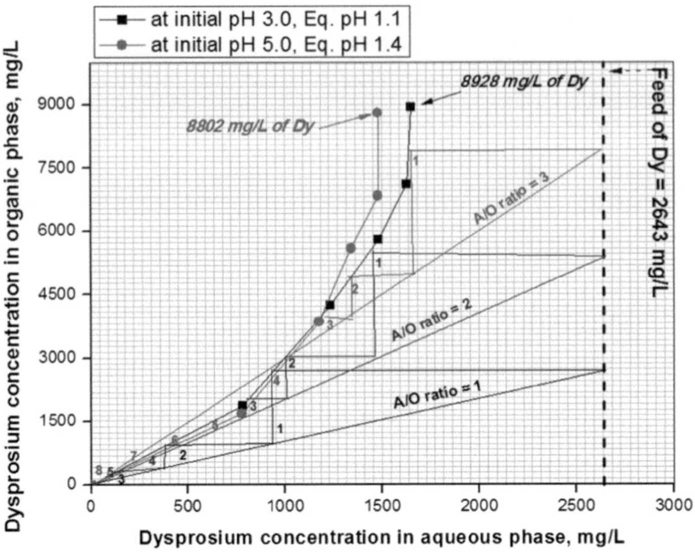

Fig. 10 Extraction isotherm (McCabe–Thiele diagram) for dysprosium

initial pH of 5, equilibrium pH of 1.29, temperature of 25 °C, and A/O of 1 and time was varied from 1 min to 120 min. The results clearly demonstrate that 30 min is enough to leach all four REEs and Fe. Thus, 30 min was selected and used in the future experiments.

Extraction Isotherm for Dysprosium

The maximum extraction ability of Cyanex 272 for Dy loading was tested by varying different A:O ratio. Based on the results, extraction isotherms were constructed (McCabe–Thiele diagram). As shown in Fig. 10, four extraction stages at A/O = 3 are required for maximum loading of Dy (extraction condition: pH 3, 8,928 mg/L of Dy; pH 5,8802 mg/L of Dy).

Conclusions

In this study, hydrometallurgical processes were developed to recovery Nd, Dy, Pr, and Tb from waste permanent magnets. In the first step, after physical separation and de-magnetization process, oxidation roasting was conducted to control iron interference during leaching process. After successful pretreatments, waste magnets were leached with HCl and solvent extraction was used to separate REEs. Solvent extraction results showed the feasibility of separating Dy and Tb at pH 3–6 using organophosphorus derivative reagents. The screening study carried out for three P-based extractants and it was found that D2EHPA and PC 88A are good for extraction, but replacement of Dy from LO was difficult with scrubbed reagent. Using Cyanex 272, loading of both metals was less and scrubbing was easy to replace Dy in place of Tb.

The phase ratio and time were optimized at 10 min and 10–15 LO/Aq ratio for the scrubbing effect.

Kinetic studies of the extraction indicate that maximum extraction of all elements is achieved at 30 min.

Acknowledgements This study was supported by the R&D Center for Valuable Recycling (Global-Top R&D Program) of the Ministry of Environment (Project No.: 2016002220005).

References

1. US DOE (US Department of Energy), Critical materials strategy. US Department of Energy, Washington, DC, December (2011), Available at http://energy.gov/sites/prod/files/DOE_CMS2011_FINAL_Full.pdf. Accessed 14 June 2014
2. Du X, Graedel TE (2011) Global rare earth in-use stocks in NdFeB permanent magnets. J Ind Ecol 15:836–843
3. Nlebedim IC, King AH (2018) Addressing criticality in rare earth elements via permanent magnets recycling. JOM 70:115–123
4. Sternberg J (2014) Howthegreatrare-earthmetalscrisisvanished. WallStreet J, 8 January. Available at http://online.wsj.com/news/articles/SB10001424052702303848104579308252845415022
5. Stegen KS (2015) Heavy rare earths, permanent magnets, and renewable energies: an imminent crisis. Energy Policy 79:1–8
6. Ni'am AC, Wang YF, Chen SW, You SJ (2019) Recovery of rare earth elements from waste permanent magnet (WPMs) via selective leaching using the Taguchi method. J Taiwan Inst of Chem Eng 97:137–145
7. Kumari A, Sinha MK, Pramanik S, Sahu SK (2018) Recovery of rare earths from spent NdFeB magnets of wind turbine: leaching and kinetic aspects. Waste Manag 75:486–498
8. Kim D, Powell LE, Delmau LH, Peterson ES, Herchenroeder J, Bhave RR (2015) Selective extraction of rare earth elements from permanent magnet scraps with membrane solvent extraction. Environ Sci Technol 49:9452–9459
9. Kumar JR, Lee HI, Lee JY, Kim JS, Sohn JS (2008) Comparison of liquid–liquid extraction studies on platinum(IV) from acidic solutions using bis(2,4,4-trimethylpentyl) monothiophosphinic acid. Sep Purifi Technol 63:184–190
10. Lee JY, Kumar JR, Kim JS, Park HK, Yoon HS (2009) Liquid–liquid extraction/separation of platinum(IV) and rhodium(III) from acidic chloride solutions using tri-iso-octylamine. J Hazard Mater 168:424–429

Part V
Poster Session

A Novel Depressant of Sodium Polyacrylate for Magnesite Flotation

Hongwei Cheng, Changmiao Liu, Dong Dong, Zihu Lv and Fei Yang

Abstract In order to inhibit magnesite effectively, a new macromolecule reagent, sodium polyacrylate (SP), is introduced. Micro-flotation results indicate that SP can strongly inhibit the flotation of magnesite in a wide pH range. When pH = 9, the flotation recovery of magnesite is 90%. After increasing the dosage of SP to 10 mg/L, the flotation recovery of magnesite declines to below 20%. Mineral crystal structure and solution chemistry show that SP easily absorbs the Mg^{2+} active sites on the surface of magnesite. Meanwhile, SP could form a bridge among the magnesite particles and lead to the flocculation and sedimentation, thus inhibiting the magnesite flotation.

Keywords Magnesite · Depressant · Sodium polyacrylate · Flotation

Introduction

Magnesite is an important industrial mineral that is widely used in metallurgy, building materials, chemical industry, as well as other industries [1], and mainly used as refractories due to the physical and chemical properties of magnesium oxide

H. Cheng · C. Liu (✉) · D. Dong · Z. Lv
Zhengzhou Institute of Multipurpose Utilization of Mineral Resources, CAGS,
Zhengzhou 450006, Henan, China
e-mail: changmiaoliu@163.com

H. Cheng · C. Liu · D. Dong · Z. Lv
Key Laboratory of Evaluation and Multipurpose Utilization of Polymetallic Ores,
Ministry of Land and Resources, Zhengzhou 450006, China

H. Cheng · C. Liu · D. Dong · Z. Lv
China National Engineering Research Center for Utilization of Industrial Minerals,
Zhengzhou 450006, Henan, China

F. Yang
Key Laboratory of Radioactive and Rare Scattered Mineral Comprehensive Utilization,
Ministry of Land and Resource, Shaoguan 512026, China

G. Azimi et al. (eds.), *Rare Metal Technology 2020*, The Minerals,
Metals & Materials Series, https://doi.org/10.1007/978-3-030-36758-9_33

such as high melting point and corrosion resistance to slag. China is a country with abundant magnesite resources of which Liaoning province has the largest reserves and output of magnesite, with a reserve of 2.91 billion tons [2]. Magnesite usually coexists with other minerals in ore deposits, such as quartz, dolomite, talc, tremolite, chlorite, charcoal materials, etc. However, the most common Mg-bearing gangue is dolomite. It is difficult to separate magnesite form dolomite due to the same Mg site on their surfaces. In practice, one of the most commonly used method for concentrating magnesite is reverse flotation, usually using fatty acid as the collector.

Effectively inhibiting magnesite is the key to reverse flotation, especially the selection of magnesite depressants [3–5]. The study of the depressant of magnesite mainly focused on small molecule depressant. Some common small molecule depressants included sodium silicate, sodium hexametaphosphate, carboxymethyl cellulose, etc. [5–9]. However, there are few reports on the macromolecule depressant for magnesite flotation.

Hence, the aim of this study is to introduce a new kind of macromolecule depressant, sodium polyacrylate (SP), for magnesite flotation. The effects of SP concentration and pH values on micro-flotation of magnesite were analyzed. Sodium oleate was used as collector. Furthermore, mineral crystal structure and solution chemistry were discussed to delineate the interaction mechanism between sodium polyacrylate and magnesite.

Materials and Methods

Pure Minerals and Reagents

The pure minerals of magnesite were obtained from Liaoning province, China. The samples were ground to powder sample of −0.074 mm. The XRD analysis result of the sample was shown in Fig. 1. The chemical analysis was carried out, and the chemical compositions of samples were listed in Table 1. Results indicated that the purities of magnesite samples were very high.

Saponified sodium oleate was used as the collector for the micro-flotation tests. NaOH and H_2SO_4 were used as pH regulator in the experiments. All the reagents were analytically pure in the tests and the water used in all the experiments was quadratic deionized water.

Micro-flotation Tests

The micro-flotation tests were carried out on the FGC-35 flotation machine, with the stirring speed fixed at 1980 rpm. Pure mineral particles (2 g) were added in a

Fig. 1 XRD analyzing figure of the clean ore of magnesite

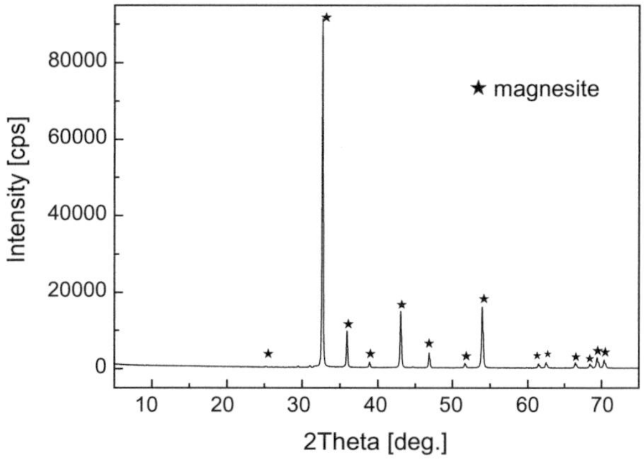

Table 1 Chemical composition of rutile ore

MgO	CaO	SiO_2	Fe_2O_3	Al_2O_3	K_2O	Na_2O	TiO_2
47.11	0.56	0.23	0.076	0.066	0.002	0.001	0.001

plexiglass cell (40 mL), which was then filled with distilled water. The pulp was continuously stirred for 1 min with pH regulator, 3 min with the depressant, and 3 min with the collector. The pH of the solution was measured before the flotation. The flotation was conducted for 3 min. The concentrate and tailing samples were collected separately and filtered, dried, and weighed to calculate the flotation recovery. All of the flotation tests were completed three times, and the average recovery was reported.

Results and Discussion

Flotation Behavior of Magnesite Using Sodium Oleate as Collector

The flotation recoveries of magnesite as a function of pH using sodium oleate as collector are shown in Fig. 2. The collector concentration remains constant at 100 mg/L. Magnesite shows good flotation in the pH region from 6 to 12, and the changes of magnesite recovery are minor in the pH, basically maintaining between 80 and 90%.

Figure 3 shows the effect of sodium oleate concentration on the flotation of magnesite at pH 5.8. The results indicate that the recovery of magnesite gradually increased with an increase in collector initial concentrate. When the dosage of sodium oleate is 100 mg/L, the recovery of magnesite is above 90%. And when the

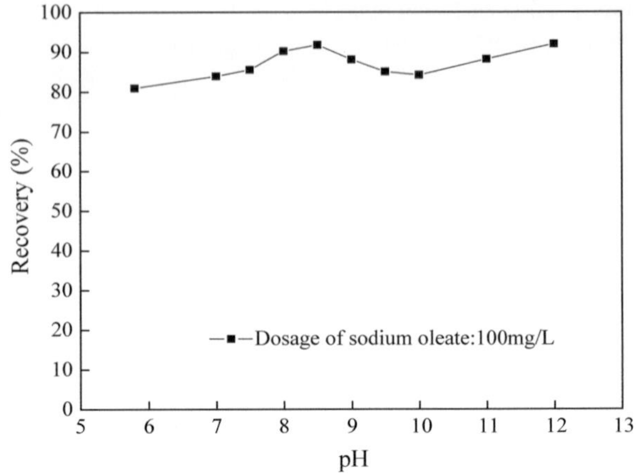

Fig. 2 Flotation recovery of magnesite as a function of pulp pH

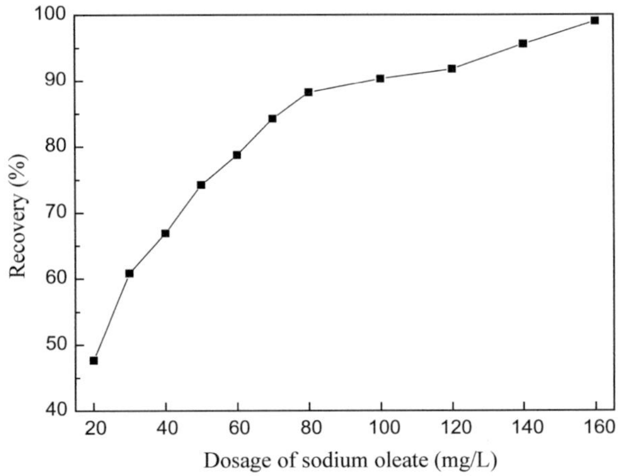

Fig. 3 Flotation recovery of magnesite as a function of collector dosage

dosage reaches 160 mg/L all the magnesite floated and thus the recovery achieves its maximum of 100%.

Depressing Effects of Sodium Polyacrylate on Magnesite

The objective of this study is to introduce a novel macromolecule depressant, sodium polyacrylate, and examine its effects on the flotation of magnesite. The flotation recoveries of magnesite as a function of pH values with 100 mg/L sodium oleate in the presence and absence of SP are shown in Fig. 4. The results in Fig. 4 indicate that the depression effect of SP is quite obvious. When pH < 10, and the recovery of magnesite decreases at least 30%, reaching to 60% at pH 9, and magnesite recovery reaching minimum of 26%. When pH > 10, the depression effect of SP becomes weak gradually [4].

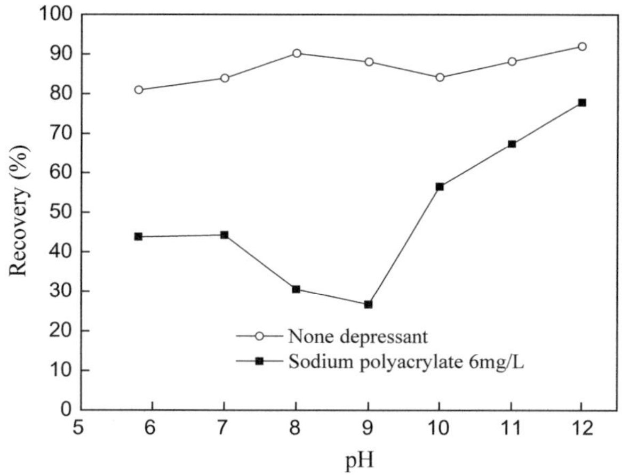

Fig. 4 Flotation recovery as a function of pulp pH in the presence and absence of SP

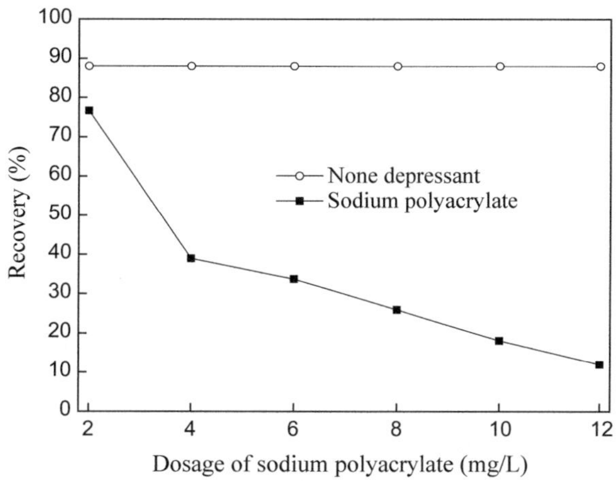

Fig. 5 Flotation recovery of magnesite as a function of the dosage of SP concentration

Figure 5 represents the flotation recoveries of magnesite as a function of SP concentration at pH 9. As can be seen in Fig. 5, the recoveries of magnesite decrease sharply as the dosage of SP increasing. When the dosage reaches to 10 mg/L, the recovery is less than 20%. The depression effect of SP on magnesite flotation is quite outstanding.

Interaction Mechanism Between Sodium Polyacrylate and Magnesite

The crystal structure of magnesite belongs to a trigonal system and rhombohedral crystal cells. One C atom and three O atoms present CO_3, and the isolated spheroid

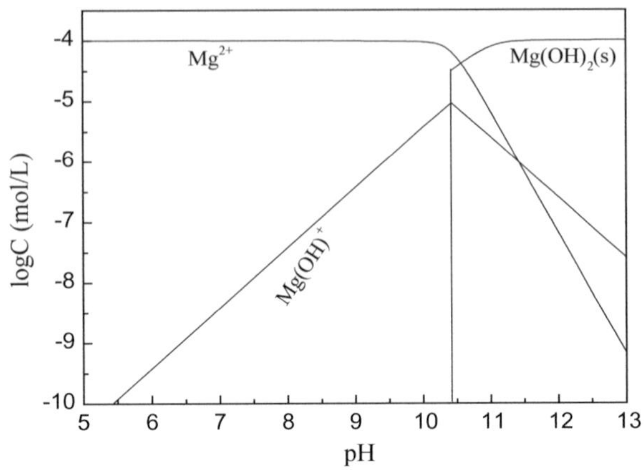

Fig. 6 The relation between the partition coefficient of the components' concentration and pH values in magnesite aqueous solution

represents Mg. In the grinding process, the Mg–O bond is more likely to break than the C–O bond, due to the strongly Connected of C atom and O atom. As a consequence, an electrostatic field with very strong surface force would keep a large amount of Mg^{2+} active spots on magnesite.

In addition, the ions on the surface of the magnesite in the pulp reacted easily with H_2O, H^+ or OH^- and dissolved into the aqueous solution. According to previous research [10, 11], magnesite had very complex dissolving chemical reactions in the aqueous solution. Many dissolved components would affect the surface property of magnesite further affecting the reaction between the magnesite surface and the flotation reagents. By detailed calculation of the dissolved components on the surface of magnesite, they finally concluded that the relation between the partition coefficient of the components and pH values, as shown in Fig. 6.

In pulp, crystal lattice cations exist in three forms: Mg^{2+}, $MgOH^+$ and $Mg(OH)_2$. Magnesium ions exist mainly in the form of Mg^{2+} when pH < 10.5, whereas magnesium ions are predominantly in the form of $Mg(OH)_{2(s)}$ when pH > 10.5. The concentration of $Mg(OH)^+$ increases at first then decreases, as the increasing of pH value, and it reaches maximum at pH = 10.5 [12]. Sodium polyacrylate belongs to low molecular weight polyacrylic homopolymer and homologous salt of formula $–(CH_2–CHCOONa)_n–$, and its molecular weight was about 8000. Sodium polyacrylate dissolves easily in pulp, bringing anions with carboxylate radicals into the pulp. Carboxylate anions can bind cations. The binding effect becomes stronger with the increase of the polyacrylic acid's dissociation degree and valence numbers of cations.

As there are a large amount of active spots of Mg^{2+} on the surface of magnesite in the solution, the sodium polyacrylate that was first added into the solution during the flotation process would strongly react with the active spots of Mg^{2+} on magnesite surface. When sodium oleate was added later during the flotation process, there was a competitive absorption between oleate anions and polyacrylic anions. The absorption of polyacrylic anions resulted in the inhibition of the magnesite flotation with the recovery rate generally under 50%. Especially in the pH range of 6–10, the positively charged Mg^{2+} and $Mg(OH)^{+}$ dominates on the magnesite surface, which would react with polyacrylic anions. The active spots of Mg on magnesite surface decreased, thus inhibiting the magnesite flotation [13].

Meanwhile, sodium polyacrylate is a kind of linear and soluble macromolecular compound. The electrostatic repulsion between carboxy groups on the molecular chain would extend the tortuous polymer chains and expose the absorptive radicals to the surface of the molecule. These functional groups can easily absorb with the surface of magnesite particles in the pulp, forming bridges between particles. These bridge functions would lead to the flocculation and sedimentation of magnesite particles, reducing the floatability of magnesite. As a result, the flotation of magnesite was depressed.

Conclusions

Sodium polyacrylate has strong depression effect on magnesite, especially with the dosage of 10 mg/L at $pH < 10$, the recovery of magnesite drops below 20%.

Sodium polyacrylate absorbed on the active spots of Mg^{2+} of the magnesite easily, forming a strong depression effect. In addition, the macromolecule structure of sodium polyacrylate could form bridges between magnesite particles. These bridges would lead to the flocculation and sedimentation of magnesite particles, reducing the floatability of magnesite.

Acknowledgements The authors acknowledge the support of the Key Scientific and Technological Project of Henan Province of China (182102310868) and Foundation of Key Laboratory of Radioactive and Rare Scattered Mineral Comprehensive Utilization, Ministry of Land and Resource (RRSM-KF2018-10).

Conflicts of Interest: The authors declare no conflict of interest.

References

1. Yan C, Xue D, Zou L, Yan X, Wang W (2005) Preparation of magnesium hydroxide nanoflowers. J Cryst Growth 282(3–4):448–454
2. Ji Z-M, Tian P-J, Chen Z, Pan K-J, Yin W-Z (2009) Research on flotation and purification of low grade magnesite. Min Metall 2:010
3. Pokrovsky OS, Schott J, Thomas F (1999) Processes at the magnesium-bearing carbonates/solution interface. I. A surface speciation model for magnesite. Geochimica et cosmochimica acta 63(6):863–880
4. Chen G, Tao D (2005) Reverse flotation of magnesite by dodecyl phosphate from dolomite in the presence of sodium silicate. Sep Sci Technol 39(2):377–390
5. Yao J, Yin W, Gong E (2016) Depressing effect of fine hydrophilic particles on magnesite reverse flotation. Int J Miner Process 149:84–93

6. Komlev A, Potapenko V (1972) Study of the role of reagents in magnesite flotation. Refractories 13(1–2):86–88
7. Ji Z, Tian P, Chen Z, Pan K-J, Yin W-Z (2008) Mechanism of research on magnesite flotation. Non-Ferr Min Metall 24:21–24
8. Shortridge P, Harris P, Bradshaw D, Koopal L (2000) The effect of chemical composition and molecular weight of polysaccharide depressants on the flotation of talc. Int J Miner Process 59 (3):215–224
9. Santana A, Peres A (2001) Reverse magnesite flotation. Miner Eng 14(1):107–111
10. Wang D-Z, Hu Y (1988) Solution chemistry of flotation. Hunan Science and Technology Press, Beijing, China, pp 235–238
11. Hongen YSZ (1990) A study on surface charges of magnesite and dolomite. Min Metall Eng 4:004
12. Chen G, Tao D (2004) Effect of solution chemistry on flotability of magnesite and dolomite. Int J Miner Process 74(1–4):343–357
13. Gence N, Özdağ H (1995) Surface properties of magnesite and surfactant adsorption mechanism. Int J Miner Process 43(1–2):37–47

Dissolution Behavior of Calcium Vanadates and Magnesium Vanadates in Sulfuric Acid

Xin Wang, Junyi Xiang, Qingyun Huang and Xuewei Lv

Abstract The dissolution efficiency of vanadium is the key to improve the recovery of vanadium from vanadium-bearing converter slag by calcification roasting-acid leaching process. An improved process, which involves roasting with calcium–magnesium salts, followed by acid leaching, was found to have a positive effect on the vanadium recovery. In this process, CaO–MgO–V_2O_5 is the basic system in the roasting step. In this study, the dissolution behavior of calcium vanadates (CaV_2O_6, $Ca_2V_2O_7$, and $Ca_3V_2O_8$) and magnesium vanadates (MgV_2O_6, $Mg_2V_2O_7$, and $Mg_3V_2O_8$) in sulfuric acid was investigated. The effect of parameters, such as stirring speed, solid/liquid ratio, and pH value, on the dissolution behavior of vanadates were determined. The results shown that both of solid/liquid ratio and pH value have significantly effects on the dissolution efficiency of vanadium, while stirring speed has a relatively minor effect under the studied conditions. Furthermore, the dissolution efficiency of magnesium vanadates has been found much higher than that of calcium vanadates.

Keywords Vanadium slag · Calcium vanadates · Magnesium vanadates
Sulfuric acid

Introduction

Vanadium, as the twelfth most abundant element in the earth's crust, has an average content of 150 g/t. The world's vanadium resources reached 6300 million tons [1]. Furthermore, the demands of vanadium in steel, chemistry, spaceflight, and aviation

X. Wang · J. Xiang (✉) · X. Lv
College of Materials Science and Engineering, Chongqing University, Chongqing 400044, China
e-mail: xiangjunyi126@126.com

Q. Huang
College of Metallurgical and Materials Engineering,
Chongqing University of Science and Technology, Chongqing 401331, China

G. Azimi et al. (eds.), *Rare Metal Technology 2020*, The Minerals, Metals & Materials Series, https://doi.org/10.1007/978-3-030-36758-9_34

industries increase rapidly in recent years [2]. The main vanadium-containing minerals are vanadium titanomagnetite, carnotite, roscoelite, and so on. Vanadium titanomagnetite ore, a naturally occurring iron ore, is regarded as one of the most important vanadium-bearing resource on earth, which supplies about 90% of the world demand for vanadium minerals [3–5]. Therefore, it is of significance to extract vanadium from vanadium titanomagnetite ore efficiently and environment-friendly for the sustainable supply of vanadium in China as well as in the world. Generally, vanadium is first reduced into hot metal in a blast furnace, then oxidized in a converter furnace, and enriched in a vanadium-bearing converter slag.

At present, vanadium slag is the main raw material to produce vanadium product. The most mainstream and commercial approaches that have been used to extract vanadium from vanadium slag are sodium roasting—water leaching and calcium roasting-acid leaching. However, sodium salt roasting will be gradually replaced due to lots of problems [6]. The disadvantage of the process is that it is not environment-friendly as the generation of harmful gases, liquid, and solid wastes [7, 8]. Calcium roasting could be an alternative with no corrosive gas emission, less harmful wastewater, and little noxious solid waste [9, 10]. Unfortunately, the existing industrial practices and fundamental research results show that sodium roasting process cannot be totally replaced by calcium roasting process because of the low vanadium recovery of calcium roasting process. In recent years, most of studies were performed in order to enhance the recovery of vanadium. Li et al. [11] have reported that when the optimal roasting conditions obtained are the $CaCO_3$ additive dosage 10%, roasting temperature 1200 °C, and roasting time 1 h, the primary vanadium-bearing phase generated is calcium metavanadate, and under the optimal conditions of pH of 0.5, mass ratio of liquid/solid of 5:1, temperature of 80 °C, and reaction time of 3 h, the leaching efficiency of vanadium is 72.1%. Zhang et al. [12] found that 91% of vanadium leaching efficiency can be achieved when leached in sulfuric acid with a pH of 2.5, L/S ratio of 3 and stirring speed of 500 rpm, and temperature of 55 °C for 30 min. Fan et al. [13] show that the maximum leaching efficiency of vanadium can reach 93.53% with high vanadium slag as raw material and 100% $CaSO_4$ as calcification agent.

In the laboratory research, it was found that the extent of recovery of vanadium can be improved by replacing calcium oxide partially with magnesium oxide during the calcification–roasting process. Furthermore, a certain amount of magnesium oxide are also contained in the vanadium slag. Therefore, $CaO–MgO–V_2O_5$ is the most basic system in the vanadium extraction process, and the dissolution efficiency of calcium vanadates and magnesium vanadates is related to the recovery efficiency of vanadium in the whole process. Since only a few kinetics studies have so far been carried out, there is still scanty data on dissolution behavior of calcium vanadates and magnesium vanadates in sulfuric acid. The objective of the present investigation is to obtain essential information on the dissolution behavior of the calcium vanadates and magnesium vanadates in sulfuric acid.

Experimental

Materials

Calcium metavanadate, calcium pyrovanadate, and calcium orthovanadate used in the present study were roasted with CaO and V_2O_5, and the mole ratio of CaO/V_2O_5 was maintained at 1:1, 2:1, and 3:1, respectively. Magnesium metavanadate, magnesium pyrovanadate, and magnesium orthovanadate were roasted with MgO and V_2O_5, and the mole ratio of MgO/V_2O_5 was maintained at 1:1, 2:1, and 3:1, respectively. The CaO, MgO, and V_2O_5 were analytical pure. The V_2O_5 and MgO were analytical grade with purity greater than 99.9%. The water used in the experiment is deionized water. Commercial sulfuric acid with a concentration of 98% H_2SO_4 from Chongqing Chuandong Chemical Co. Ltd was also used in this study. The X-ray diffraction patterns of calcium vanadates and magnesium vanadates are shown in Fig. 1. Comparing the diffraction lines of the prepared samples with the standard cards, it can be found that the most diffraction peaks match with the standard cards, and only a very few impurity peaks exist.

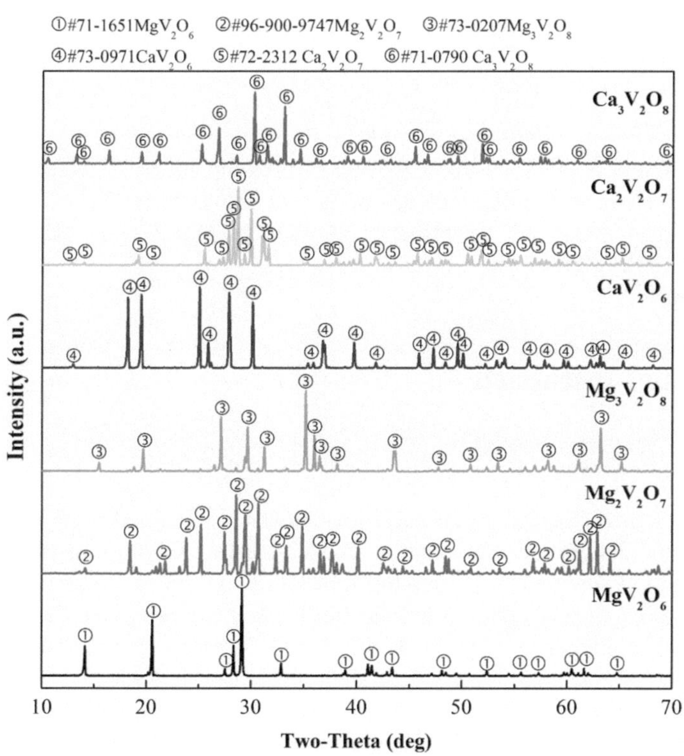

Fig. 1 X-ray diffraction patterns of calcium vanadates and magnesium vanadates

Procedure

All the dissolution experiments were conducted in 500 mL three-necked round-bottomed flask equipped with a condenser, a thermometer, a magnetic stirrer, and a pH meter. After the solution in flask was heated to the desired temperature, a measured amount of the sample was charged into the reactor. Then, the mixture of the reactants was dissolved under designed stirring speed and fixed pH meter. The pH value of mixed solution was controlled by adding sulfuric acid through the all dissolution process. At the end of each experiment, the slurry was filtered, and the obtained solutions and residues were kept separately.

The dissolution efficiency is a calculated with the following formula:

$$\eta(\%) = \frac{C_1 \times \mathrm{V}}{C_2 \times \mathrm{m}} \times 100\%$$

where η is the dissolution efficiency, V is the leaching liquor volume (mL), C_1 is the concentration of vanadium in the liquor (g/mL), C_2 is the concentration of vanadium in the solid sample (g/mL), and m is the quality of the solid sample (g).

Characterization

The phase of calcium vanadates and magnesium vanadates was analyzed by X-ray powder diffraction (XRD) using a PANalytical Company of the Netherlands. The vanadium content of the obtained solutions was identified by inductively coupled plasma optical emission spectrometry (ICP-OES) (ICAP 6000, Thermo Fisher Scientific, USA).

Results and Discussion

Effect of Stirring Speed

The effect of stirring speed on the sulfuric acid was investigated in 100 mL solution at a constant dissolution temperature of 50 °C using solid/liquid ratio of 1:150 (g/ml) and pH value of 2.5, by treating four kinds of stirring speeds, namely, 100, 200, 300, 400 rpm. The results are given in Fig. 2. It can been seen from Fig. 2 that the dissolution efficiency of CaV_2O_6 and MgV_2O_6 increases with the increase of speed from 100 rpm to 300 rpm, then slight decreases at 400 rpm. Obviously, the dissolution efficiency of MgV_2O_6 is significantly higher than that of CaV_2O_6 at the same dissolution conditions. The dissolution efficiency of MgV_2O_6 is above 50%, and that of CaV_2O_6 is around 35%. On the basis of the data, subsequent experiments were carried out at a stirring speed of 300 rpm to ensure that calcium vanadates and magnesium vanadates have a better dissolution efficiency.

Effect of Solid/Liquid Ratio

The effect of solid/liquid ratio on the dissolution process was tested by several experiments, using solid/liquid ratio varying from 1:50 to 1:200 (g/ml). The other dissolution parameters were fixed at temperature of 50 °C, pH value of 2.5, stirring speed of 300 rpm, and reaction time of 20 min. The obtained results are plotted in Fig. 3.

From the obtained results, it is evident that the dissolution efficiency of CaV_2O_6 and MgV_2O_6 is strongly dependent on the solid/liquid ratio. The dissolution efficiency increases first and then decreases, reaching the maximum value at solid/liquid ratio of 1:150 (g/ml). This could be because of the fact the area of sample and sulfuric acid solution increases with the solid/liquid ratio increasing, resulting more sufficient of the reaction. As a result, the dissolution efficiency is higher. Therefore, the solid–liquid ratio was determined to be 1:150 (g/ml) in the follow-up experiment process.

Effect of pH Value

In order to study the effect of pH value on calcium vanadates (CaV_2O_6, $Ca_2V_2O_7$, and $Ca_3V_2O_8$) and magnesium vanadates (MgV_2O_6, $Mg_2V_2O_7$, and $Mg_3V_2O_8$) in sulfuric acid, several dissolution experiments were performed using various pH values controlled by pH meter with sulfuric acid solution, while the other

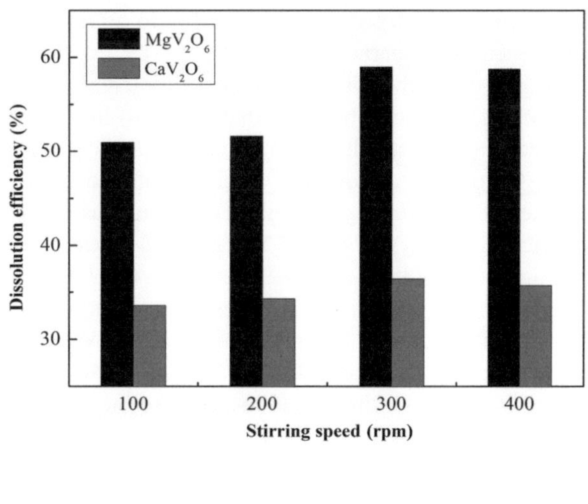

Fig. 2 Effect of stirring speed on the dissolution of CaV_2O_6 and MgV_2O_6

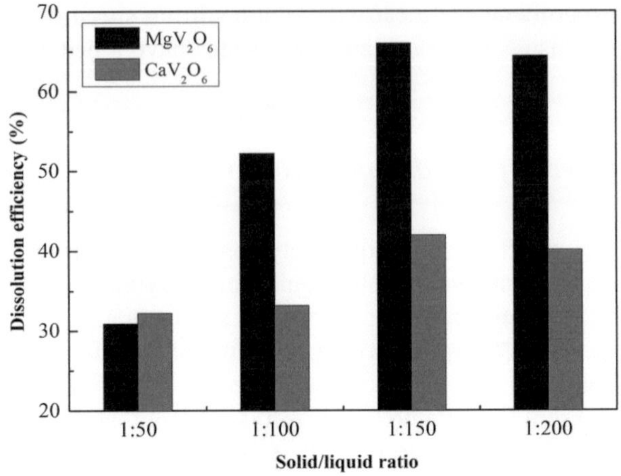

Fig. 3 Effect of solid/liquid ratio on the dissolution of CaV_2O_6 and MgV_2O_6

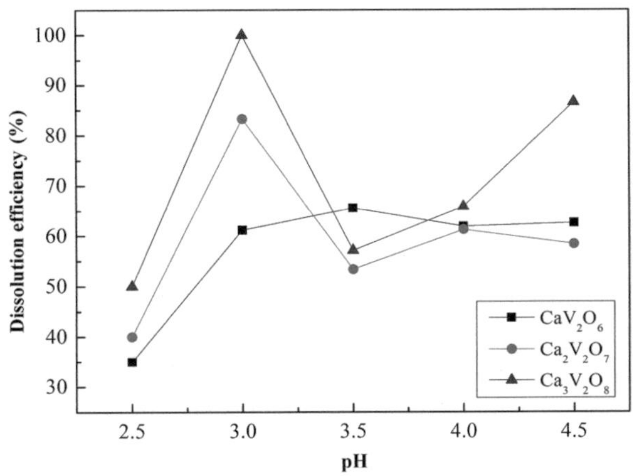

Fig. 4 Effect of pH value on calcium vanadates dissolution process

parameters were fixed at solid/liquid ratio of 1:150 (g/ml), reaction time of 20 min, and reaction temperature of 50 °C.

Figure 4 shows the dissolution efficiency of calcium vanadates at different temperatures, it can be seen that the dissolution efficiency of calcium vanadates reaches the lowest value when the pH value is 2.5. The pH value has a little effect to CaV_2O_6 on the dissolution efficiency in range of 3.0–4.5, with the dissolution efficiency changing only slightly with increasing pH value, while the dissolution efficiency of $Ca_2V_2O_7$ and $Ca_3V_2O_8$ has a complex effect on the pH value of the sulfuric acid solution. But both $Ca_2V_2O_7$ and $Ca_3V_2O_8$ have the highest dissolution efficiency at a pH value of 3.0, especially $Ca_3V_2O_8$, almost 100%.

The dissolution efficiency of magnesium vanadates at different temperatures is shown in Fig. 5. The results in Fig. 5 show that the dissolution efficiency of MgV_2O_6 decreases first, then has a sharp increase and reaches the maximum value

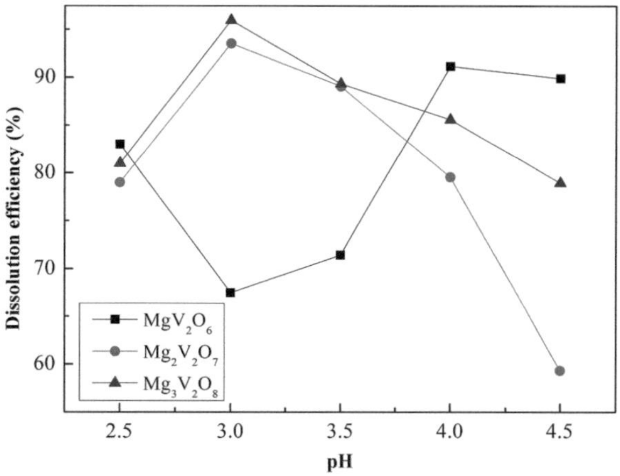

Fig. 5 Effect of pH value on magnesium vanadates dissolution process

at pH of 4.0, then slightly decreases with further increasing the pH value. It is clear that the dissolution efficiency of $Mg_2V_2O_7$ and $Mg_3V_2O_8$ increases when the pH goes from 2.5 to 3. When pH value is in the range of 3.0–4.0, the dissolution efficiency of $Mg_2V_2O_7$ and $Mg_3V_2O_8$ is inversely proportional to pH value. Under the current experiments conditions, the dissolution efficiency reaches maximum when pH value is equal to 3.0.

Conclusions

The following conclusions can be drawn from this study:

(1) The solid/liquid ratio has a significant effect on the efficiency of CaV_2O_6 and MgV_2O_6, while stirring speed has relatively minor effect. And the dissolution efficiency of MgV_2O_6 has been found much higher than that of CaV_2O_6. It is found that CaV_2O_6 and MgV_2O_6 have a best dissolution efficiency under the studied conditions, when the solid/liquid ratio and the stirring speed are 1:150 (g/ml) and 300 rpm, respectively.

(2) The results indicate that pH value has relatively complex effect on the dissolution efficiency of vanadium. Under the dissolution condition of solid/liquid ratio of 1:150, reaction time of 20 min, and temperature of 50 °C, the dissolution efficiency of CaV_2O_6 is higher at pH of 3.5, and $Ca_2V_2O_7$ and $Ca_3V_2O_8$ have the best dissolution efficiency when pH value is equal to 3. Under the same dissolution condition of calcium vanadates, MgV_2O_6, $Mg_2V_2O_7$, and $Mg_3V_2O_8$ have the best dissolution efficiency when pH value is equal to 4.0, 3.0, and 3.0, respectively.

Acknowledgements This work was supported by the Project funded by China Postdoctoral Science Foundation (grant numbers 2018M640898), the National Key Research and Development Program of China (2018YFC1900501), and the open project founded by State Key Laboratory of Vanadium and Titanium Resources Comprehensive Utilization.

References

1. Moskalyk R, Alfantazi A (2006) Processing of vanadium: a review. Miner Eng 16(9):793–805
2. Baker TN (2014) Processes, microstructure and properties of vanadium microalloyed steels. Metal Sci J 25(9):1083–1107
3. Speer JG, Michael JR, Hansen SS (1987) Carbonitride precipitation in niobium/vanadium microalloyed steels. Metall Trans A 18(2):211–222
4. Deng J, Xue X, Liu GG (2007) Current situation and development of comprehensive utilization of vanadium-bearing titanomagnetite at Pangang. J Mater Metall 6(2):83–86
5. Tan QY, Chen B, Zhang YU, Long YB, Yang YH (2011) Characteristics and current situation of comprehensive utilization of vanadium titano-magnetite resources in Panxi region. Multipurp Utilization Miner Resour 6:6–10
6. Li XS, Xie B (2012) Extraction of vanadium from high calcium vanadium slag using direct roasting and soda leaching. J Miner Metall Mater 19(7):595–601
7. Lasheen TA (2008) Soda ash roasting of titania slag product from Rosetta ilmenite. Hydrometallurgy 93(3):124–128
8. Liu ZH, Yan L, Chen ML, Nueraihemaiti A, Du J, Xing F, Tao CY (2016) Enhanced leaching of vanadium slag in acidic solution by electro-oxidation. Hydrometallurgy 159:1–5
9. Chen HS (1992) Study on extraction technology of V2O5 by calcining vanadium slag lime. Steel Iron Vanadium Titanium 3:3–11
10. Fu ZB (2011) Development process and trends of vanadium extraction from vanadium-titanium magnetite ore. Chin J Nonferrous Metals 40(6):29–33
11. Li LJ, Zhang L, Zheng SL, Lou TP, Zhang Y, Chen DH, Zhang Y (2011) Acid leaching of calcined vanadium titanomagnetite with calcium compounds for extraction of vanadium. Chin J Process Eng 11(4):573–578
12. Zhang JH, Zhang W, Zhang L, Gu SQ (2014) Effect of acid leaching on the vanadium leaching rate in process of vanadium extraction using calcium roasting. J Northeast Univ (Natural Sci) 35(11):1574–1578
13. Fan K, Li ZC, Li ZS, Wang WP, Zhang WL, Zheng HY, Shen FM (2015) Effect of different calcification agents on vanadium extraction from high-vanadium slag by calcified roasting-acid leaching. J Chongqing Univ 5:151–156

PRICE—PRocess Industries in the Circular Economy

Dag Øistein Eriksen

Abstract PRICE is an acronym for PRocess Industries in the Circular Economy. The main objective of the project is to prepare the collaborating industrial partners to the circular economy. This competence building project gathers industrial partners, research institutes, and the two largest universities in Norway. The participants are: Boliden Odda, Glencore Nikkelverk, KA Rasmussen, NOAH, Solberg Industri, Yara International, SINTEF industry, University of Oslo, and the Norwegian University of Science and Technology (NTNU). To reach this ambitious goal, there are sub-objectives: increased recirculation and recovery of metals and minerals—in general and between the participants in PRICE; understanding the behaviour (speciation) of and recovery or removal of valuable elements and toxic components present at low concentrations in process solutions; use of electrochemical means to enhance separation of elements present in trace concentrations; and assessment to evaluate the impacts of circular economy value-chains through environmental modelling (LCA) and techno/economic optimization.

Keywords Circular economy · Recycling · Hydrometallurgy · Knowledge building

Introduction

The green economy requires reduced waste, recycling of materials, and more efficient consumption of energy and materials. An important prerequisite for a successful transition to the process industry's value-chains and, ultimately, a circular economy, is a better understanding of the economic and environmental impacts from changing current practices. Evaluating new business concepts rely on a multidisciplinary, system-perspective approach integrating leading competence on process technology; innovation and new business development; and economic,

D. Ø. Eriksen (✉)
Department of Chemistry, University of Oslo, Oslo, Norway
e-mail: d.o.eriksen@kjemi.uio.no

G. Azimi et al. (eds.), *Rare Metal Technology 2020*, The Minerals, Metals & Materials Series, https://doi.org/10.1007/978-3-030-36758-9_35

environmental, and technical analyses. This means that the technological challenges and opportunities must be understood and seen in relation to the economic and environmental impacts. In Norway, the process industries based on aqueous solutions cover most of the requirements for sustainable productions:

- Clean energy from hydroelectric power,
- Production plants in coastal areas making transportation cheap and easy by ships,
- Chemical infrastructure comprising the availability of acids and bases,
- Strong environmental legislation and control,
- Safe handling of waste materials,
- Know-how, and
- Capital.

However, all these topics have room for improvements and must also be developed further to face the needs of the future.

PRICE is a *knowledge building* project supported by the Research Council of Norway and Norwegian Process industrial companies. The education and research are conducted by SINTEF industry, University of Oslo (UiO), and the Norwegian University of Science and Technology (NTNU). The industrial companies are: Boliden Odda AS, Glencore Nikkelverk AS, K.A. Rasmussen AS, NOAH, Solberg Industri AS, and Yara International ASA [1]. Four Ph.D. grantees are engaged by the project, two at NTNU and two at UiO. UiO provides the project management. Figure 1 presents the circular economy in a generic mode, but our industrial contributors and sponsors all fit into parts of this figure.

Knowledge needs for the industry

It is a well-known fact that many of the earth's mineral resources are being depleted or the ores' production has peaked. Studies performed by acknowledged research institutes [2, 3] conclude that over 30 elements are in the critical zone. Some of these will become very critical in the future. Some elements are only found in few areas of the world, and often subject to conflicts between different interest groups, both corporative and on a national level. It has therefore become an increasingly important issue to recover and recycle as much as possible of the metals and minerals in use, so-called circular economy.

For the process industry, there is always a question of economy in any activity to be implemented. The challenges facing metal recycling are several:

- The elements of interest are always mixed with other unwanted elements.
- Existing methods for separating the elements are often complex and of multi-step nature.
- Existing methods of separation/refining are often sensitive to variations in the composition and conditions.
- If the level of dilution has come too far, the cost of recycling with current technologies is larger than the value of recycled materials.
- Unwanted elements from one industry might be an interesting raw material for another.

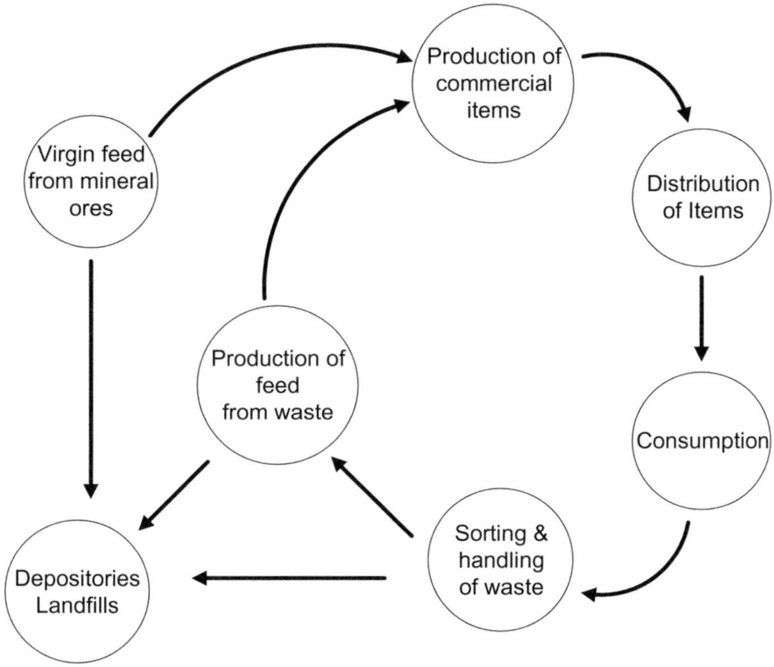

Fig. 1 Circular economy including disposal of waste, utilisation of waste as new feed and input of new, virgin material from minerals

- The volumes at one site may be too small for economical treatment, but economy of scale at one central site may improve the situation.
- Coordination and interchange of residual elements are lacking.

Dividing the hydrometallurgical industry into areas

In this context, we use the term "hydrometallurgy" as any process where aqueous solution is essential and at least one of the traditional unit operations, such as mixing, stirring, leaching, crystallization, filtration, centrifugation, distillation, solvent extraction, ion exchange, membrane separations, and electrolyzes, is involved.

We can divide the participating process industry utilizing aqueous solutions into five areas:

a. Processing minerals, e.g., Yara International.
b. Processing concentrates and making pure metals, e.g., Glencore Nikkelverk and Boliden Odda.
c. Recycling of metals, e.g., K.A. Rasmussen and Glencore Nikkelverk.
d. Treatment of waste streams, e.g., NOAH and Solberg Industri, Glencore Nikkelverk.

e. Handling of wastes, e.g., NOAH, but also common to most of the companies mentioned.

In PRICE, however, we divide the companies into different areas of interest or knowledge needs:

a. Recovery of heavy metals: Yara, NOAH, KA Rasmussen, Solberg Industri, Boliden Odda, and Glencore Nikkelverk.
b. Removal of halogenides: Yara, Boliden Odda, NOAH, and Glencore Nikkelverk.
c. Electrochemical methods: KA Rasmussen, Boliden Odda, Glencore Nikkelverk.
d. Crystallization, precipitation, and filtering: NOAH, Yara, Boliden Odda, Glencore Nikkelverk, Solberg Industri.

Four Ph.D. students are involved in this project. They all have to perform high-quality research. The different research topics will be related to topics and challenges the industrial participants have and need to solve. PRICE is therefore a research program, and we aim at locating bottlenecks and problem areas for each individual industry, to develop new knowledge and coordinate scientific results found, making tailor-made solutions for each industry as well as trying to solve the coordination problem: waste from one company may be raw material for another. The research program will evaluate the different existing methods as well as exploring the effect recent advances within chemistry/metallurgy can have in combination with each other, and in combination with other disciplines (pyrometallurgy, material science).

It is our common goal that the new knowledge gained will give us robust and versatile processes, allowing the recycling of the metals at low cost and low environmental impact, feasible implementation of technology to industrial scale, and in addition, building up centers of competence at the research institutes. What is therefore needed is robust technology with a high selectivity. This is a very demanding goal.

The project was launched on May 1, 2019; the Ph.D. grantees were selected in June and are all starting in the Fall of 2019. The project will be finished in 2024.

Acknowledgements The author is grateful for the economic support by the Research Council of Norway and the PRICE collaboration for permission to present the project.

References

1. www.ka-rasmussen.no/om-k-a-rasmussen/, www.nikkelverk.no/en/Pages/home.aspx, www.solbergindustri.no/en.noah.no/, www.yara.com/ www.boliden.com/operations/smelters/boliden-odda
2. Marscheider-Weidemann F, Langkau S, Hummen T, Erdmann L, Tercero Espinoza L, Angerer G, Marwede M, Benecke S (2016) Rohstoffe fur Zukunftstechnologien 2016.—DERA Rohstoffinformastionen 28:353 S., Berlin
3. Swedish Agency for Growth Policy Analysis, Studentplan 3, SE-831 40 Östersund, Sweden Ref. no.: 2016/227

Reductive Leaching of Indium-Bearing Zinc Leaching Residue in Sulfuric Acid and Sulfur Dioxide

Zhi-gan Deng, Guang Fan, Chang Wei, Gang Fan, Min-ting Li, Xing-bin Li and Cun-xiong Li

Abstract The leaching residue in zinc smelting system is taken as the research object, and the experiment of leaching residue in reduction leaching under sulphuric acid–sulfur dioxide system is carried out. The effects of reaction temperature, liquid–solid ratio, initial sulfuric acid concentration, partial pressure of SO_2, and stirring speed on leaching rate of medium leaching residue under sulfuric acid-sulfur dioxide reduction conditions were studied. The leaching rates of zinc 95.24%, iron 98.66%, indium 95.04%, and copper 0.036% in medium leaching residue can be obtained at the conditions of temperature of 105 °C, time of 4 h, liquid–solid ratio of 8 ml/g, initial sulfuric acid concentration of 120 g/L, SO_2 partial pressure of 200 kPa, and stirring speed of 500 r/min.

Keywords Zinc leaching residue · Reductive leaching · Hydrometallurgical zinc Sulfur dioxide

Introduction

Marmatite is an important resource of zinc, which exist in forms of high iron and high indium [1, 2]. The conventional zinc recovery process is the roast-leaching-electrowinning process [3, 4]. In this process, a significant part of zinc ferrite is obtained when zinc concentrate was roasted. The indium what is associated with zinc concentrate enter into the crystal lattice of zinc ferrite through substituted iron to form indium-bearing zinc ferrite [5, 6]. In the hydrometallurgical zinc process to treat zinc concentrate, the majority of indium comes into the leach residue which

Z. Deng (✉) · G. Fan · C. Wei · G. Fan (✉) · M. Li · X. Li · C. Li
Faculty of Metallurgical and Energy Engineering, Kunming University of Science and Technology, 253 Xuefu Road, Kunming 650093, Yunnan, China
e-mail: dengzhigan83@163.com

G. Fan
e-mail: kgfangang2008@163.com

G. Azimi et al. (eds.), *Rare Metal Technology 2020*, The Minerals, Metals & Materials Series, https://doi.org/10.1007/978-3-030-36758-9_36

has a significant part of indium-bearing zinc ferrite. Unfortunately, the indium-bearing zinc ferrite is stable and difficult to dissolve at low acid concentrations. Hot acid leaching is the main way to dissolute zinc ferrite [7–9]. There is much ferric ion in lixivium of the hot acid leaching. Sphalerite concentrate can successfully be used as a reductant in acidic solution [10–14]. The reduction of ferric to ferrous, and then oxidation and precipitation of ferrous ion as hematite would solve the problem with losses of zinc in jarosite process; and also overcome the problems in the conversion process, where several valuable metals (Ag, Au, Pb) are lost in the precipitation. It is also good for separations of indium and zinc when the ferric ion is converted to the ferrous ion [15, 16].

Using sulfur dioxide as reductant in reductive leaching, the majority of the ferric ion present in the leach process is converted to the ferrous ion. The main reactions are as follows:

$$ZnO \cdot Fe_2O_3 + 4H_2SO_4 = ZnSO_4 + Fe_2(SO_4)_3 + 4H_2O \quad (1)$$

$$ZnO \cdot Fe_2O_3 + SO_2 + 2H_2SO_4 + 2H_2O = 2FeSO_4 + ZnSO_4 \quad (2)$$

In order to recycle the valuable metal from these ores, the test work of reductive leaching of indium-bearing zinc leaching residue in sulfuric acid and sulfur dioxide was conducted. The aim of this research was to state the reductive leaching process of zinc residue has a significant part of zinc ferrite by sulfuric acid in the presence of sulfur dioxide as reductant. The extraction of indium and zinc from residue was demonstrated, and the effects of SO_2 partial pressure, initial sulfuric acid concentration, reaction time, reaction temperature, and rotational speed on the leaching behavior and leaching rate of metal were investigated.

Experimental

Materials

The zinc leaching residue used in the present was from Yunnan Province, China. Analytical grade sulfuric acid was purchased from AR, Chengdu Union Institute Chemical & Reagent. SO_2 within high-pressure gas bottles was from Kunming MeiSaiEr.

Experimental Procedure

The zinc leaching residue was crushed and ball milled to 85% less than 0.074 mm and dried at 65 °C in an oven for 24 h. Leaching experiments were carried out in a 2 L Polyclave autoclave (Büchi Glas Uster, Switzerland). The autoclave was made of

CrNiMoTi and equipped with a sampling system, a proportional–integral–differential (PID) temperature controller, a heating circulation bath, data logging software for a personal computer, and a stirrer with a rotation speed of 0–1500 r/min. The working temperature of the autoclave was 0–300 °C.

At the end of the experiment, the ore pulp was withdrawn and filtered. The solid residue and solution were chemically analyzed. The extraction rates of Zn, In, Cu, and Fe were calculated.

Analytical Methods

Zinc was analyzed by complex titration with EDTA. Indium was analyzed by atomic absorption spectrometry (Shanghai Morning Field Equipment Co., Ltd). Fe was analyzed by complex titration with potassium bichromate. X-ray powder diffraction (XRD) was carried out using Rigaku D/MAX 2500v diffractometer (Japan). The potential of solution was measured using a PHSJ-4F digital pH meter by a platinum electrode with an Ag/AgCl electrode used as the reference electrode (Shanghai Leici Instrument Co., Ltd).

Results and Discussion

Characterization of the Materials

The zinc leaching residue used in the present study was from Hualian Zinc and Indium Co., Ltd. in Yunnan Province, China. The compositions of the zinc leaching residue are listed in Table 1. X-ray diffraction of zinc leaching residue identified zinc ferrite ($ZnFe_2O_4$) as the main mineral components as shown in Fig. 1. No special indium bearing mineral was identified. It can be seen in the SEM image shown in Fig. 2 that the zinc leaching residue was dense.

Effect of SO_2 Partial Pressure

Addition of SO_2 into the reactor provided a reductant to reduce Fe^{3+} to Fe^{2+}, there by dissolving iron into the solution. Figure 3 presents the effect of SO_2 on the

Table 1 The main chemical component of zinc leaching residue (wt%)

Zn	Fe	Cu	In	S	Pb	Cl	As
21.90	26.16	1.04	827.9 g/t	6.54	1.53	<0.01	0.42

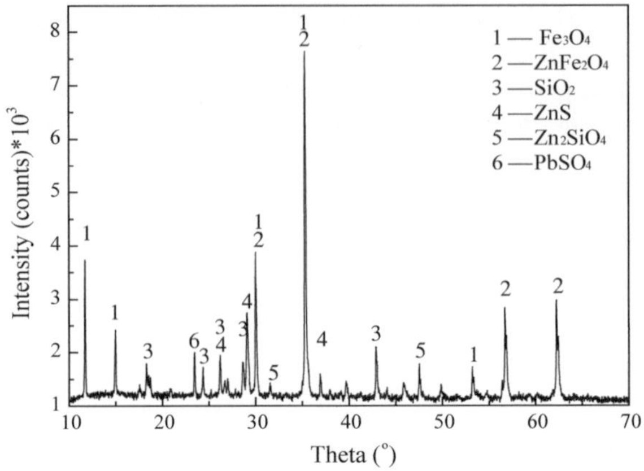

Fig. 1 XRD spectrum of the zinc leaching residue

Fig. 2 SEM images of the zinc leaching residue

extraction of Zn. As shown in Fig. 3, the extraction zinc was about 65% after 4 h without SO_2 addition. With the addition of 200 kPa of SO_2, the extraction of Zn was increased significantly to around 96%.

Figure 4 presents the effect of SO_2 on the concentrations of Fe^{3+} and Fe^{2+} and redox potential in lixivium of the reductive leaching. It can be seen in Fig. 4 that the redox potential was decreased with increasing the partial pressure of SO_2.

It can be seen in Fig. 4 that the Fe^{3+} content was decreased and the Fe^{2+} content was increased with increasing the SO_2 amount. If Fe^{2+} is present in the lixivium, in the absence of other reductants it determines the redox potential. With increasing the Fe^{2+} amount, the redox potential decreased.

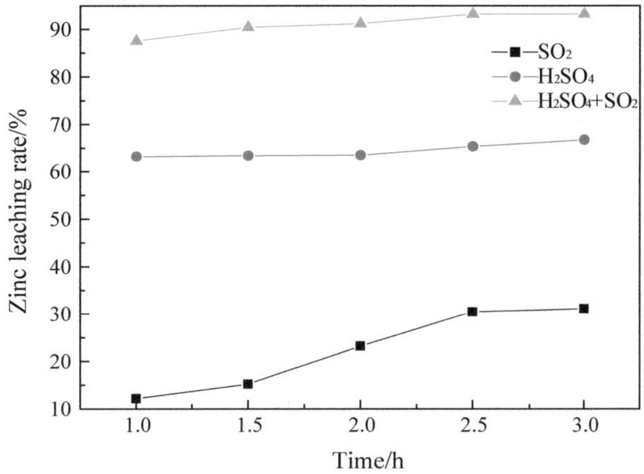

Fig. 3 Effect of SO_2 addition on extraction of Zn. (105 °C, liquid–solid ratio 8 ml/g, initial H_2SO_4 120 g/L, stirring speed 500 r/min, and time 4 h)

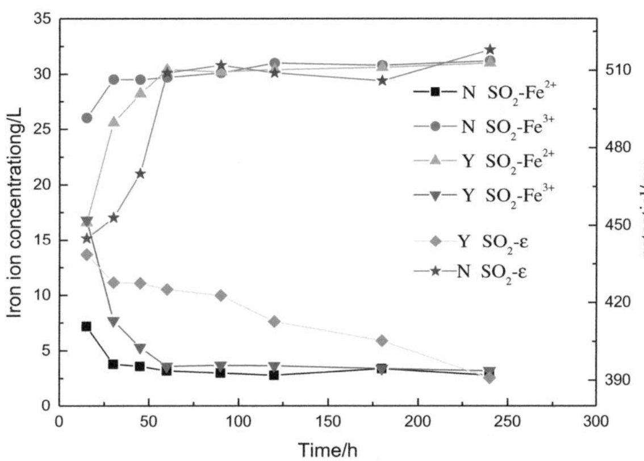

Fig. 4 Effect of SO_2 addition on the concentration of Fe^{3+} and Fe^{2+} and redox potential of lixivium. (105 °C, liquid–solid ratio 8 ml/g, initial H_2SO_4 120 g/L, stirring speed 500 r/min, and time 4 h) (N: No SO_2 addition; Y: With SO_2 addition)

Figure 5 presents the effect of SO_2 partial pressure on the extraction of Zn, Fe, In, and Cu. With the addition of 200 kPa SO_2, the extraction of Zn, Fe, and In was increased significantly to ~96%, ~93%, and ~95%, respectively. However, more than 90% copper was precipitated into the reduction leaching residue.

Effect of Initial Sulfuric Acid Concentration

In the reductive leaching process, the reacted acid of the lixivium improved with increasing sulfuric acid concentration, and selecting the appropriate concentration of sulfuric acid was necessary. The effect of sulfuric acid concentration on the leaching of metals was studied in the range of 100–140 g/L. Figure 6 shows the extraction of Zn, Fe, In, and Cu as a function of sulfuric acid concentration in the leaching solution. It can be seen in Fig. 6 that the curves of leaching efficiencies of

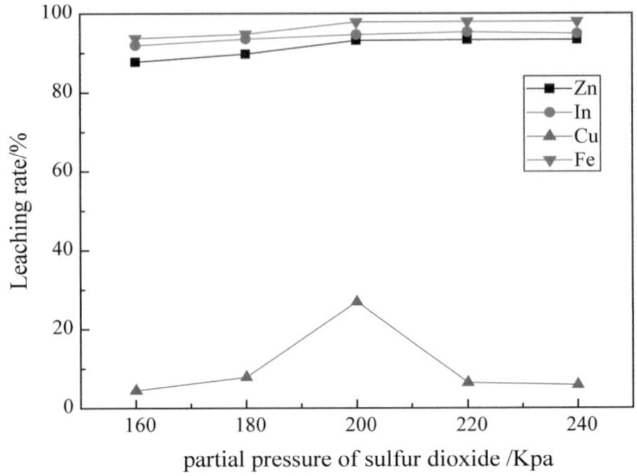

Fig. 5 Effect of SO_2 partial pressure on the extraction of Zn, Fe, In, and Cu. (105 °C, liquid–solid ratio 8 ml/g, initial H_2SO_4 120 g/L, stirring speed 500 r/min, and time 4 h)

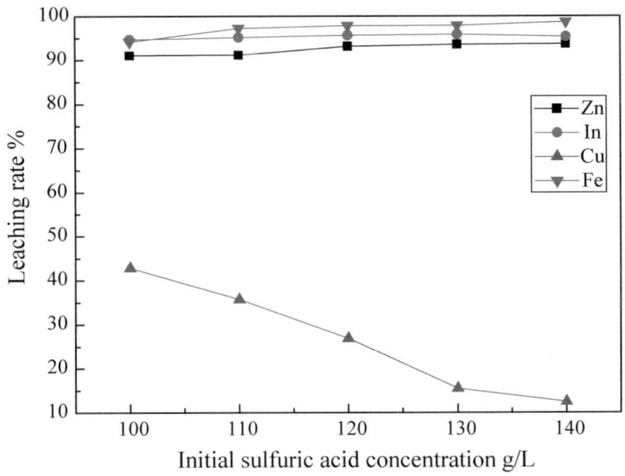

Fig. 6 Effect of sulfuric acid concentration on the extraction of Zn, Fe, In, and Cu. (the SO_2 partial pressure 200 kPa, 105 °C, liquid–solid ratio 8 ml/g, stirring speed 500 r/min, and time 4 h)

Zn, Fe, and In increased faster with increasing the sulfuric acid concentration from 100 to 140 g/L, but the leaching efficiency of Cu was slightly decreased. The leaching efficiencies of Zn, Fe, and In reached ~93%, ~96%, and ~95%, respectively, when sulfuric acid concentration was 130 g/L, and more than 92% copper was precipitated into the reduction leaching residue.

Effect of Reactive Time

The effect of reaction time on the extraction of metals was studied in the range of 60–300 min. As shown in Fig. 7, the extraction of Zn, Fe, and In increased from ~83%, ~65% and 87% to ~96%, ~98%, and ~95%, respectively, when the

Fig. 7 Effect of time on extraction of Zn, Fe, In, and Cu. (the SO_2 partial pressure 200 kPa, 105 °C, initial H_2SO_4 120 g/L, stirring speed 500 r/min, and 4 h)

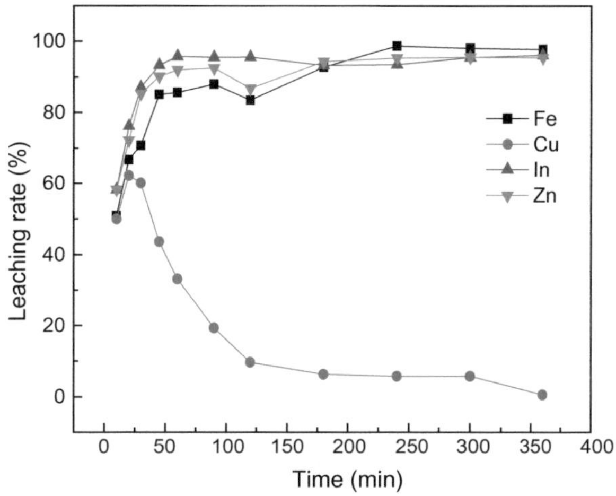

time increased from 50 min to 250 min. But the extraction of copper slightly decreased with increasing reaction time, and more than 90% copper was precipitated into the reduction leaching residue at 250 min.

Effect of Reaction Temperature

The extraction rate of metals was also affected by leaching temperature. High temperature was beneficial for the dissolutions of zinc ferrite. However, high temperatures had disadvantages, such as higher energy consumption and higher evaporation. A medium temperature range from 85 to 125 °C was selected and its effect on the metal extraction was studied. Figure 8 shows the effect of temperature on the extraction of metals. The extraction of metal increased with increasing of temperature. The extraction of Zn, Fe, and In increased when the temperature increased from 85 to 125 °C, but a further increase in temperature to 105 °C had no significant effect. Maximum amount of copper precipitated into the reduction leaching residue at 105 °C.

Effect of Liquid–Solid Ration

The effect of liquid–solid ratio on the extraction of Zn, Fe, In, and Cu is shown in Fig. 9. As can be seen, the extraction rate of Zn, Fe, and In does not increase significantly with increasing the liquid–solid ratio. The effect of liquid–solid ratio on the extraction of Cu was significant at lower liquid–solid ratio.

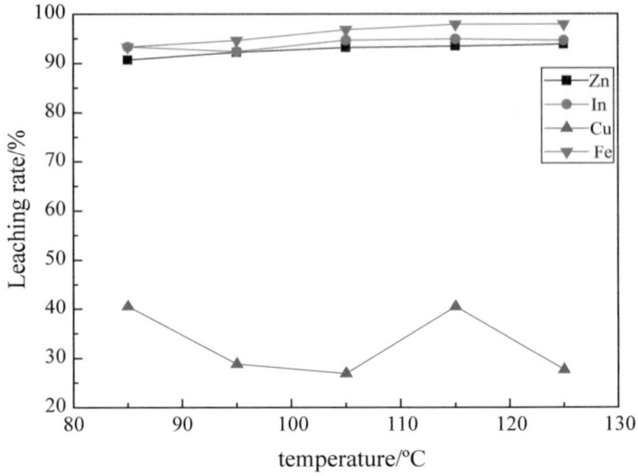

Fig. 8 Effect of temperature on the extraction of Zn, Fe, In, and Cu. (the SO_2 partial pressure 200 kPa, liquid–solid ratio 8 ml/g, initial H_2SO_4 120 g/L, stirring speed 500 r/min, and time 4 h)

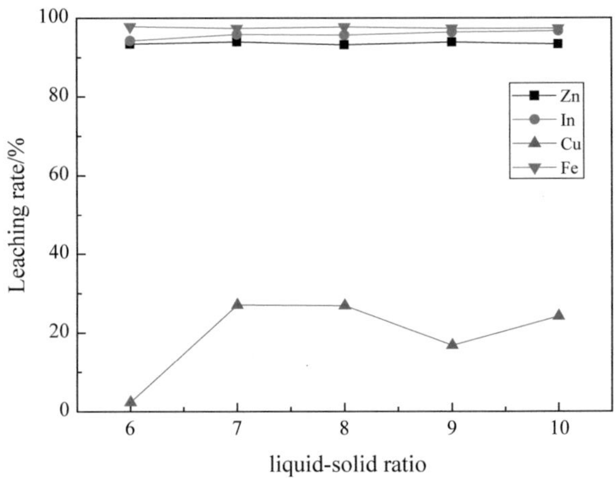

Fig. 9 Effect of liquid–solid ration on extraction of Zn, Fe, In, and Cu. (the SO_2 partial pressure 200 kPa, 105 °C, liquid–solid ratio 8 ml/g, initial H_2SO_4 120 g/L, and time 4 h)

Characterization of the Reductive Leaching Residue

Reductive leach residue obtained under conditions of 200 kPa SO_2 partial pressure, 105 °C, liquid–solid ratio of 8 ml/g, initial H_2SO_4 concentration of 120 g/L, stirring speed of 500 r/min, and time of 4 h was characterized. The results indicated that under the conditions employed up to 95.24% Zn, 98.66% Fe, and 95.04% In, and only 0.036% Cu extraction was achieved. The reductive leaching residue was chemically analyzed, and the results are presented in Table 2. As can be seen, the

Table 2 The main chemical components of reductive leaching residue (wt%)

Zn	Fe	Cu	In	S	Pb
5.27	3.92	4.63	256 g/t	4.21	1.37

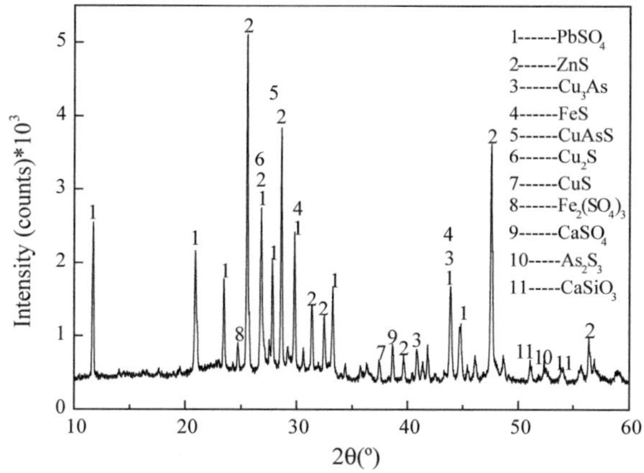

Fig. 10 XRD pattern of reductive leaching residue

reductive leaching residue became an rich material with high copper content. There were small amounts of Zn, Fe, and In present.

A typical XRD pattern of the reductive leaching residue is shown in Fig. 10. It can be seen clearly that constituent phases of the reductive leaching residue changed significantly compared with the zinc leaching residue. The main phases of the reductive leaching residue are sulphur, anglesite ($PbSO_4$), sphalerite ((Zn, Fe)S), and copper sulfide (Cu_2S). The zinc ferrite was almost entirely dissolved. There is a small amount of sphalerite in the leaching residue.

Conclusions

Experimental results indicated that under the conditions with 200 kPa SO_2 partial pressure with initial sulphuric acid concentration of 120 g/L, stirring speed of 500 r/min, at leaching temperature of 105 °C and 240 min time, up to 95.24% Zn, 98.66% and 95.04% In, and only 0.036% Cu extraction is achieved. Under optimal conditions, the reductive leaching residue became a rich material with high copper. Zinc ferrite was almost entirely leached. The reductive leaching of zinc ferrite in the presence of high-iron sphalerite as reductant is a viable process that effectively extracts zinc and indium and converts Fe^{3+} into Fe^{2+} in the solution.

Acknowledgements This work was supported financially by the National Natural Science Foundation of China (Grant Nos. 51804146, 51664030, and 51564030), National Key Research Development Program of China (Grant No. 2018YFC1900402), and the Analysis and Testing Center of Kunming University of Science and Technology, China.

References

1. Yanjuan Z, Xuanhai L, Liuping P, Xinyuan L, Xueping L (2010) Studies on the kinetics of zinc and indium extraction from indium-bearingzinc ferrite. Hydrometallurgy 100:172–176
2. Ismael MRC, Carvalho JMR (2003) Iron recovery from sulphate leaches liquors in zinc hydrometallurgy. Miner Eng 16:31–39
3. Xuelan WU, Shunke WU, Wenqing QIN, Xihong MA, Yinjian NIU, Shaoshi LAI, Congren YANG, Fen JIAO, Liuyi REN (2012) Reductive leaching of gallium from zinc residue. Hydrometallurgy 113–114:195–199
4. Cunxiong LI, Chang WEI, Hongsheng XU, Minting LI, Xingbin LI, Zhigan DENG, Gang FAN (2010) Oxidative pressure leaching of sphalerite concentrate with high indium and iron content in sulfuric acid medium. Hydrometallurgy 102:91–94
5. Zhifeng X, Qingzheng J, Chengyan W (2013) Atmospheric oxygen-rich direct leaching behavior of zinc sulphide concentrate. Trans Nonferr Metals Soc China 23(12):3780–3787
6. Parvatheeswara Rao B, Rao KH (2005) Distribution of In^{3+} ions in indium-substituted Ni–Zn–Ti ferrites. J Magn Magn Mater 292:44–48
7. Ramachandra Sarma VN, Deo K, Biswas AK (1976) Dissolution of zinc ferrite samples in acids. Hydrometallurgy 2:171–184
8. Leclerc Nathalie, Meux Eric, Lecuire Jean-Marie (2003) Hydrometallurgical extraction of zinc from zinc ferrites. Hydrometallurgy 70:175–183
9. Filippou D, Demopoulos GP (1992) A reaction kinetic model for the leaching of industrial zinc ferrite particulates in sulphuric acid media. Can Metall Q 31(1):41–54
10. Han JW, Li W, Qin WQ, Chai LY, Zheng YX, Yang K (2016) Thermodynamic and kinetic studies for intensifying selective decomposition of Zinc Ferrite. JOM 68(9):2543–2550
11. Zhang F, Wei C, Deng ZG, Li CX, Li XB, Li MT (2016) Reductive leaching of zinc and indium from industrial zinc ferrite particulates in sulphuric acid media. Trans Nonferr Metals Soc China 26(9):2495–2501
12. Aydogan S, Aras A, Canbazoglu M (2005) Dissolution kinetics of sphalerite in acidic ferric chloride leaching. Chem Eng J 114:67–72
13. Markus H, Fugleberg S, Valtakari D, Salmia T, Murzina DYu, Lahtinen M (2004). Reduction of ferric to ferrous with sphalerite concentrate kinetic modeling. Hydrometallurgy 73:269–282
14. Yangyang F, Yan L, Liping N, Tingle J, Weiguang Z, Tingan Z (2019) Reductive leaching of indium-bearing zinc ferrite in sulfuric acid usingsulfur dioxide as a reductant. Hydrometallurgy 186:192–199
15. Shiqing L, Motang T, Jing HE, Shenghai Y, Chaobo T, Yongming C (2006) Extraction of indium from indium-zinc concentrate. Trans Nonferr Metals Soc China 16:1448–1454
16. Fan Z, Chang W, Zhigan D, Xingbin L, Cunxiong L, Minting L (2016) Reductive leaching of indium-bearing zinc residue in sulfuric acid using sphalerite concentrate as reductant. Hydrometallurgy 161:102–106

Author Index

G. Azimi et al. (eds.), *Rare Metal Technology 2020*, The Minerals, Metals & Materials Series, https://doi.org/10.1007/978-3-030-36758-9

Subject Index

G. Azimi et al. (eds.), *Rare Metal Technology 2020*, The Minerals, Metals & Materials Series, https://doi.org/10.1007/978-3-030-36758-9

MIX
Papier aus verantwortungsvollen Quellen
Paper from responsible sources
FSC® C105338

If you have any concerns about our products,
you can contact us on
ProductSafety@springernature.com

In case Publisher is established outside the EU,
the EU authorized representative is:
Springer Nature Customer Service Center GmbH
Europaplatz 3, 69115 Heidelberg, Germany

Printed by Libri Plureos GmbH
in Hamburg, Germany